Structural-Chemical Systematics of Minerals

Alexander A. Godovikov ·
Svetlana N. Nenasheva

Structural-Chemical Systematics of Minerals

Third Edition

 Springer

Alexander A. Godovikov
Moscow, Russia

Svetlana N. Nenasheva
Fersman Mineralogical Museum
Russian Academy of Science
Moscow, Russia

Translated by I. A. Godovikov

ISBN 978-3-319-72876-6   ISBN 978-3-319-72877-3   (eBook)
https://doi.org/10.1007/978-3-319-72877-3

Library of Congress Control Number: 2019934785

This Springer imprint is published by the registered company Springer Nature Switzerland AG
The registered company address is: Gewerbestrasse 11, 6330 Cham, Switzerland

Contents

4. TYPE: MINERALS WITH PRINCIPAL COVALENT-IONIC AND IONIC BOND - HALOGEN COMPOUNDS: HALOGENIDES (ISODESMICAL) → HALOGENOSALTS (ANISODESMICAL) 243

5. TYPE: CARBON, ITS COMPOUNDS (WITHOUT CARBONATES) AND RELATED SUBSTANCES. 250

UDK 549 + 548.3

Structural-Chemical systematic of minerals. A.A.Godovikov, S.N.Nenasheva. – 3 nd updated edition. – M., 2016. – n p. Ill. 3. Bibliography: 69 ref.

The classification tables are complemented by the new mineral species, that were discoveried at 2007-2016 years [4], [50].The formulae of some mineral species were corrected and some mineral species were transfered to another taxons on account of the appearance of new data about the chemical composition or the crystal structures.

Structural-chemical systematic of minerals is representative of recent data on connection between chemical composition and crystal structures and properties of minerals, conditions of mineral formations, paragenesis. The chemical signs are the basis of structural-chemical systematic. The crystal structures of mineral consider on the middle and low-level taxons, but not on the high-level taxons, because the mineral structure is depend on chemical composition and physical-chemical parameters of mineralforming systems.

The classification tables are given, which include near by 5000 mineral species. This enables to use developed classification for scientific and practical purposes.

This book designed for the wide circle of mineralogists, petrographers, geochemists, students of geolodical institutions and colleges.

COVENTIONAL ABBREVIATIONS:
CN – coordinate number
HPC – hexagonal close-packed
CPC – cubic close-packed
CP – close-packed
FC – force characteristic (γ)

$$\gamma_{orb} = F/r_{orb}$$
$$\gamma_{orb}{}^{n+} = I_n/r^{n+}$$
$$\gamma_i = I_n/r_i$$

F – affinity of electron
I_n – ionization potential of n –th electron
r_{orb} – atom´s orbital radius
$r_{orb}{}^{n+}$ – ion´s orbital radius
r_i – ionic radius

A.A. Godovikov introduced unnamed taxons. The names for these taxons was proposed by S.N.Nenasheva. These taxons marked by asterisk on the right from taxon. For example: quasitype*, quasisubtype*, overclass*.

The new taxons and new mineral species was introduced by S.N.Nenasheva, marked by asterisk on the left. For example: *3.2.5.3.6.3. Oxido-thiosulpfates,*museumite, *telyushenkoite.

Formulas of mineral species are given in modern reference books: [4], [42], [46], [50].

Introduction

Inspite of the existing opinions, especially of young scientists, that there is no sense to explore the systematic (in all forms) and that it is more important to get quite definite physical data, there are hundreds works which are dedicated to the mineral systematic. Some well-known scientists worked on them, such as M.Lomonosov, J. Bercelius, V. Severgin, J.D. Dan, V. Vernadskiy. H. Strunz, A. Povarennykh, I. Kostov and other ones, and dedicate their investigations to systematic of minerals.

This can be explained by the following reasons:

1) There is a necessity to systematize different and numerous information on individual minerals. Without this it is not possible to get slim and scientific description. The comprehension of it usually comes with the age of scientist and with spreading of the scientific range of interests.

2) There are a lot of mineral properties which are used in their descriptions. We need to know them not only for mineral diagnostics, but also for clarification of their searching features, forming conditions of their paragenetic associations, capabilities of mineral utilization by humans.

3) By belonging of minerals to the complete different chemical compound types; by differences and complexity of their composition.

4) Among minerals there are substances which were formed in completely, even in interexcluded physical-chemical conditions.

Any of mineral systematics appears to be multidimentioned, because it should consider all the multitude of different mineral features. The systematic which is expressed in most usual table appearance could mirror, on essence, only one of all possible sections of multidementioned area of mineral properties. The best of all modern systematic's appearances is systematic in computer databases on diskettes or on compact disks which are considered maximum number of different mineral properties. They allow to get different concrete sections of multidementioned area of mineral properties, to clarify all correlations and queries.

Historically earlier were those of mineral systematics which were based on chemical features. Unfortunately all these variants are based on highly limited number of general chemical features, on very general ideas about connection between properties of elements and their position in Periodic system. These ideas could not mirror in necessary measure all varieties and regularities of chemical compounds and features of minerals, real complexities of connection between electronic structure of elements and structure, properties of substances they forming. Big troubles of chemical systematics are determined by unsatisfactory condition of nomenclature and classification of inorganic crystalline substances, including the majority of mineral species. All this brings to considerable drawbacks of these systematics, impossibility of their wide use. This is complicated by that usually authors of chemical systematics are not stopped on basing of their works, satisfying either on schemes of chemical features which were used in classification tables or on table of itself.

The next time group can be represented by crystallochemical, rather structural-geometrical, systematics which are widely spread. Appealing feature in all these systematics is that the concrete peculiarities of crystalline structures of minerals which can be obtained experimentally are in their basis. However, all similar systematics which can excellently demonstrate regularities of connections of mineral properties with geometrical peculiarities of their crystalline structures, be found practically weak in

© Springer Nature Switzerland AG 2020
A. A. Godovikov and S. N. Nenasheva, *Structural-Chemical Systematics of Minerals*, https://doi.org/10.1007/978-3-319-72877-3_1

solution questions on mineral genesis and paragenesis. This is clear, because the structure of minerals appears to be the secondary feature on comparison with their chemical features, as long as the structure itself is determined by chemical composition of mineral, physical-chemical parameters of system in which it forms and stays in.

Inspite of widely spreading of such systematics, what is most likely the time tribute, they can not satisfy mineralogists, because of the shown reasons. This is most brightly expressed in whole series of works by I. Kostov, who has written in one of them: "In modern crystalline chemical classifications paragenetic connections between minerals are ignored, therefore we have formal, although slim in some cases distribution [39]. Unsatisfaction of mineralogists in this kind of systematics gives as a result the appearance of geochemical systematic of minerals by I. Kostov [40] in which by division of minerals in subclasses the element triads which characterize definite mineral associations set as a leading feature. Obviously, this cause the appearance of extremely unsuccessful and voluntary "chemical" systematic by E. Semenov [60] and very detailed, but without descriptive principles and often inconsequent systematic by A. Hoelzel [34] and A. Clark [12]. Speciesforming cations have the supreme position in class systematic in these works.

The luck of satisfying mineral systematic, obviously played its role in that the last time more and more reference books comes which have alphabetic mineral classification [16]; [2]; [57]; [17]; [51].

It is apparent that mineralogist will be satisfied, if systematic has on its basis chemical features which could enable to understand their connection with mineral structure and properties, mineralforming conditions and paragenesis. Features which are characterizing the structure of minerals should have not the highest taxon positions, but rather middle or even inferior taxon positions, because they are in straight dependence on chemical composition and physical-chemical parameters of mineralforming systems. It is important to aim for selecting in systematic such taxons which:

a) could unite possibly greater number of mineral species, this should ease their general characteristic and save mineralogy from numerous repetitions which appear in such cases, when mineralogy, because of the luck of really well-developed systematic, is stated as descriptive science;

b) could show gradual transformations from one taxons to other, their natural and numerous interconnections.

Maximally all these features could be represented in systematic which based on:

a) detaily developed ideas about the connection between properties of elements and their electronic structure, their position in Periodic table;

b) numerous chemical features which are often not considered in necessary measure and from which the chemical properties of minerals, mineralforming conditions and paragenesis are dependent;

c) clarified regularities in connection between fundamental properties of elements and structure of chemical compound, their forming;

d) regularities in structure and mineral properties variance on dependence on physical-chemical parameters of systems in which they have formed and stated.

Developing any systematic of natural objects including minerals, two approaches which are contrary on essence should be considered. One of them, the simple one, brings us for putting all considered objects in developed scheme, basing on "strong" logical requests. Inspite of bribing scientific vividness this way is impossible for natural objects and it could not bring to creation of their really natural systematic which considers all numerous natural connections a transformations among them. And exactly natural systematic is rather necessary and acquittal for them. This could explain that V. Vernadkiy [65], who complained on natural mineral systematic's absence, felt respect to it. The Mendeleev's success in creation of Periodic low and developing the table variant of Periodic system, could be explained simultaneously; when he instead of following the dependence of properties of elements on their atomic weights, proclaimed by himself, did three deviations, changing position in the following systems of element pairs: Co and Ni, Se and Br, Te and I which has been found its explanation later.

Such complexities which appeared on the simplest level of chemical objects - elements risen many times in conversion to their compounds, when atomic electronic structure, energy levels of electronic orbitals revealed not impurely, but rather through the chemical bond, crystalline structure of substance which extremely complicate the creation of natural systematic of minerals.

However, the situation is not completely lost and the exit could be found, if in systematic development consider not only the formal features, but also real mineral associations, their paragenesises, differences in physical chemical parameters of mineralforming systems, giving favor in mineral order in systematic exactly to the last features. This way could be vendicated by two reasons. First, it is necessary to consider that natural mineral associations, sequential transformations in mineralforming processes are not accidental, but appear to be the results of selecting, sometimes long and complex, of individual elements by their properties in natural associations in whole. Thus, this feature could be considered as criteria of naturality of mineral systematic. More than, its utilization allows to unseal deep connections in properties of individual elements, motives of their definite behavior in mineralforming processes. Second, coordination of systematic not only with properties of minerals, their structure, but also with mineralforming conditions, their paragenesis, transformation conditions from one taxon to other one either on the same level or by its consequent deepening and detailing, has to be the most important demand of the mineralogy. Without it the scientific under-standing and statement and finally the creation of natural mineral systematic is impossible. The numerous examples of what was said are mentioned below.

Let's concentrate on selecting principles of highest taxon systematic, their sequence, connections; however, remember, that definitions of lower taxons - mineral species, subspecies, seria (genus), varieties, groups, families which were given earlier [22], [25] are not considered by us. To the highest taxons we refer all taxons up to classes, quasiclasses and subclasses inclusive, although not all authors give them special names, tracing their sequence by numbers (first of all roman, and then arabic) and letters (first of all capital, and then small ones) or by consequent rising numeric indexes. We use for this purpose only arabic numbers and primary small letters of alphabet.

Before we go to consideration of taxons themselves it is necessary to stop on precise definition of the number of basic terms and positions.

We should begin with systematic of elementary substances[1], compounds in dependence on prevalence of one or another definite type of chemical bond in them -

---

[1] In majority chemistry courses, especially primary ones, elementary substances which include substances, consisting of atoms of one sort called simple substances. Mineralogists, who however, did not pay attention that this term is completely identical to term which defines simple binary compounds in contrast to complex or binary compounds, have borrowed this term form them.

geochemical systematic of elements and cations which is used in developed natural mineral systematic.

Division of elementary substances on metals, semi-metals and nonmetals

Special atomic properties of individual elements are most visible in elementary substances, because they contain no other atoms, except the atoms of given element. But even in this case there is no simple connection between the fundamental properties of elements, for example values of their force characteristics, order numbers, and physical or chemical properties of elementary substances as far as this dependence is complicated by chemical bond between atoms, by its type, by crystalline structure of given elementary substance which is stable in definite intervals of temperature and pressure, by other physical chemical parameters[2].

Elementary substances at all times are divided on: 1) metals, 2) semimetals or metalloids and 3) nonmetals. In Russia, however, the tendency of refusal from terms "semimetals" and "metalloids" which are exclude from chemical reference books and textbooks [1], has been strongly become apparent. In USA in latest reference books on chemistry they are sometimes even selected in tables by special color [4], [60]. Unfortunately, there are no strong term definitions given, that is why different authors have completely different boundaries between these groups of substances.

For example, in P.W. Atkins's and J.A. Beran's book [2] we can find the following definitions:

"A **metal** is a substance that conducts electricity, has a metallic luster, and malleable and ductile" (p.43).

"A **nonmetal** is a substance that does not conduct electricity and is neither malleable nor ductile" (p.43).

"A **metalloid** has the physical appearance and properties of a metal but behaves chemically like a nonmetal" (p.44).

Inspite of their lapidarity and simplicity these definitions have insufficiency which brings to the different number of elementary substances, included in every group by different authors. First of all it is remarkable, that all used features are nonadequate. So, in definition of metal and nonmetal - it is electroconductivity, malleability, plasticity which are some of the physical properties. At the same time, there is no allusion about how these substances look (by the way it is not clear enough, what authors are consider by that - luster, color or something else), not rather about their chemical properties, but it is not clear which of them, although exactly some inconcrete physical magnitudes, including external look, and chemical properties are put in the term definition basis of "metalloid". It is not amazing, that such situation brings to contradictions in relating some elementary substances to each of three stressed groups by different authors. Following that tables of comparing physical and chemical properties of metals and nonmetals from which, by incomprehensible reasons, the semimetals are completely excluded, are not improve the situation. Such situation is not confusing for majority of

---

[2] The narrow connection between elements properties - their electronic structure, their position in Periodic table and properties of elementary substances clarifying clearly on curve, called "atomic volumes curve" of elements which was used for explaining of Periodic low by D.I. Mendeleev [48], J.L. Meyer [49] for basing of elements division on atmo-, litho, chalco- and siderophylic by V.M. Goldschmidt [32] and many of the followers. However, usually scientists forget that this curve shows the dependence of specific element volumes of elementary substances , not the elements, on ordinal (atomic) numbers or atomic masses of elements. This could be explained by intersubstitution of terms "element" and "elementary (simple) substance", what is completely impossible as noted earlier [21]

authors and is in accordance with dialectic logic, when metalloids or semimetals are defined as substances which have to be found on boundary between metals and nonmetals and which have properties both of metals and nonmetals [67].

As a result K.W. Whitten and K.D Gailey [67] build boundary metals ↔ nonmetals strongly through diagonal B → Si → As → Te → At, and all mentioned elements are placed to the right from this boundary. They did not select semimetals concretely, although the indistinct definition for semimetals is included, and their selection is considered to be individual for every author. P.W. Atkins and J.A. Beran [3] as semimetals pick out Si, Ge, As, Sb, Te and Po, as far as Al, Ga, Sn and Bi are considered to be rather metals and B, C, P, Se, I and At are refered to nonmetals.

Considered definitions from native encyclopedic reference books are not more successful.

The most strong of them is definition of metals which is given in PED [54]. According to it, elementary substances which "have some typical properties: high electroconductivity and thermal conductivity, negative thermal electroconductivity coefficient[3], ability to reflect electromagnetic waves (luster and untransparency), plasticity[4] at usual conditions", considered as metals (p. 409).

Metallic conductivity type usually explained by presence of free electrons in metals which appear as a result of valent and conductivity zone overlapping, because the width of the forbidden zone in metals (DE_0) is 0.

Definitions of "semimetal" term which are given by chemists are completely unsuccessful. Thus, in CED [11] we can find: "**Semimetals** (semimetalic elements, "fragile metals") - elements, occupying places on boundary between metals and nonmetals (how this boundary comes - it is not mentioned, although by different authors it is completely different; meaning of "on boundary" is not clear enough - A.G.) in Periodic system of elements by Mendeleev (here is the evident mixing of terms "elementary substance" and "element" - A.G.). They characterize by covalent crystalline lattice (this is not right, because "lattices" of elementary As, Sb and Te which are practically refered to semimetals by all authors are quasimolecular - A.G.) along with metallic conductivity (As, Sb, Te have semiconductive type of conductivity - A.G.). To semimetals are refered Sb, Bi, Po, sometimes (when ? - A.G.) also Ge, As, Te, although they are conductors by conductivity type (not conductors, but semiconductors ! - A.G.), and by chemical properties (which of them ? - A.G.) - nonmetals, and Sn which has simiconductor modification" (p. 472). In addition to numerous rough mistakes in this definition it can be noted, that as basis of referring of element (as a matter of fact simple substance) to semimetals the existence of polymorphs with semiconductor properties (case of Sn which has lowtemperature modification - "gray tin" with diamond structure and semiconductor properties) have been chosen. This present to be absolutely impossible, because polymorphs, including those of them which have semiconductor properties are known for other elementary metallic substances and vice-versa, some nonmetals and semimetals have polymorphs with metallic conductivity (for example, polymorphs of Ge and Si which are stable at very high pressures).

More clear definition is given in PED [54], where we can find: "**semimetals**" - substances, occupying by electrical properties intermediate position between *metals* and

---

[3] Metallic Al has unlike the other metals positive thermal electroconductivity coefficient.
[4] Not all of the metals are plastic and have low hardness. The most bright examples of fragile metals which differ by perfect cleavage and high hardness, are Cr and Os. The cleavage can be also found in monoctystalls of iron which has malleability properties in polycrystallic blanks, and other metals. As it was shown earlier [18], such properties of metals could be explained by peculiarities of theirs chemical bond, when the covalent component of chemical bond is appeared equally with its metallic component.

semiconductors; … from one side semimetals remain to be conductors up to absolute temperature naught, but from another side - with minor (comparing to metals) concentration of current carriers i. With temperature rising the number of carriers increased, electroconductivity raised too" (p. 563). Thus, semimetals characterized although by high conductivity, but of semiconductor type.

In connection with it, it is expedient to give both more strong definition of boundaries between metals, semimetals and nonmetals, and list of elementary substances, refered to each of these groups.

Earlier [21] was shown, that the boundary between metals from one side and nonmetals and semimetals from the other one in expanded variant of Periodic system comes among s-elements through H and He, refered to nonmetals and Li - Be, refered to metals. Further this boundary comes through p-elements as broken line B → Si → Ge → Sb → Bi. To the left from this boundary are the elements, forming only metallic elementary substances which completely suited for definition from PED [54], mentioned earlier. To the right of this boundary there are elements, forming elementary substances which have the properties of nonmetals and semimetals.

Indicated boundary is determined by diagonal similarity in properties of element pairs B - Si and Ge - Sb; and belonging of pairs Si -Ge and Sb- Bi to shrink-analogies after *d* and *f* pressing respectively.

Comparing the values of forbidden zone width of elementary substances of p elements (DE_0, eV):

B	C(diamond)	N	O	F	Ne
1,5	5,2	-	-	-	-
Al	Si	P(bl.)	S	Cl	Ar
0	1,21	0,33	2,6	-	-
Ga	Ge	As	Se	Br	Kr
0	0,78	1,2	2,1	-	-
In	Sn	Sb	Te	I	Xe
0	0	0,12	0,32	1,35	-
Tl	Pb	Bi	Po	At	Rn
0	0	0	0	?	-

and the values of their specific resistancy (p, om cm; at 20^0C): (see the next page),

B	C(diamond)	N	O	F	Ne
$4·10^6-10^7$	$>10^{16}$	-	-	-	-
Al	Si	P(bl.)	S	Cl	Ar
$2,5·10^{-6}$	$2,3·10^5$	$3,1·10^6$	$2,0·10^{16}$	-	-
Ga	Ge	As	Se	Br	Kr
$1,4·10^{-5}$	48-60	$3,5·10^{-5}$	1,2.10	-	-
In	Sn	Sb	Te	I	Xe
$8,2·10^{-6}$	1,0 10-5	$4,3·10^{-5}$	$3-5·10^{-1}$	$1,3·10^9$	-
Tl	Pb	Bi	Po	At	Rn
$1,8·10^{-5}$	$1,9·10^{-5}$	$1,0·10^{-4}$	$42-10·10^{-3}$	?	-

we can see, that by the value of ΔE_0 not only elementary substances which could be found to the left of the boundary metal $\leftrightarrow$ nonmetal - Al, Ga, In, Tl, Sn, Pb, but also Bi and Po which laying to the right of this boundary, should be refered to metals. By the value of specific resistancy the number of substances which could be refered to metals is even greater. Thus, besides mentioned ones, there are As, Sb and with small strain Te among them.

Another picture is drawn by comparing structures with physical properties, connected with them, of considered elementary substances. So, elementary Al, Ga, In, Tl, Sn and Pb which are found to the left from the boundary metal $\leftrightarrow$ nonmetal, have specific for metals crystalline structures (except Ga) and malleability, plasticity. Quite another matter - elementary As, Sb, Bi and Te, found to the right from this boundary which characterized by quasimolecular crystalline structures, typical for nonmetals, and as an effect - perfect cleavage, high fragility, not usual for metals.

Such properties of elementary As, Sb, Bi and Te, satisfied both for metals and for nonmetals from the other side, give us opportunity to mark out these substances as semimetals.

Besides As, Sb, Bi and Te, rare radioactive Po and At should be refered to semimetals. Because of the rarity and small quantities in which these substances known, we can consider on properties on whole about their belonging to semimetals.

Thus, on definition of Po in SCE [59] we could find, that elementary Po slowly dissolves in HCl with formation of ion Po^{2+} which oxidize by radioactivity itself up to Po^{4+}. Hydrogen sulfide sediments sulfide PoS from solutions of Po salts. For Po the following substances are well known: oxides - PoO and PoO_2, halogenides PoX_4, sulfates $Po[SO_4]_2$ and $PoO[SO_4]$, halogensalts $M[PoX_6]$. By these and many other features Po - is typical metallic element. Considerably rarely Po reveals features of nonmetals, come forward with Po^{2-} in hydride H_2Po and in polonides.

In definition of At from CE [9] we could find: "For At it is typical to combine properties of both nonmetals (halogens) and metals (Po, Pb, and so on). Thus, analogically to iodine, At could be easily dissolved and extracted by organic solvents; by fugitiveness it gives way to I_2, but it could also be distilled away. By acting on At solution of hydrogen, gaseous HAt forming in reaction moment. Simultaneously to I_2, At in water solution is reduced by SO_2 and oxidized by Br_2. However, like metals, At could be sedimenated from hydrochloric acid solutions by hydrogen. In presence of oxidants in water acid solutions At exists in single charged cation which probably has the following structure: $[At(H_2O)_x]^+$, where x = 1 or 2" (p. 397).

Especially we should concentrate on elementary Si and Ge which are referred by some authors [3] to semimetals, although, according to mentioned earlier values of width of forbidden zone and their specific resistancy values, they should be refered to nonmetals. Their referring to semimetals most likely is a result of misunderstanding. This could be connected with that elementary Si and Ge in private life and trade received names "metallic" silicon and germanium respectively because of their strong not metallic, but rather semimetalic lustre which is a result of pretty high concentration of current bearers at normal temperature in these semiconductors with diamond type structure. In accordance with it, it is necessary to remember that simultaneous names are used for some other nonmetals, for example, "metallic selenium" and "metallic iodine" which are characterized in elementary state by semimetalic lustre too.

All that was said shows, that elementary substances should be divided on: metals, semimetals and nonmetals.

To the last ones the elementary B; C (different allotropes), Si, Ge, N, P (different allotropes), O_2 and so on allotropes of O, S_8 and other allotropes of S, Se_8 and other allotropes of Se; F_2, Cl_2, Br_2, I_2, and also H_2 and noble gases - He, Ne, Ar, Kr, Xe and Rn refered. Among them diamond and other allotropes of C, native Se, Ge, S (different allotropes), Se are known as minerals.

Boundaries between considered groups of elementary substances, in that way, could be presented by following:

B	**C**	**N**	**O**	**F**	**Ne**
Al	**Si**	**P**	**S**	**Cl**	**Ar**
Ga	**Ge**	*As*	**Se**	**Br**	**Kr**
In	Sn	*Sb*	*Te*	**I**	**Xe**
Tl	Pb	*Bi*	*Po*	*At*	**Rn**

Here is the boundary between metals which are typed in normal style, and nonmetals which are typed in semibold style, and semimetals, selected by italic and underlined, is shown by semibold broken line; thin broken line - boundary between semimetals and nonmetals.

Basic types of chemical substances, selected by primary type of chemical bond

The majority of elementary substances are refered to metals. Considerably less of them are semimetals. Nonmetals are not so numerous too. There are incomprehensibly more compounds of different elements with each other. The chemical bond type in them, their crystalline structures and properties are dependent on differences in electronegativeness between substanceforming elements (atoms, ions), their atomic (ordering) numbers. Changing of chemical bond type, properties of compounds in dependence on changing of fundamental properties of atoms, forming them, could be happened either consequently or sporadically. Earlier [18], [21], [27] it was mentioned, that chemical bond should be considered as metallic-covalent-ionic-residuum one in whole; its character in compound is defined not only by extent of covalency (ionicity), in general case by proportional difference in force characteristics (electronegativenesses), but also by extent of metallicity, determined by ordering number of element or by average ordering number of electropositive elements in compound. The extent of its "residuality" could in some cases be one more measurement of chemical bond which although could not be easily expressed through the mentioned fundamental properties of elements and atoms. Despite that the nature of chemical bond in all its demonstrations is the same, usually four extreme types of chemical bond are selected: metallic, covalent, ionic and residuum bond which are expressed in different compounds in different extent.

Metals which elements occupied left greatest part of extended variant of Periodic system, by interaction with each other are forming substances with typical metallic properties - first of all with high conductivity, strong metallic lustre, determined by metallic type of bond or simply by metallic bond. Such substances called metallides or intermetallides. Unfortunately these terms have lost the most part of their definition up today, so now it is necessary to concentrate on them.

Thus, in CED [11] we could find: "**Metallides** (intermetallides), chemical compounds of two or several metals. Compounds of transitive metals with more electropositive nonmetals (H, B, C, N and so on), characterized predominantly by metallic type of chemical bond, are also frequently refered to metallides (but may be in vain ! - A.G.)" (p. 325). We can see, that this "definition" of metallides which have been given by S.S. Kiparisov, not so much appears to be really definition, as noncritical exposition of different ways to it. When different authors are unrestricted in referring various compounds to metallides by their matter of taste, but not by the basis requests of their definition.

Definition of metallides, given in SCE [58] is not much better. "**Metallical compounds** (metallides, metalsimilar compounds, intermediate phases in alloys) - chemical compounds which have metallic properties (which of them ? - A.G.) The majority of metallical compounds are formed by interacting of several metals (intermetallic compounds), but they could also include (in which appearance, in which quantities ? - A.G.) C, N, B, Si, H and other nonmetals (which of them ? - A.G.). In accordance with it metalsimilar carbides, nitrides, borides, silicides, hydrides and so on are refered to metallical (?!! - A.G.) compounds" (p. 139). But where should we refer numerous nonmetalsimilar carbides, nitrides, borides, silicides with covalent (muassonite) or even ionic chemical bond, compounds with polyradicals as $Ca[C_2]$ and what means "metalsimilar" - is not clear. In this way, this definition, belonging to B.K. Wolf, can not sustain any criticism too.

More resent definition from CE [10] is not also clarify this situation. Where we can find: "**Metallical compounds** (metallides) have metallic properties, in particular electrical conductivity which caused by metallic character of chemical bond. To metallical compounds refer compounds of metals with each other - *intermetallides* and more other compounds of metals (in general of transitive ones) with nonmetals. Metallic properties are usually strongly clarifying in rich-metal compounds - lowest carbides, nitrides, sulfides, oxides and so on" (p. 42).

The wide treatment of notion "metallides" (intermetallides) to which refered not only compounds of metals with each other, but also compounds with semimetals, H (hydrides), light nonmetals with from 1 up to 3 p-electrons - B, C, Si, N and P (borides, carbides, silicides, nitrides, phosphides), accepted by W. Pearson [53], P.I. Kripyakevich [41]. And what is more, P.I. Kripyakevich as intermetallides considers some selenides, sulfides, sulfosalts and even oxides, especially those of them which are subcompounds or substances with cluster [5] structure.

At the same time, in reference book on chemical nomenclature by A.I. Busev and I.P.Efimov [8] we could find much more narrow treatment of these terms : "**Metallides** (intermetallic compounds) - chemical compounds of metals with each other. The metallic chemical bond is the prevalent one in such compounds. Metallides are not submitting by regularities of composition permanency and simple divisible ratios." (p. 109). The last statement is not, however, correct. Thus, among metals from one side there are compounds which are not submitting by simple divisible ratios, by usual valent states of elements - in composition of such compounds electropositive elements sharply predominate upon electronegative ones, and they called *subcompounds*. The electronic structure of such substances, their abilities of isomorphism are usually analyzed with consideration of so-called electronic concentration [33]. From the other side among metals there are a lot of compounds, in which elements have their usual valences and characterized by simple integer ratios. At the same time there are many compounds with variable in definite limits composition which are refered to metallides - bertollides, ordered alloies - Kurnakov's phases.

Considering all mentioned contradictions in terminology, we will limit term metalloids only by compounds of metals with each other, - elements which could be found to the left from mentioned earlier boundary metal ↔ nonmetal in extended variant of Periodic system. Term "intermetallides" like all its synonyms, should be recognized as indefinite, obsoleted and should be excluded from usage. Metallic conductivity, frequently metallic lustre and malleability are typical for metallides. These properties are defined by metallic type of chemical bond which is typical for these substances.

Simultaneously to metallides it is expedient to select also *semimetallides* to which we could refer compounds of metals with semimetals - arsenides, antimonides, bismuthides and tellurides (polonides and astatides which are not known in mineral forms, could be refered to them also). Here are the substances with high conductivity, frequently of semiconductive type, with metallic lustre, fragile, in some cases with cleavage, up to perfect one (tetradimite mineral family). These minerals are characterized by chemical bond with predominance of its metallic component.

After that, the association of compounds of electropositive (metallical) and semimetallical elements with the rest of nonmetallical elements in *nonmetallides* is coming real by itself. Overwhelming number of compounds from chalcogen up to halogen with pretty various properties, defined by consequent decreasing of metallicity extent and covalency extent of chemical bond and increasing of its ionicity extent in mentioned direction, are refered to nonmetallides.

Among nonmetallides IVa-nonmetallides are distinguished by their properties, what is defined by special - "passing" - place of IVa-elements in Periodic system. This clarifies in what to the left to IVa group practically only metallic elements could be found, but to the right from it - rather nonmetallic and semimetallic. There is only one exception - B which is refered to nonmetallic elements, placed in IIIa-group; the last two elements of IVa group - An and Pb are refered to metallic elements too. Exactly the same peculiar placement of IVa elements in Periodic system brings E. Zintl and H. Kaiser [69] to confinement, that this group could be considered as boundary between metallides and nonmetallides. Later this boundary was called Zintl boundary, although it was unclear, but rather diffusual, what was stressed by T.V. Massalskyi [47] and what is presented to be natural, because of that was said.

By special placement of IVa group in Periodic system such specific properties of IVa-nonmetals as their ability to form entire variety of organic compounds, including hydrocarbons, siliciumhydrogens (silanes, silones and so on), germaniumorganic compounds which are more rare, but rather close by structure and properties to them, could be explained. At the same time carbides, silicides and germanides could be considered as derivatives of respective hydrogen compounds. Depending on electropositive element, they could have completely different type of chemical bond, including metallic, covalent, residual or ionic one. By this the variety in properties of such compounds, including differences in crystalline structure, difficulty of determining their place in total sequence of taxon changing in systematic, is defined. Thus, it was already mentioned, that some authors include them in "intermetallides". However, it is uncomfortable not only because of the specific properties of these substances, but also therefore, that such their position splits natural connections and transforms in united series: metallides → semimetallides → chalcogen compounds → oxygen compunds → halogen compounds which is presented to be natural changing of chemical bond type: metallic → metallic-covalent → ionic-covalent → covalent-ionic → ionic bond.

Considering what was said, the most correct way is to select from total series of different compounds those substances which are formed by IVa-nonmetallic elements, excluding compounds like carbonates which should be considered with the rest of oxygen compounds by particularities of their composition and properties. Compounds of IVa-nonmetals, including native minerals, carbides, silicides and organic compounds, should be considered as concluding ones in the systematic. Such compounds of Vanonmetals (N and P) as nitrides and phosphides which are close by their properties, structures, mineralforming conditions to carbides and silicides, and which have such common features with IVa-nonmetals, as for example that they belonging to the light (containing from 1 up to 3 p-electrons) typical p-elements (2 and 3 periods), are also refered here. It is important, that in this case organic compounds which are concluding the mineralogical systematic, are found their legal place.

Simple compounds, binary and more complex compounds, salts.

Such terms as simple compounds, binary and more complex compounds, salts, are very important in order to understand the basis of mineral systematic. Compounds of two elements which are consist of simple cations and anions are usually refered to *simple* substances[5]. For example of simple compounds we could mention sphalerite ZnS, chalcosine Cu_2S, antimonite Sb_2S_3, hematite Fe_2O_3, halite $NaCl$ and so on minerals. By their properties they could be refered to bases, amphotericcompounds, anhydrydes. Some of them, for example, sphalerite ZnS, chalcosine Cu_2S, antimonite Sb_2S_3, halite $NaCl$ are, at the same time, salts of nonoxygen acids, such as H_2S, HCl. Minerals which contain simple complex ions, for example, ammonium chloride NH_4Cl or simple complex anions, for example, pyrite $Fe[S_2]$, arsenopyrite $Fe[AsS]$, scutterudite $Co_4[As_4]_3$ could be refered to simple compounds too.

To *binary* or more complex compounds referred minerals which contain more than one cation. For example chalcopyrite $CuFeS_2$, stannite $CuFeSnS_4$, perovskite $CaTiO_3$, ferberite $FeWO_4$, gagarinite $NaTRCaF_6$, carnallite $KMg(H_2O)_6Cl_3$.

Significantly more complex is term "salts". Thus, in book by A.I. Busev and I. P. Efimov [8] we could find: "Salts - the class (what is class is not defined at all and in this case this term hasn't any special taxonomic assignment - A.G.) of chemical compounds, ctrystalline substances which have ionic structure (not always! - A.G.). By dissociation in aqueous solutions salts give positive charging ions of metals and negative charging ions of acid residues (sometimes also ions of hydrogen or hydroxo-group)" (p. 419).

In CED [11] the following definition is given: "Salts - are the products of substitution of H atoms in acids instead of metal or of OH group in bases instead of acid residuum. By the complete substitution normal or average salts are forming, for example $NaCl$, K_2SO_4, $(C_{17}H_{35}COO)_3Al$. Incomplete substitution of H atoms brings to acid salts (for example ammonium hydrosulfate NH_4HSO_4) incomplete substitution of OH groups - to basis salts, for example aluminum dihydroxostearate $(C_{17}H_{35}COO)Al(OH)_2$ ... Salts usually have ionic crystalline structure[6] and characterized by relatively high temperature values of melting and boiling points. Many salts are soluble in water[7], with complete dissociation on ions" (p. 533).

The most essential drawback of all these definitions is that they are limited by oxygen and often water-soluble compounds. Only in acid definition which is given in

---

[5] Electropositive components of compound called cations without any dependence on chemical bond type. That is why cations in compounds with different type of chemical bond are playing different role, approaching to idealized positive charged particle by increasing of chemical bond ionic extent. Thus, cations in the majority of sulfates, nitrates, chlorides (except minerals of Ag), halogen salts are close to ideal cations. By this time, in compounds with preferred metallic and covalent chemical bond, so-called cations takes their part only as electropositive components of compound. We could say the same thing about anions, as electronegative components of compound.

[6] The chemical bond between cation and acid residuum in chalcophylic element salts, especially of weak acids, often is covalent.

[7] Silicium acid salts - silicates, the most numerous among minerals, in majority hardly soluble in water; low solubility of compounds is one of the most important conditions of their long preservation in form of minerals in nature.

CED, it is mentioned, that the substitution of O atoms instead of S atoms brings to formation of tioacids. At the same time, salts of tioacids - tiosalts[8] - are extremely numerous among minerals and have been considered since V.I. Vernadkiyi as products of reaction between tiobases (Na_2S, FeS, Cu_2S and etc.) and tioanhydrides (As_2S_3, Sb_2S_3, Bi_2S_3 and so on). The fact of existence of halogenanhydrides, halogen acids[9], halogenbases and product of their reacting - halogensalts[10] which are also numerous in nature [24] is practically ignored. Talking about bases, anhydrides, acids, salts, it is necessary, in this way, to distinguish chalco- oxy- and halogenbases, chalco- oxy- and halogenanhydrides, chalco-, oxy- and halogenacids, chalco-, oxy and halogensalts [26], [27].

Their forming could be presented by the following reactions:

a) for bases:
$2Na + H_2S \rightarrow H_2 + Na_2S$ (tiobase),
$2Na + H_2O \rightarrow H_2 + 2NaOH$ (hydroxybase),
$2Na + 2HCl \rightarrow H_2 + 2NaCl$ (halogenbase);
b) for acids:
$As_2S_5 + 3H_2S \rightarrow 2H_3AsS_4$ (tioacid),
$SO_3 + H_2O \rightarrow H_2SO_4$ (oxyacid),
$BF_3 + HF \rightarrow HBF_4$ (halogenacid);
c) for salts:
$3Na_2S + As_2S_5 \rightarrow 2Na_3AsS_4$ (tiosalt),
$Na_2O + SO_3 \rightarrow Na_2O_4$ (oxysalt),
$NaF + BF_3 \rightarrow NaBF_4$ (halogensalt).

Every of the enumerated groups of substances has special properties. Thus, typical bases are characterized by ionic cation-ligand bond, ionic homodesmic coordinate structure, and in limit, the most typical cases - good solubility in water and dissociation in solutions by the reactions like $Na_2S \rightarrow 2Na^+ + S^{2-}$, $NaOH \rightarrow Na^+ + OH^-$, $NaCl \rightarrow Na^+ + Cl^-$. Baseses are characterized by low valency (W) of cations, relatively high temperatures of melting point, low vapor resiliency. By increasing of covalent extent of bond cation-ligand, all mentioned properties become less definite.

Typical anhydrides, for example As_2S_5, As_2S_3, SO_3, P_2O_5, BF_3, SiF_4 are characterized by molecular (SO_3, P_2O_5, BF_3, SiF_4) or quasimolecular[11] (As_2S_5, As_2S_3)

---

[8] Tiosalts are referred to chalcosalts, including tio- and seleniumsalts; here and further we use prefix "chalco" not as uniting term, covering compounds of all IVa-elements - O, S, Se, Te, Po, what is unsuccessfully were recommended by "IUPAK chemical nomenclature rules " [52], but rather just as referring to compounds of S and Se one.

[9] In CED [11] halogenacids are referred to so-called beyongacids (superacids, magical acids), defined as "complex nonwater (more correct - nonoxygen - A.G.) mineral acids, acidity of which is higher, then of 100% H_2SO_4" (p. 517); among their examples the products of reaction of halogenanhydrides (AsF_5, SbF_5, BF_3) with protoncontaining nonoxygen acids (HF and so on) are noted.

[10] It should be noted, that chalcobases and halogenbases are at the same time the salts of nonoxygen acids such as sulfurhydrogen, hydrogen chloride and so on, differing by all usual for them properties, particularly by that they easily dissolve in water, dissociating on cations Na^+ and anions S^{2-} (in case of Na_2S) and Cl^- (in case of NaCl).

[11] Term "quasimolecular structure", "quasimolecular crystall" are introduced for solid state substances, intermediate between lowmolecular or simply molecular (SO_3, P_2O_5, BF_3, SiF_4 - in their structures there are small discreet molecules which are connected with each other by residual bond) and polymeric highmolecular compounds. There are indefinite in one (chains and ribbons), two (layers) or even three dimensions (frameworks) in structure of quasimolecular compounds. Quasimoleculas are connected in structure by residuum, covalent-residual and so on bonds. From polymeric moleculas, quasimoleculas are

structure with covalent bond inside molecules (and quasimolecules) and residual bond between them. That is why the most typical of such substances are not dissolve in water or react with it intensively, they have low temperatures of melting points, high vapor resiliency. Highcovalent cations-anionformers (W usually 4, rare 3 like B) are extremely usual for them.

Chalco-, oxy- or halogensalts could be defined as products of reaction of chalco-, oxy- or halogenbases with chalco-, oxy- or halogenanhydrides which are in most typical cases have heterodesmical bond - ionic one between cations and acid residues and ionic-covalent one between anionformer and ligands. Such salts are good dissolving in water with formation of positive charged cations and negative charged complex anions, preserving in solutions the same form as in crystalline solids, although they are often hydrotated in one measure or another. Increasing of the covalent or metallicity extent of bond cation - acid residuum (anionic radical) brings to decreasing in salt solubility[12].

From the crystallochemical point if view the essential feature of salts is heterodesmicity of their structures for which both simple (Na^+, Ca^{2+} and so on) and complex ($[Mg(H_2O)_6]^{2+}$ and so on), including polymer ($[Na(H_2O)_4]^{3+}$ and so on) cations and in obligatory order complex, mono- or heteronuclear, mono- or polymeric, for example, $[SO_4]^{2-}$, $[SiO_4]^{4-}$, $[Si_2O_6]^{4-}$, $[AlSiO_4]^-$, $[B_2O_5]^{4-}$ and so on anions are typical.

Belonging of given substance to simple, binary compound or to salt is determined by fundamental properties of substanceforming atoms - their FC and Z, their ratios, by affiliation of substance to chalcogen, oxygen or halogen compounds. General regularity is that transformation: simple compound $\rightarrow$ binary compound $\rightarrow$ salt, has come by increasing of difference in FC of substanceforming cations, i.e. by increasing of ionic extent of chemical bond; as far as decreasing of their total or average Z, i.e. decreasing of metallicity extent of chemical bond could be favorable for transformation either to one or to another direction.

All what was said could be illustrated by transformation simple compound $\rightarrow$ binary substance $\rightarrow$ salt in sequence of oxygen compounds AO_2 (simple cation A oxide) $\rightarrow ABO_4$ (complex cations A and B oxide) $\rightarrow M[TO_4]$ (oxysalt of M^{n+} cation and tetrahedral anionic radical $[TO_4]^{n-}$ with anionformer T) with dependence on difference in FC between cations (A and B; M and T) and total value of Z, in which cations A(B) and M(T) respect to s- and d-cations (Fig. 1). On considered figure simple oxides AO_2 with rutile structure and binary oxides ABO_4 with dirutile structure placed in the left part with minimal values of differences in FC and with oscillating in wide limits values ΣZ (area I). To the right and upward from this area there is area II - complex oxides with ferberite structure $^{(6)}Fe^{(6)}WO_4$, changing with increasing of differences in FC by area III - oxysalts with scheelite structure $^{(8)}Ca[WO_4]$. Further increasing of differences in FC brings to increasing of cation CN in oxysalts $M[TO_4]$ (changing of areas IV $\rightarrow$ V $\rightarrow$ VI) along with preserving their oxysalt nature.

---

differed by simple structure, saving of their forms in crystallic solid (in solid polymers macromoleculas laid in packages or globulas). Crystallic solids which include quasimoleculas, have not only near, but also far order, as far as in solid polymers far order is absent, that is why they have renthenoamorphic properties.

[12] To salts, as it was already said, referred also substances which could be considered as products of reaction between oxybases and nonoxygen acids like H_2S, HF and so on, for example, Na_2S, NaF and so on which in typical caseses have isodesmical structures, ionic bond, high solubility, dissociating on simple cations and anions and which are, at the same time, chalco- and halogenbases which should be considered, in order to avoid the confusion and because of the matives, mentioned earlier, as such, with distinguishing from other salts.

Analogical regularities were found for great number not only of other oxygen, but also of chalcogen and halogen compounds [26], [27].

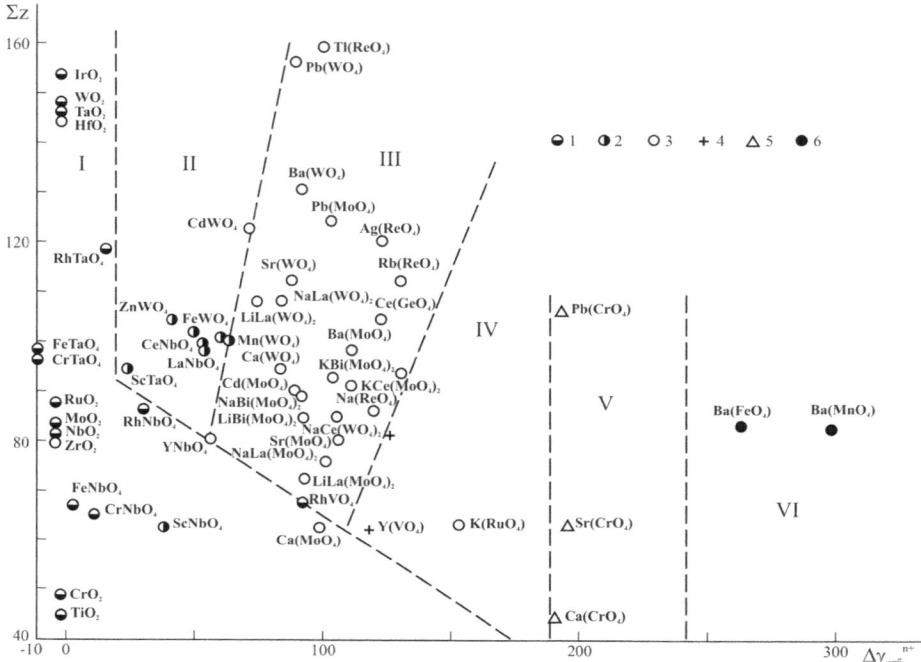

Fig.1 The dependence between the total value of atomic number (ΣZ) of atoms A(B), M(T), difference of their force characteristics ($\Delta\gamma_{opb.}^{n+}$) and the structure of compounds formed by them in sequence of oxygen compounds $2AO_2 \rightarrow ABO_4 \rightarrow M[TO_4]$.

(T)= d- cations; I – area of crystals with rutile (TiO_2 and so on.) or dirutile ($FeNbO_4$ and so on) structures, II – area of crystals with structure of ferberite $FeWO_4$, III – of scheelite $Ca[WO_4]$ type, IV – of type of $Y[VO_4]$, V – of type of $Pb[CrO_4]$, VI – of type of $Ba[FeO_4]$; the type of structure: 1 – TiO_2, 2 – $FeWO_4$, 3 – $Ca[WO_4]$, 4 – $Y[VO_4]$, 5 – $Pb[CrO_4]$, 6 – $Ba[FeO_4]$.

All what was said tells not only about considerable differences in crystalline structures and properties of simple compounds, complex compounds, salts, but also about their natural change in mentioned order as long with changing of FC and Z, and creates the basis for classification of substances in limits of chalcogen, oxygen and halogen compounds.

Geochemical classification of elements and cations

The first highly successful geochemical systematic of elements which haven't lost its significance by now was element separating on geochemical (genetic) groups, offered 70 years ago by V.M. Goldschmidt [32]. This systematic he has based from one side on division of elements along with fusion of metals on metallic melt - sulfide melt, harden as matte and silicate melt, harden as slag, taking this process as a model of Earth substance differentiation in development process. From the other side he has paid the attention that elements which are typical for mentioned metallurgy products, have taken not accidental, but rather defined position in curve of dependence of atomic volumes of elementary substances upon atomic (ordered) element number[13] (Fig. 2).

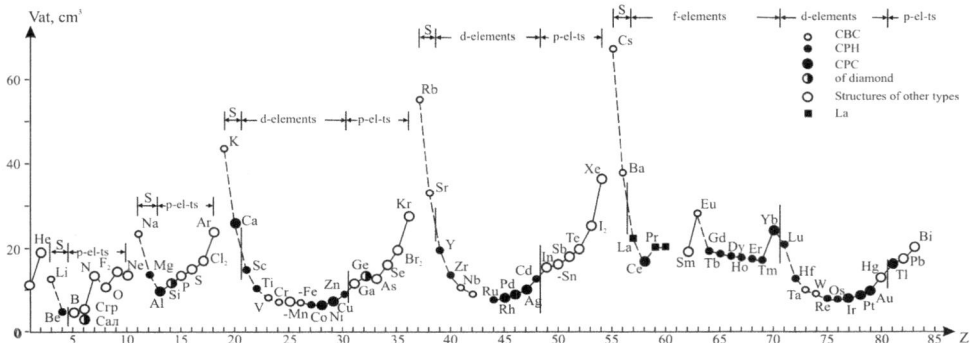

Fig. 2 The variation of atomic volume of elementary substances (V) by increasing of atomic number (Z) of corresponding elements.1 – cubic body-centered, 2 – close-packing hexagonal, 3 – close-packeing cubic, 4 – of diamond, 5 – ionic type structure, 6 – structures of La.

On this basis V.M. Goldschmidt has divided all elements on 4 groups:

1. *Siderophylic* elements, placed in minimal parts of curves of dependence of atomic volumes upon atomic number. To them he refered C, P, Fe, Co, Ni, Tc(Ma), Mo, Ru, Rh, Re, Os, Ir and Pt.
2. *Chalcophylic* elements, placed in ascending parts of curves of dependence of atomic volumes upon atomic number - S, Cu, Zn, Ga, Ge, As, Se, Pd (more correctly it should be refered to siderophylic elements), Ag, Cd, In, Sn, Sb, Te, Au, Hg, Tl, Pb, Bi и Po.
3. *Lithophylic* elements, placed in descending parts of curves of dependence of atomic volumes upon atomic number - Li, Be, B, Na, Al, Si, K, Ca, Sc, Ti, V, Cr, Mn, Rb, Sr, Y, Zr, Nb, Cs, Ba, Ln, Hf, Ta, W, Fr (it has not been placed yet on V.M.Goldschmidt's curve), Ra, Ac, Th, Pa, U; here he refered also O, F, Cl, Br, I, placed on ascending parts of considered curves, taking into account commonness of these elements in minerals of rocks and doing deviation from logical-formal in favor to natural element classification.

---

[13] This curve, firstly used by D.I. Mendeleev [48] and L. Mayer [49], usually because of substitution of terms element and elementary (simple) substance, incorrectly called "curve of atomic volumes of elements". Along with it, it is forgotten that in its basis have been put only one property of elements - their ordered numbers, as far as values of atomic volumes are referred to elementary substances, but not to elements and their utilization with clarifying of element properties is not correct enough, that was discussed in detail before [18], [19].

4. *Atmophylic* elements, placed in upward parts of curves of dependence of atomic volumes upon atomic number - He, N, Ne, Ar, Kr, Xe, Rn; here he refered H too.

Not disclaiming successfulness of offered by V.M. Goldschmidt terms which are correspond to objective existing regular connections of element structures and their behavior in natural processes, expressed, in particular, on mentioned figure, what was determined their vitality, it is necessary to stress, that such general conclusion of element properties inevitably brings us to disputable places of some of them, especially those ones which could be refered equally to different groups. We could note more contradictions which have roots in fact that curve of atomic volumes, used for systematic, was built with using only one element property - their order numbers. All other used values - atomic volumes - are refered not to elements - totality of atoms of one sort in free state, but rather to properties of elementary substances, depending upon their crystalline structure, physical-chemical parameters which are defined the stability of one or another concrete polymorphic modification.

Definite discrepancies of his systematic of elements to actual material V.M. Goldschmidt has felt himself soon. In accordance with it in 1928 year [33] he has already noted, that element comparing on atomic volumes is correct only if they are stated at the same valency condition and have the same CN. The same thing was noted by A.E. Fersman [14].

Definite limited nature of element classification offered by material V.M. Goldschmidt has been token into account by V.V. Shcherbina [61], who stressed the importance of magnetic properties of elementary substances, type of cations, formed by given element for classification of elements. In connection with it he wrote: "… being of certain element in one or another group (considering geochemical groups of elements by Goldschmidt - A.G.) not disturbs it to clarify its peculiarities which could be typical for other groups. Thus, for iron it could be said, that it possess both lithophylic and chalcophylic and siderophylic properties, but, taking in mind, that world consists of metallic iron, that iron has a lot of properties of siderophylic elements, it could be refered to siderophylic elements. Tin, owing to paramagnetic properties, could be found mainly in form of cassiterite in association with lithophylic elements in nature, but by several of its properties (atomic structure, formation of complex sulfides and so on) it could be refered to chalcophylic elements. Presence of small quantities of tin in many sulfide minerals confirms this point of view.

Arsenic, when it plays cation part, i.e. in form of As_2S_3 and its compounds, undoubtedly is chalcophylic. Vice versa, anionic arsenic in speryllite $PtAs_2$, smaltite $CoAs_2$ and so on characterize the siderophylic nature of these elements. Halloids, laid on ascending curve, refered to the typical lithopylic elements.

Palladium, owing to the sequence of its properties, is refered to chalcophylic elements by many authors, although, perhaps, more correctly to refer it to siderophylic group.

Molybdenum, inspite of its primary place in nature exclusively in form of sulfide - molybdenite[14] - nevertheless could not be refered to chalcophylic elements, because molybdenum is paramagnetic (this is about elementary substance, but not about element - A.G.), not forming ion with 18 outer-shell electrons, not forming complex sulfides like $R_2S\,MoS_2$ or $RS\,MoS_2$; molybdenite frequently occurs without escort of other chalcophylic elements in nature. At least, litho-, chalco- and siderophylic properties,

---

[14] V.V. Shcherbina failed to take in consideration the existence of different complex oxides of Mo and W, molybdates and tungstenates in nature. Along with it in some tungstenates, for example in scheelite, the quantity of isomorphous dashes of Mo achieves industrial significant values.

based on ionic structure, are periodical properties too. Lithophylic elements changed by siderophylic, and siderophylic - by chalcophylic, that is why element, placed between litho- and siderophylic elements, could be refered to chalcophylic elements hardly. As for its affinity with sulfur, it could be explained by nontypical placement of outer-shell electrons in molybdenum atom." (p. 32-33).

Unsatisfaction in V.M. Goldschmidt's systematic brings to development of personal geochemical systematics by several scientists. Particular interesting were those of them which were offered by A.E. Fersman [14] and A.N. Zavaritskyi [68].

In Fersman's systematic we should stress the selection of typical elements (i.e. elements with cenosymmetrical outer-shell and subouter-shell electrons). This were done by separating of these elements from the rest by double horizontal line. In upward part of table there are all typical elements from first three periods, and elements from K up to Ni including too. By this the special properties of majority of elements, refered to first seria of d-elements which refered, as it has been established later [62], to cenosymmetricals too, were pointed out.

Geochemical classification by A.N. Zavaritskyi differs by the greatest consequence and detailness. It is based, from one side, on element properties which are reflected in extended variant of Periodic system, from the other side - on enormous personal experience of outstanding petrographier and petrologer. He selected in limits of Periodic system 11 element areas:

1. Hydrogen area - H, stressing its special role in minerals.
2. Noble gas area - He, Ne, Ar, Kr, Xr, Rn.
3. Rock elements area - Li, Na, K, Rb, Cs, Be, Mg, Ca, Al, Si.
4. Area of elements of magnetic emanations - B, C, N, O, P, S, F, Cl.
5. Iron group elements area - Ti, V, Cr, Mn, Fe, Co, Ni. Touching this group A.N.Zavaritskyi noted, that "geological processes of iron ore forming are much closer to processes of rock forming, then to processes of forming of other element ores. Such ores are magmatic iron ores and ores of sediment genesis" (p. 16).
6. Rare element area - Sc, Y, Ln, Zr, Hf, Nb, Ta, as that A.N.Zavaritskyi noted definite convention in uniting of elements in this group.
7. Radioactive elements area - Fr, Ra, Ac, Th, Pa, U.
8. Platinum group elements area - Ru, Rh, Pd, Os, Ir, Pt.
9. Metallic (colour) elements area - Cu, Ag, Au, Zn, Cd, Hg, Ga, In, Tl, Ge, Sn, Pb.
10. Metalloid metalogenic elements (elements "sulfoacids") area - As, Sb, Bi, Se, Te, Po.
11. Heavy halloids area - I, Br, At.

Mo, W, Tc and Re remained without selected areas in A.N. Zavaritskyi's systematic.

Separately we need to note, that systematic by A.N. Zavaritskyi, as he noted himself, is by many features close to technical classification of elements by G.Berg, developed on the basis of exceptionally industrial (technical) features, what is consolidates the importance of both of them.

Our mineralogy-crystallochemical systematic of elements is prove to be close to considered earlier. This systematic developed with taking into account recent data on peculiarities of electronic structure of elements - studies about ceno- and noncenosymmetrical electrons, elements, different extents of closeness (analogy) in properties of elements as a dependence on their electronegativeness, expressed by force characteristics - FC and order number of element Z [5], [26], [31], although it has some essential differences. It is comfortable to combine selected in this systematic 13 areas of elements into more important groups, using terminology of V.M. Goldschmidt, changing, however, groups filling with considering of mentioned data. In this case we will have the following element systematic:

1. Hydrogen - H. Selecting H into separated group is corresponding to systematic by A.N. Zavaritskyi.
2. Lithophylic elements with low FC.
    2.1 Alkaline and alkaline-earth elements - Li, Na, K, Rb, Cs, Fr; Mg, Ca, Sr, Ba.
    2.2 Rare earth and radioactive elements - Sc, Y, Ln (La - Yb), Th, U.
    2.3 Amphoteric elements - Be, Al, (Ga)
    2.4 Cenosymmetrical d'-elements - Ti, V, Cr, Mn, Fe, Co, Ni - group of elements fully analogical to iron group of A.N. Zavaritskyi; Ti connected by diagonal similarity with Nb and Ta, refered to the next group 3.1. The isomorphism between them and commonness of Ti dashes in Nb and Ta minerals and vice versa could be explained by this.
3. Lithophylic elements with middle FC
    3.1 Noncenosymmetrical d'-complexformers - Zr, Hf, Nb, Ta.
    3.2 Mo and W.
4. Noble-metallic (siderophylic) elements - Ru, Rh, Pd, Ag, Os, Ir, Pt, Au; among them Ag and Au often are found in chalcophylic minerals and associations.
5. Chalcophylic elements.
    5.1 Chalcophylic elements with low FC - Cu, Zn, Cd, Hg, (Ga), In, Tl, Pb.
    5.2 Chalcophylic elements with middle FC - Ge, Sn, As, Sb, Bi, Se, Te. Groups 5.1 and 5.2 are close to groups 9 - metallic (colour) elements and 10 - metalloid metalogenic elements (elements "sulfoacids") of A.N. Zavaritskyi respectively, with the exception of Ge and Sn which he refered to group 9.
6. Light anionformers - B, C, Si, N, P, O, S, F, Cl - group of elements fully analogical to group of elements of magmatic emanations by A.N. Zavaritskyi, included in form of anionformers (B, C, Si, N, P, S) or anions (O, F, Cl) into lithophylic minerals, and only S plays, along with it, anion role in chalcophylic minerals.
7. Heavy anionformers - Br and I.
8. Noblegases elements - He, Ne, Ar, Kr, Xe, Rn.

This mentioned systematic, however, needs in additional specifications, because, as it was said earlier [26], [27], the majority of elements are amphoteric and their acid-basic properties are defined by properties of other elements, included in compound, their ratios, physical-chemical parameters of systems in which this mineral was formed or placed. It was shown, that the considerations about amphoterity in use for crtystallic substances, should be based on values of CN of cations. General regularity here is that increasing of cation's CN brings to strengthening of its basic properties, and decreasing of its CN, opposite, - to increasing of its acid properties. Thus, the cations of typical siderophylic elements could in crystallochemical relation be the analogies of typical lithophylic elements, and cations of typical chalcophylic elements could be the analogies of typical lithophylic elements. In this connection, we could point-out Mn^{2+}, Fe^{2+}, Co^{2+}, Ni^{2+} which along with CN = 6 are the crystallochemical analogies for Mg^{2+}, what is clarifying in wide isomorphism among them, formation not only isoformular, but also isostructural compound. Simultaneously, Pb^{2+} with CN 12 is crystallochemical analogue of Ba^{2+}, K^+ in minerals like hollandite, Ca^{2+} - in makedonite ($PbTiO_3$) which is close to perovskite structure ($CaTiO_3$); Tl^+ with CN = 12 is analogue of typical lithophylic alkaline K^+ in isostructural pair djerfisherite - thalfenisite. There are a lot of such examples, what is difficulties using of generalized systematic of elements (cations) like considered before.

Differences in acid-alkaline properties of cations with different CN is comfortable to express by set values of FC - their references to CN - FC/CN which along with the electronic type of cation (s-, f-, d-, p-), its order number allow to come to very important conclusions. Let's see it in detail in order, replying for transformation from s-, f- through d- to p-elements.

Elements of s- and f-type in majority give cations with so low values of FC/CN, that almost all of them could be refered to cations with alkaline or clear basic properties (Fig.3).

Particular place among s-cations occupies Li^+, for which with CN = 4 FC/CN = 7,1, Mg^{2+} which has FC/CN with CN = 6 equal to 10,2, and with CN = 4 - 15,3; and Be^{2+} which FC/CN with CN = 6 equal to 21,8, and with CN = 4 - 32,8. All mentioned peculiarities of these cations clarify in their special behavior in compounds. Thus, Li^+ is not isomorphous to other s-cations of Ia-group with lower FC/CN. At the same time, its compounds by their properties and mineralforming conditions are clearly differ even from isostructural minerals. For example, we could cite holmquistite, lepidolite and spodumene which are in this connection "rara avises" in families of amphiboles, micas and pyroxenes respectively.

The role of $^{(6)}Mg^{2+}$ in minerals is also original. In many of them it shows isomorphysm up to full one with cations like $^{(6)}Fe^{2+}$, $^{(6)}Mn^{2+}$, but not with other IIa-cations, in sequence of minerals (for example, in micas, chlorites) it is isomorphous in wide limits to $^{(6)}Al^{3+}$. But cations of $^{(4)}Mg^{2+}$ could be isomorphous even to cations like $^{(4)}Al^{3+}$ or they could play crystallochemical role, simultaneous to last ones which is appeared, for example, in oxyspinellides. By this way, in crystallochemical relation in Mg^{2+} the transformation to amphoteric cations is clarified.

The typical amphoteric s-cation is Be^{2+}, for which in minerals CN = 4 is usual (FC/CN = 32,8). The formation of minerals like beryllosilicates, characterized by that they contain heteronuclear beryllium-silicium radicals, beryllophosphates with mixed heteronuclear beryllum-phoshatoes radicals and so on minerals, is directly connected with this.

The U^{6+} for which more then 180 mineral species are known with U^{6+} in form of ion uranyl $(UO_2)^{2+}$, has special position among considering f-cations. Among them there are small number of respectively simple compounds, as far as the majority of them has properties of salts of weak acids, in which uranile ion is included in complex anion radicals. This has been established in uranates and in numerous salts of uranyl-acids. Because by now there is no precise values of FC of U^{6+}, in order to determine the FC/CN ratio we have to take two extreme values of FC. Along with it, it is important, that even with respectively high values of CN (8 = 2 + 6 and 6 = 2 + 4) they are noticeable higher then analogue values of almost all considered s- and f-cations, except $^{(4)}Be^{2+}$. Thus, all considered cations could be divided on two types 1)lithophylic s- and 2)lithophylic f-cations which are clearly isolated on mentioned Fig. 3.

As it was known, two d-elements - Sc and Y are close by their properties to lantanides. Their closeness is so great, that they are usually situated in the same minerals in form of isomorphous mixtures and they could be hardly separated by chemical ways from

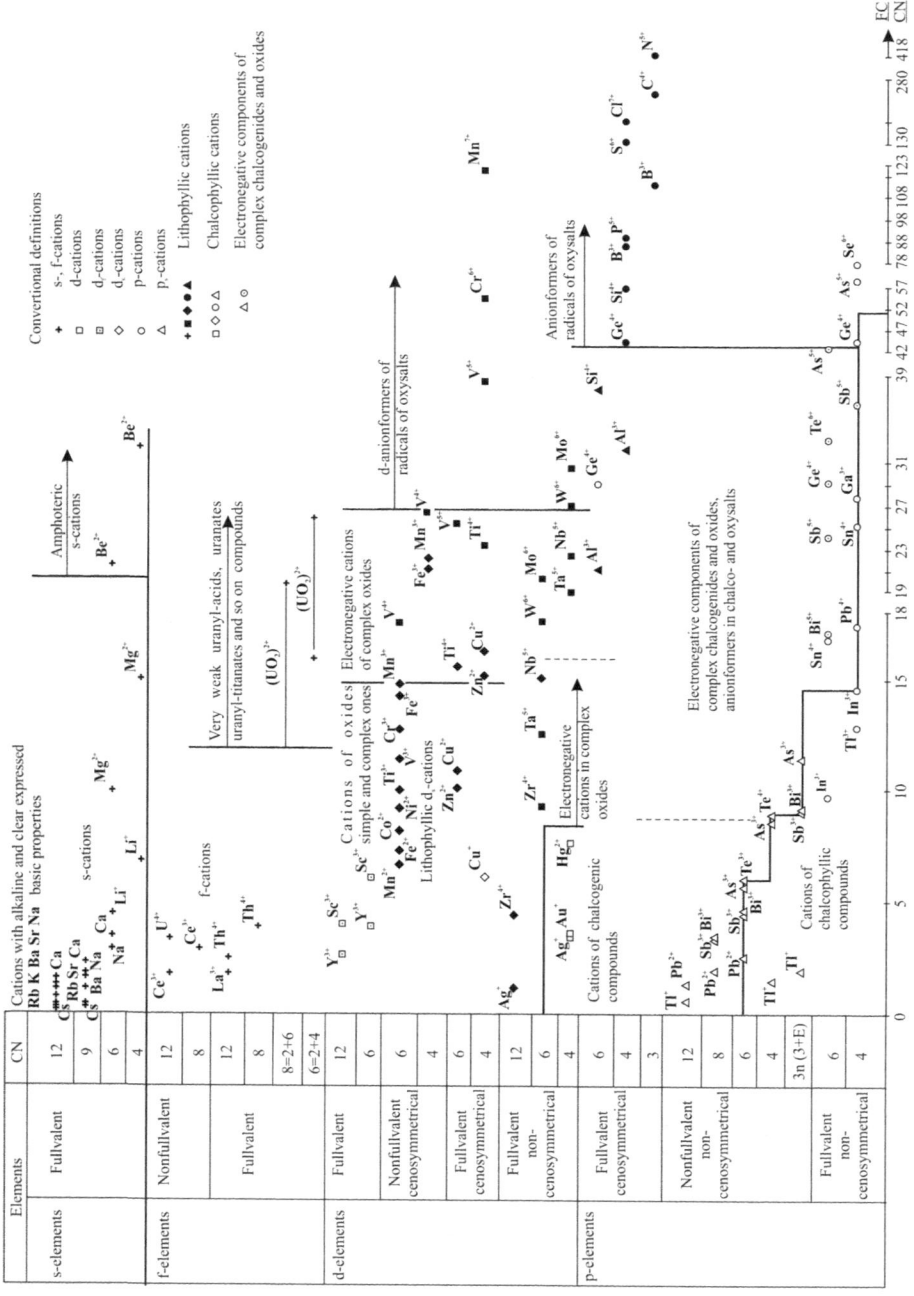

Fig. 3 The connection of reduced force characteristics (FC/CN) with acid-basic properties of cations.

each other. All this bring to association of all considered elements to rare-earth's group, for designation of which symbol TR was accepted. The comparison of values of FC/CN of Sc^{3+} and Y^{3+} with these values of lantanides cations which was done in the same Fig.3, shows their pretty beg closeness. Considering all that was said, these cations are designated as lithophylic d_f-cations. It shouldn't be forgotten, that maximal closeness with f-cations they have only with high CN, as long as with CN = 6, for example, Sc^{3+} could play crystallochemical role of $^{(6)}Al^{3+}$ which is presented in bazzite $(Sc,Al)_2$ Be_3 $[Si_6O_{18}]$ which is crystallochemical analogue of beryl Al_2 Be_3 $[Si_6O_{18}]$.

Significantly more complex are the connections of values of FC/CN with acid-basic properties in other d-cations (Fig.3) (platinoides are excepted from consideration, because of their original properties and differences in behavior in minerals from majority of other elements). This is determined either by great variety of their valences and CN or by their belonging not only to lithophylic, but also to chalcophylic elements (cations).

Here first of all let's stop on nonfullvalent cations of cenosymmetrical d-elements which have been associated before by A.N. Zavaritskyi to group of iron, and by us - to group of cenosymmetrical d'-elements. In Fig. 3 it is clear, that by valency and value of FC/CN such cations as $^{(6)}Mn^{2+}$ (6,7), $^{(6)}Fe^{2+}$ (7,4), $^{(6)}Co^{2+}$ (8,3), $^{(6)}Ni^{2+}$ (9,3) are close to $^{(6)}Mg^{2+}$ (10,2). This is presented in isomorphism up to full one between these cations in many of minerals. Close values of FC/CN have also threevalent $^{(6)}Ti^{3+}$ (10,1), $^{(6)}V^{3+}$ (11,5), $^{(6)}Cr^{3+}$ (12,9), $^{(6)}Fe^{3+}$ (14,4), $^{(6)}Mn^{3+}$ (14,9), for which wide isomorphism both with each other and with $^{(6)}Al^{3+}$ (21,4) (oxyspinellides, pyroxenes, garnets and so on) is also noticed. It should be pointed-out, that in many of minerals there is a wide isomorphism between considered two- and threevalent cations (especially in micas, chlorites) too.

By value of FC/CN threevalent $^{(6)}Ti^{3+}$, $^{(6)}V^{3+}$, $^{(6)}Cr^{3+}$, $^{(6)}Fe^{3+}$, $^{(6)}Mn^{3+}$ are also close to $^{(4)}Mg^{2+}$ (15,3), what is explaining their close crystallochemical role in minerals like oxyspinellides. $^{(6)}Ti^{4+}$ (15,8), $^{(6)}V^{4+}$ (17,8), $^{(4)}Fe^{3+}$ (21,6), $^{(4)}Mn^{3+}$ (22,4), $^{(4)}Ti^{4+}$ (23,7), $^{(6)}V^{5+}$ (25,6), $^{(4)}V^{4+}$ (26,7) are playing different crystallochemical role. They are presented in complex oxides, for example, as electronegative amphoteric cations with very weak acid properties, weak anionformers in minerals like ferrisilicates getting closer in this relation to $^{(4)}Be^{2+}$ (32,8) и $^{(4)}Al^{3+}$ (32,2).

Here we should note, that by value of FC/CN among cations of Ti even $^{(4)}Ti^{4+}$ (23,7) is far from typical anionformers. That is why the respective substances which have not been known as minerals yet should be considered as complex oxides. Use of accepted in chemistry name titanates, considering salts of titanium acids, is very probational.

Considering mentioned closeness of these cations with such s-cations as $^{(6)}Mg^{2+}$, $^{(4)}Mg^{2+}$, $^{(4)}Be^{2+}$ and belonging of respective elements to lithophylls, all of them are designated by us as d_s-cations.

To d_s-cations also refered $^{(6)}Zn^{2+}$ (10,2) and $^{(6)}Cu^{2+}$ (11,0) which are close by FC/CN to $^{(6)}Mg^{2+}$ (10,2) and $^{(4)}Mg^{2+}$ (15,3), by what their close crystallochemical role, for example in oxyspinellides, pyroxenes and some other minerals is explained.

Special position on Fig.3 have cations of elements, selected by us as noncenosymmetrical d'-complexformers - Zr, Hf, Nb and Ta. Inspite of low values of FC/CN, these cations rarely play role of cations of simple compounds in minerals, for example, in simple oxides, zircon and so on. The role of electronegative cations in complex compounds like zirkonates, tantaloniobates, where they, however, are not reaching the role of typical anionformers, are more characterizing for them. Their inclination for complexforming is clarifying in formation of mixed anionic radicals of zircono- and tantalosilicates. By these features Ti^{4+} is very close to them. It is also forming minerals like titanosilicates, in which isomorphous substitutions between Ti and Nb, Ta are presented, what is reverberation of diagonal similarity Ti with Nb and Ta which was mentioned before.

By this way, Ti is presented as element with dual crystallochemical peculiarities - from one side it is one of the cenosymmetrical d'-elements, from another one - it comes forward as equitable member of noncenosymmetrical d'-complexformers.

The role of electronegative cations in complex oxides, rushing for role of anionformers in oxysalts is more clearly presented in Mo^{6+} , W^{6+}, especially with CN = 4, responding to molybdates and tungstenates. At the same time it should be stressed, that anhydrides of respective acids are solid substances with high melting temperatures, they are hardly dissolve in water by what they are completely differ from anhydrides of typical oxyacids. They have considerable lower values of FC/CN (see Fig. 3). All this shows their natural connection with complex oxides and unnaturalness of their removing to principle different taxons, how it, for example, was done by A.S. Povarennykh [55], who considered minerals like tungstenates with CN of Mo^{6+} и W^6 equal to 6 among complex oxides, and minerals like scheelite with CN of Mo^{6+} и W^{6+} equal to 4 among complete different place in systematic - among oxysalts which are close to sulfates, with which they have not any genetic connections. They are associated with them only by formal features - similarity in CN and valences of anionformers. The same thing we could say about dragging off the vanadates with CN V^{5+} = 4 from other closely connected with them minerals of V^{5+} in which it has CN = 5 or 6 and consideration of them in common or closely with minerals of phosphates class only by formal features - isoformularity of respective acids, same valency of anionformers, similarity of their CN.

Thus, cenosymmetrical $^{(6)}Ti^{4+}$, $^{(6)}V^{4+}$, $^{(4)}Fe^{3+}$, $^{(4)}Mn^{3+}$, $^{(4)}Ti^{4+}$, $^{(6)}V^{5+}$, $^{(4)}V^{4+}$ and noncenosymmetrical $^{(6)}Zr^{4+}$, $^{(6)}Ta^{5+}$, $^{(6)}Nb^{5+}$, $^{(6)}W^{6+}$, $^{(4)}Ta^{5+}$, $^{(6)}Mo^{6+}$, $^{(4)}Nb^{5+}$, $^{(4)}W^{6+}$, $^{(4)}Mo^{6+}$ are present the group of amphoteric cations with properties varying from weak basic ones with CN = 6 up to weak acid ones with CN = 4, ending on $^{(4)}V^{5+}$ which shows the most acid properties. At the same time, fullvalent $^{(4)}Cr^{6+}$ and $^{(4)}Mn^{7+}$ have strong enough acid properties, what is expressed in their role as anionformers of strong chromic and manganic acids and their salts.

Considering what was said, it is expedient to divide all mentioned d-cations on three groups by their crystallochemical role in minerals:

a) Lithophylic d-cations with   low FC - $^{(4)}Fe^{3+}$, $^{(4)}Mn^{3+}$. It is convenient to unite them   with s-cations, considering them as d_s-cations, because they are crystallochemically closely connected with other three- and bivalent d_s-cations, what is presented in paragenesis of respective minerals, placed in association or appropriately changing each other by changing of oxidizing potential Eh. At the same time it should not be forgotten about significant role of $^{(4)}Fe^{3+}$ in chalcogen minerals.

b) Lithophylic d-cations with middle FC - $^{(6)}Ti^{4+}$, $^{(6)}V^{4+}$, $^{(4)}Ti^{4+}$, $^{(6)}V^{5+}$, $^{(4)}V^{4+}$, $^{(6)}Zr^{4+}$, $^{(6)}Ta^{5+}$, $^{(6)}Nb^{5+}$, $^{(6)}W^{6+}$, $^{(4)}Ta^{5+}$, $^{(6)}Mo^{6+}$, $^{(4)}Nb^{5+}$, $^{(4)}W^{6+}$, $^{(4)}Mo^{6+}$, $^{(4)}V^{5+}$. They form both simple and complex oxides which by their composition, sequence of properties and especially by their genesis are closely connected with oxysalts, ending this sequence. Dragging off the last ones by high constants of respective acids feature and consideration of these classes of minerals from this basis with oxysalts of p-anionformers (phosphates and sulfates), as it was usually done by logical-formal motives, is absolutely unnatural. Middle values of Eh, needed for formation of such minerals, their low solubility, character of paragenesis are in good agreement with it.

c) Lithophylic d-cations with high FC - $^{(4)}Cr^{6+}$ и $^{(4)}Mn^{7+}$ which are d-anionformers of strong acids, consideration of which should end the oxygen compounds of lithophylic d-cations.

In addition to what was said let's cite the sequence of considered classes of oxysalts by increasing of pKa of respective acids:

Class (4) - Vanadates (3,74; $H_3[VO_4]$)
Class (4) - Molybdates (2,54; $H_2[MoO_4]$)
Class (4) - Tungstenates (2,20; $H_2[WO_4]$)
Class (4) - Chromates (0,80; $H_2[CrO_4]$)

As d_s-cation in some minerals could be $^{(12)}Ag^+$ (1,2), playing, for example, in argentojarosite role of $^{(12)}K^+$ (0,61), $^{(4)}Zn$ (15,3) и $^{(4)}Cu$ (16,5) which are usually typical chalcophylic cations and just sometimes play role of d_s-cations, for example, in oxyspinellides.

The rest of d-cations with CN = 4 and lower are typical chalcophylic.

The same features are in the basis of p-cations systematic (Fig. 3) which could be united by their role in minerals in 5 groups:

a) p_s-cations of lithophylic elements - $^{(6)}Al^{3+}$ и $^{(4)}Al^{3+}$ which are close crystallochemically, as it have already been pointed-out, to $^{(6)}Mg^{2+}$ and especially to $^{(4)}Mg^{2+}$ and $^{(4)}Be^{2+}$. Here by value of FC/CN we could refer $^{(6)}Ge^{4+}$ and $^{(6)}Si^{4+}$. However, $^{(6)}Si^{4+}$ is stable only at very high pressures (stishovite with structure of rutile) and it appears extremely rarely in composition of minerals, formed in Earth crust. From them we could point only on one certainly established mineral - thaumasite with formulae $Ca_3Si(OH)_6 [CO_3][SO_4] 12H_2O$. And if we take in consideration narrow genetical connection of minerals $^{(6)}Ge^{4+}$ and $^{(6)}Si^{4+}$ with minerals $^{(4)}Ge^{4+}$ и$^{(4)}Si^{4+}$, refered to very weak anhydrides of silicium and germanium acids and to their salts, it is expedient, because of the same purposes which were discussed previously, to consider the minerals of $^{(6)}Ge^{4+}$ и $^{(6)}Si^{4+}$ with minerals of $^{(4)}Ge^{4+}$ и $^{(4)}Si^{4+}$, separating them from those of $^{(6)}Al^{3+}$ и $^{(4)}Al^{3+}$. We should not forget about definite crystallochemical closeness of $^{(4)}Al^{3+}$ and $^{(4)}Si^{4+}$, presented in fact of existence of numerous alumosilicates with heteronuclear alumosilicium polymeric radicals.

b) Lithophylic anionformers of oxyacids, to which fullvalent cations of p-elements, separated previously as light anionformers - $^{(4)}Si^{4+}$, $^{(4)}B^{3+}$, $^{(4)}P^{5+}$, $^{(3)}B^{3+}$, $^{(4)}S^{6+}$, $^{(4)}Cl^{7+}$, $^{(4)}C^{4+}$, $^{(4)}N^{5+}$ are refered. Here especially should be noted difference in CN which could have B, owing to its properties as anionformer are considerably changing - $^{(4)}B^{3+}$ by these features is placed between $^{(4)}Si^{4+}$ и $^{(4)}P^{5+}$, as far as $^{(3)}B^{3+}$ - between $^{(4)}P^{5+}$, and $^{(4)}S^{6+}$. Besides that, there are many minerals, consisting of polymeric radicals with $^{(4)}B^{3+}$ and $^{(3)}B^{3+}$ at the same time and which are considerably differ by their properties and forming conditions, paragenesis from other borates. To the same group we could refer the salts of $^{(4)}S^{4+}$ - sulfites, because inspite of low value of FC/CN of this anionformer (17.8), sulfuric acid is refered to strong acids (pKa=1.85). And its salts could be considered in connection with sulfates, from which they differ first of all by lower value of Eh, needed for their formation.

c) p_s-cations of chalcophylic elements - $^{(12)}Tl^+$, $^{(12)}Pb^{2+}$, $^{(8)}Pb^{2+}$,$^{(8)}Sb^{3+}$ и $^{(8)}Bi^{3+}$. Among them $^{(12)}Tl^+$ is, as it have already been said, crystallochemical analogue of $^{(12)}K^+$ in pair djerfisherite - thalfenisite and so on minerals; $^{(12)}Pb^{2+}$- $^{(12)}Ca^{2+}$ in macedonite; $^{(12)}Ba^{2+}, ^{(12)}K^+$ - in minerals like hollandite; $^{(8)}Pb^{2+}$, $^{(8)}Sb^{3+}$ and $^{(8)}Bi^{3+}$ - $^{(8)}Ca^{2+}$, $^{(8)}TR^{3+}$ in minerals with structure of pyroclore. The same cations are included in composition of some other lithophylic minerals.

d) Chalcophylic p-cations with low FC, - $^{(4)}Tl^+$, $^{(3)}Tl^+$, $^{(6)}Pb^{2+}$, $^{(6)}Sb^{3+}$, $^{(6)}Bi^{3+}$, $^{(6)}As^{3+}$, $^{(6)}Te^{4+}$, $^{(4)}As^{3+}$, $^{(4)}Te^{4+}$, $^{(6)}In^{3+}$, $^{(4)}Tl^{3+}$, $^{(4)}In^{3+}$, $^{(6)}Sn^{4+}$, $^{(4)}Pb^{4+}$, $^{(4)}Sn^{4+}$, $^{(4)}Ga^{4+}$, coming forward as electropositive components in simple and complex chalcogen compounds, chalcosalts, in which CN of many cations is increasing, for example, for Pb^{2+} up to 7 and even 8, for Cu^+ up to 6 an so on; and also in oxygen minerals, refered to arsenites, ..., tellurites, tellurates.

e) Chalcophylic p-cations with middle FC - $^{(3+E)}Sb^{3+}$, $^{(3+E)}Bi^{3+}$, $^{(3+E)}As^{3+}$, $^{(6)}Bi^{5+}$, $^{(6)}Sb^{5+}, ^{(6)}Te^{6+}, ^{(4)}Sb^{5+}, ^{(6)}As^{3+}$, coming forward as electronegative components in complex sulfides and oxides, anionformers of chalcosalts → oxysalts like arsenites, ..., tellurites and tellurates. Along with it by their properties and conditions of formation the last minerals are rather close to respective complex compounds, then to typical oxysalts like phosphates and sulfates, among which they usually are considered without any sense and basis. In this relation they are resembling considered earlier tantaloniobates, molybdates, tungstenates, vanadates.

f) Chalcophylic anionformers of oxysalts radicals - $^{(4)}Ge^{4+}, ^{(4)}As^{5+}, ^{(4)}Se^{6+}$. Oxysalts, responding to such acids because of crystallochemical logical-formal similarity usually are refered in systematic just behind silicates, phosphates and sulfates respectively. Conditionally selenites are inserted here too which inspite of that anionformer $^{(4)}Se^{4+}$ has low value of FC/CN are the salts of strong enough acids, resembling in this relation sulfites with anionformer $^{(4)}S^{4+}$. However, taking in mind the conditions of formation of these minerals mainly because of oxidizing the chalcogen minerals, narrow their chemical and paragenetic connection with minerals, formed by elements of two previous groups, the most natural is to place this group at the end of our consideration of chalcophylic elements. Along with it, the following sequence of mineral classes depending on strength of their acids should be also contemplated:

Class (6) - Arsenites (9,23; H_3AsO_3)
Class (4) - Germanates (9,10; $H_2[GeO_4]$) - are considered just behind silicates
Class (6) - Tellurates (7,61; H_6TeO_6 or $Te(OH)_6$)
Class (4) - Tellurites (2,57; H_2TeO_3)
Class - Selenites (2,75; H_2SeO_3)
Class (4) - Arsenates (2,25; $H_3[AsO_4]$)
Class (4) - Selenates (1,92; $H_2[SeO_4]$)

Resuming and taking into account some well-known facts, the systematic of the most important ions, including in minerals could be presented in the following form (the values of FC/CN are pointed-out in brackets):

1. Hydrogen and hydrogencontaining ions, neutral molecules - H^+, $(H_3O)^+$, NH^+_4; $(OH)^-$; H_2O.
2. Cations of lithophylic elements with low FC, including:
   a). s-Cations - $^{(12)}Cs^+$ (0,35), $^{(12)}Rb^+$ (0,47), $^{(9)}Cs^+$ (0,47), $^{(12)}K^+$ (0,61), $^{(9)}Rb^+$ (0,63), $^{(12)}Ba^{2+}$ (0,96), $^{(12)}Sr^{2+}$ (1,3), $^{(9)}Ba^{2+}$ (1,3), $^{(12)}Na^+$ (1,5), $^{(12)}Ca^{2+}$ (1,8), $^{(9)}Sr^{2+}$ (1,8), $^{(9)}Na^+$ (2,0), $^{(9)}Ca^{2+}$ (2,5), $^{(6)}Na^+$ (3,1), $^{(6)}Ca^{2+}$ (3,7), $^{(6)}Li^+$ (3,7), $^{(4)}Li^+$ (7,1), $^{(6)}Mg^{2+}$ (10,2), $^{(4)}Mg^{2+}$ (15,3), $^{(6)}Be^{2+}$ (21,8), $^{(4)}Be^{2+}$ (32,8).

b). f-Cations – $^{(12)}Ce^{3+}$ (2,0), $^{(12)}La^{3+}$ (2,0), $^{(12)}Th^{4+}$ (2,7), $^{(8)}Ce^{3+}$ (3,1), $^{(12)}U^{4+}$ (3,65), $^{(8)}Th^{4+}$ (4,1); special properties has U^{6+} which is found in minerals mostly in form of ion of uranile $(UO_2)^{2+}$, forming weak uranyl acids, uranates and numerous uranyl-oxysalts.

c). d_f-Cations - $^{(12)}Y^{3+}$ (2,7), $^{(6)}Y^{3+}$ (4,0), $^{(12)}Sc^{3+}$ (4,2), $^{(6)}Sc^{3+}$ (6,3).

d). d_s-Cations - $^{(12)}Ag^+$ (1,2), $^{(6)}Mn^{2+}$ (6,7), $^{(6)}Fe^{2+}$ (7,4), $^{(6)}Co^{2+}$ (8,3), $^{(6)}Ni^{2+}$ (9,3), $^{(6)}Ti^{3+}$ (10,10), $^{(6)}Zn^{2+}$ (10,2), $^{(6)}Cu^{2+}$ (11,0), $^{(6)}V^{3+}$ (11,5), $^{(6)}Cr^{3+}$ (12,9), $^{(6)}Fe^{3+}$ (14,4), $^{(6)}Mn^{3+}$ (14,9), $^{(4)}Fe^{3+}$ (21,6), $^{(4)}Mn^{3+}$ (22,4); partially the same role is played by $^{(4)}Zn^{2+}$ (15,3) and $^{(4)}Cu^{2+}$ (16,5) which are more characteric for chalcophylic minerals; cations $^{(6)}Mn^{2+}$ (6,7), $^{(6)}Fe^{2+}$ (7,4), $^{(6)}Co^{2+}$ (8,3), $^{(6)}Ni^{2+}$ (9,3), $^{(6)}Cr^{3+}$ (12,9), $^{(6)}Fe^{3+}$ (14,4), $^{(6)}Mn^{3+}$ (14,9), $^{(4)}Fe^{3+}$ (21,6), $^{(4)}Mn^{3+}$ (22,4), especially those which are selected with bold, are coming forward also as chalcophylic cations with low FC (Group. 4.d).

e). p_s-Cations of lithophylic elements - $^{(6)}Al^{3+}$ (21,4), $^{(6)}Ge^{4+}$ (29,2), $^{(4)}Al^{3+}$ (32,2), $^{(6)}Si^{4+}$ (37,6).

f). p_s-Cations of chalcophylic elements - $^{(12)}Tl^+$ (0,48), $^{(12)}Pb^{2+}$ (1,3), $^{(8)}Pb^{2+}$ (1,9), $^{(8)}Sb^{3+}$ (3,3), $^{(8)}Bi^{3+}$ (3,4), $^{(6)}Pb^{2+}$ (2,5), $^{(6)}Sb^{3+}$ (4,4), $^{(6)}Bi^{3+}$ (4,6), $^{(6)}As^{3+}$ (5,7), $^{(6)}Te^{4+}$(6,0); cations $^{(6)}Pb^{2+}$ (2,5), $^{(6)}Sb^{3+}$ (4,4), $^{(6)}Bi^{3+}$ (4,6), $^{(6)}As^{3+}$ (5,7), $^{(6)}Te^{4+}$(6,0) are presented as cations also in some chalcosalts.

3. Cations - complexformers of lithopyllic elements with low - middle FC, including:

   a). Cations of noncenosymmetrical d'-complexformers with low - middle FC - $^{(12)}Zr^{4+}$ (4,5), $^{(6)}Zr^{4+}$ (9,4), $^{(6)}Ta^{5+}$ (12,7), $^{(6)}Nb^{5+}$ (15,2), $^{(6)}W^{6+}$ (17,8), $^{(4)}Ta^{5+}$ (19,1), $^{(6)}Mo^{6+}$(20,6), $^{(4)}Nb^{5+}$(22,7), $^{(4)}W^{6+}$(26,8), $^{(4)}Mo^{6+}$(30,9).

   b). Cations of cenosymmetrical d'-complexformers with middle FC - $^{(6)}Ti^{4+}$ (15,8), $^{(6)}V^{4+}$(17,8), $^{(4)}Ti^{4+}$ (23,7), $^{(6)}V^{5+}$ (25,6), $^{(4)}V^{4+}$ (26,7), $^{(4)}V^{5+}$ (38,4), $^{(4)}Cr^{6+}$ (54,4), $^{(4)}Mn^{7+}$ (121,3)).

4. Chalcophylic cations with low FC, including:

   a). Cations of chalcophylic elements with low FC - $^{(4)}Tl^+$ (1,45), $^{(3)}Tl^+$ (1,9), $^{(6)}Pb^{2+}$ (2,5), $^{(4)}Ag^+$ (3,5), $^{(4)}Au^+$ (3,6), $^{(6)}Sb^{3+}$ (4,4), $^{(6)}Bi^{3+}$ (4,6), $^{(6)}As^{3+}$ (5,7), $^{(6)}Te^{4+}$ (6,0), $^{(4)}Cu^+$ (6,2), $^{(4)}Hg^{2+}$ (7,7), $^{(4)}As^{3+}$ (8,6), $^{(4)}Te^{4+}$ (8,9), $^{(3+E)}Sb^{3+}$ (8,9), $^{(3+E)}Bi^{3+}$ (9,1), $^{(6)}In^{3+}$ (9,7), $^{(6)}Zn^{2+}$ (10,2), $^{(6)}Cu^{2+}$ (11,0), $^{(3+E)}As^{3+}$ (11,4), $^{(4)}Tl^{3+}$ (12,8), $^{(4)}In^{3+}$ (14,6), $^{(4)}Zn^{2+}$ (15,3), $^{(4)}Cu^{2+}$ (16,5), $^{(6)}Sn^{4+}$ (16,9), $^{(4)}Pb^{4+}$ (17,5), $^{(4)}Sn^{4+}$ (25,3), $^{(4)}Ga^{3+}$ (37,8) and cations of the same elements with lower CN; cations $^{(6)}Pb^{2+}$ (2,5), $^{(6)}Sb^{3+}$ (4,4), $^{(6)}Bi^{3+}$ (4,6), $^{(6)}As^{3+}$ (5,7), $^{(6)}Te^{4+}$ (6,0) are coming forward as cations in some oxides too (see group. 2.f), $^{(6)}Sb^{3+}$ (4,4), $^{(6)}Bi^{3+}$ (4,6), $^{(6)}As^{3+}$ (5,7), $^{(6)}Te^{4+}$ (6,0), $^{(4)}As^{3+}$ (8,6), $^{(4)}Te^{4+}$ (8,9), $^{(3+E)}Sb^{3+}$ (8,9), $^{(3+E)}Bi^{3+}$ (9,1), $^{(3+E)}As^{3+}$ (11,4), $^{(6)}Sn^{4+}$ (16,9) - as complexformers in sequence of complex chalcogen and oxygen compounds (group 5).

   b). Some. firstly selected with bold, cations of cenosymmetrical d'-elements - $^{(6)}Mn^{2+}$ (6,7), $^{(6)}Fe^{2+}$ (7,4), $^{(6)}Co^{2+}$ (8,3), $^{(6)}Ni^{2+}$ (9,3), $^{(6)}Cr^{3+}$ (12,9), $^{(6)}Fe^{3+}$ (14,4), $^{(6)}Mn^{3+}$ (14,9), $^{(4)}Fe^{3+}$ (21,6), $^{(4)}Mn^{3+}$ (22,4), widely known in minerals as d_s-cations (see group 2.f).

5. Chalcophylic cations-complexformers (in general cations of chalcophylic elements with middle FC) - $^{(6)}Sb^{3+}$ (4,4), $^{(6)}Bi^{3+}$ (4,6), $^{(6)}As^{3+}$ (5,7), $^{(6)}Te^{4+}$ (6,0), $^{(4)}As^{3+}$ (8,6), $^{(4)}Te^{4+}$ (8,9), $^{(3+E)}Sb^{3+}$ (8,9), $^{(3+E)}Bi^{3+}$ (9,1), $^{(3+E)}As^{3+}$ (11,4), $^{(6)}Sn^{4+}$ (16,9), $^{(6)}Bi^{5+}$ (17,2), $^{(6)}Sb^{5+}$ (24,3), $^{(6)}Ge^{4+}$ (29,2), $^{(6)}Te^{6+}$ (33,1), $^{(4)}Sb^{5+}$ (36,4), $^{(6)}As^{5+}$ (42,4), $^{(4)}Ge^{4+}$ (43,6), $^{(4)}As^{5+}$ (63,7), $^{(4)}Se^{6+}$ (78,0); cations $^{(6)}Sb^{3+}$ (4,4), $^{(6)}Bi^{3+}$ (4,6), $^{(6)}As^{3+}$ (5,7), $^{(6)}Te^{4+}$ (6,0), $^{(4)}As^{3+}$ (8,6), $^{(4)}Te^{4+}$ (8,9), $^{(3+E)}Sb^{3+}$ (8,9),

$^{(3+E)}Bi^{3+}$ (9,1), $^{(3+E)}As^{3+}$ (11,4), $^{(6)}Sn^{4+}$ (16,9) are coming forward similar to cations of chalcophylic elements with low FC (group 4.a).

6. p-Anionformers (cations of light and heavy anionformers) - $^{(4)}Si^{4+}$(56,4), $^{(4)}B^{3+}$ (86,2), $^{(4)}P^{5+}$(90,3), $^{(3)}B^{3+}$ (114,9), $^{(4)}S^{6+}$ (130,2), $^{(4)}Cl^{7+}$ (181,8), $^{(3)}C^{4+}$(238,8), $^{(3)}N^{5+}$(418,3); $^{(4)}S^{4+}$ (17,8), $^{(3)}I^{5+}$ (23,7), $^{(3)}Br^{5+}$ (27,3).

The presented systematic of cations has been placed by us in the basis of subsequent mineral systematic, especially in limits of classes, selected by character of anions. At the same time it shows that the same cations could play cation role in minerals of different types, for example, in lithopylic and chalcophylic, come forward in some minerals as cations, presenting their basic properties, and in other ones - as anionformers, presenting their acid properties. The value of FC/CN predetermines the role of cations in compounds in total, but it could not be used as stricted formal criteria, because the properties of cations mostly depend also on delicacies in atomic structure, in particular on belonging to cenosymmetricals or noncenosymmetricals, on values of their Z. At the same time, in many cases it allows to consider not only the role of cation in compounds, but also about possibility of isomorphism between it and other cations, especially if they are close by other features.

Highest taxons of structural-chemical systematic of minerals, preceding classes

All what was said shows, that the principal type of chemical bond in compounds should be considered as major feature of the most important taxons of mineralogical systematic. By this feature all minerals could be united in five types:

1. Type: Minerals with principal metallic and metallic-covalent bond - native metals and semimetals, metallides and semimetallides.
2. Type: Minerals with principal metallic-covalent and ionic-covalent, rare van der Waals forces - chalcogen compounds and native VIa-nonmetals.
3. Type: Minerals with principal ionic-covalent and covalent-ionic - nonmetallides of light (typical, cenosymmetrical) VIa-element (O) - oxygen compounds.
4. Type: Minerals with principal covalent-ionic and ionic bond - halogen compounds.
5. Type: Carbon, its compounds (without carbonates) and related substances.

Touching the selected types, it should be noted, that type "elementary (simple) compounds" which is usually started the majority of well-known systematics is disappeared from their number. From our point of view this is not only justified, but also is necessary action because of the following purposes:

a). In type of elementary substances usually unite the substance with completely different types of chemical bond - metals, semimetals, nonmetals, semiconductors, molecular substances which have only one in common - every of these substances consists of atoms of one sort. At the same time the properties, formation conditions, paragenesis of elementary (native) substances with different type of chemical bond have nothing common with each other.

b). Because of what was said it is very difficult to build systematic of elementary substance so, that we could get natural transformation from them to minerals of the following taxon.

Considering what was pointed out and necessity of association of the most close to each other by properties, genesis and paragenesis minerals, in natural systematic of minerals, it is more expedient to specify native metals, metallides, semimetals, semimetallides from selected earlier type of "elementary (simple) substances" to independent type, what we have been done previously. Along with it native metals are removing in front of respective metallides, and semimetals - in front of respective semimetallides. Simultaneously, native nonmetals - S and Se, are considered in front of chalcogen compounds, at the expense of which they are often formed.

The considerable peculiarity of classification of minerals, as far as inorganic crystalline substances at all, is impossibility of utilizing the strict system approach in its development. This have been already clarified in special position of type V among all mentioned types, in specific place of native minerals in systematic, what was discussed above. The same thing should be token into account with selection of further taxons in every type, when in the basis of their selection we should put different features, determined by complexity of their composition, peculiarities of chemical bond, crystalline structures, properties and conditions of formation of respective substances.

Thus type 1 divided on two subtypes: 1.1 - metals and metallides, 1.2 - semimetals and semimetallides, reflecting the considerable differences in chemical bond, structures, properties of both these and those ones. The last one, in its turn is divided on 2 quasisubtypes[*]: 1.2.1 - semimetals and semimetallides of Va-semimetals, 1.2.2 - semimetals and semimetallides of VIa-semimetals.

Type 2 firstly is dividing on two quasitypes[*], uniting completely different by type of chemical bond substances: 2a. - elementary (native) VIa-nonmetals which are characterized by residuum chemical bond, and 2b. - chalcogen compounds which have principal metallic-covalent and ionic-covalent chemical bond. Further quasitype[*] 2b. is dividing on two subtypes: 2b.1. - chalcogen substances of sidero- and chalcophylic cations, and 2b.2 - chalcogen compounds of lithophylic cations, reflecting considerable differences in composition, stability in natural conditions, properties, conditions of formation and paragenesis of minerals from every subtype. In this case the requirement of consequent transformation from one taxons to other ones, is observed, because some minerals of the last subtype are close by composition, properties, conditions of formation to minerals, beginning the next type 3. Subtype 2b.1 in its turn is divided on two quasisubtypes[*]: 2b.1a. - sulfides and sulfosalts of sidero- and chalcophylic cations, 2b.1b. - selenides and selenosalts of sidero- and chalcophylic cations, reverberating differences in properties, conditions of formation and paragenesis of natural chalcogen compounds of S and Se, although many of them are connected by continuous isomorphous transformations. It should be token in mind, that in minerals, refered to chalcogen compounds the consequent transformation from simple chalcogenides through complex ones to chalcosalts is often observed. Exactly by this it is expressed, that one authors consider all chalcosalts as complex chalcogenides, as far as other ones persistently isolate them in separate taxons as chalcosalts which are the natural end of sequences like simple chalcogenides → complex (binary and so on) chalcogenides → chalcosalts.

---

[*] The titles of taxons, selected by "*", are suggested by S.N. Nenasheva for comparing convenience. A.A. Godovikov leave these taxons without any titles.

Type 3 is divided by belonging of minerals to iso- or anisodesmical compounds first of all on two subtypes: 3.1 - oxides and hydroxides (isodesmical), 3.2. - oxysalts (anisodesmical). In dependence on belonging of mineral cations of first of them to one or another type it is divided further on six consequently changing each other quasisubtypes*, what is responding for transformation from cations with low FC to cations with high FC, from lithophylic cations through chalcophylic ones to cations of nonmetallic elements with highest FC:

3.1a - oxides and hydroxides of lithophylic cations with low FC,
3.1b - oxides and hydroxides of lithophylic cations with middle FC,
3.1c - oxides and hydroxides of chalcophylic cations (except Va- and VIa-cations),
3.1d - oxides and hydroxides of Va-cations (As, Sb, Bi),
3.1e - oxides and hydroxides of VIa-cations (Te),
3.1f - oxides and hydroxides of nonmetallic (lithophylic) cations.

Subtype 3.2 is also divided further immediate on classes, what is considered below.

Type 4 initially is divided by the same features on two subtypes: 4.1 - halogenides (isodesmical) and 4.2 - halogensalts (anisodesmical). The last one by dependence on belonging of anionformers to d- or p-cations in its turn is divided on two quasisubtypes*: 4.2a. - halogensalts with d-anionformers and 4.2b. - halogensalts with p-anionformers. Here it is necessary to pay attention on, that the majority of authors, however, because of the tradition, taking roots in last century, by this time are considering halogensalts as complex halogenides. By this the fact of clear isolation of radical groups in their structures, for example, tetrahedral $[BF_4]^-$, $[BeF_4]^{2-}$, octahedral - $[SiF_6]^{2-}$, $[AlF_6]^{3-}$, polymeric - $[BeF_3]^-$, $[B_2F_7]^-$, $[Al_3F_{14}]^{2-}$ and other, many of which are stable in solutions, and high solubility of many of halogensalts, is completely ignored. Some halogenacids, for example, $H[BF_4]$, as it have been already noted, are so strong, that they have in chemistry name superacids or magical acids. All this monosemantically testifies about salt nature of halogensalts [24]. Misunderstanding of it carried on, for example H. Strunz [63] to consideration of halogensalts not only as complex halogenides, but also as isodesmical (!) compounds, to associating by this feature of all halogensalts with oxides. Rather exactly with this, the place, leading up to present time in mineralogical systematic to halogen compounds between chalcogen and oxygen compounds [51], [34], or between oxides and oxysalts [12], is connected. Although by chemical bond type, existence of halogensalts among halogen compounds, they should be considered behind oxygen compounds.

The sequence of types: chalcogen → oxygen → halogen compounds - responds, as it was stressed earlier, for regular changing in their representatives of principal chemical bond type in direction: metallic-covalent → ionic-covalent → covalent-ionic → ionic bond.

The systematic of type 5 is building significantly more difficulty because of purposes, showed before. Initially it is divided on two quasitypes*, completely different by number of representatives, first of which - 5a. associating inorganic compounds, including native IVa-nonmetals, second one - 5b. - organic compounds. First of them is further divided on two subtypes: 5a.1. - elementary (native) IVa-nonmetals, 5a.2. - minerals with principal covalent and metallic-covalent chemical bond - carbides, and compounds which are close to them - silicides, nitrides and phosphides. Organic compounds in which the most important role is played by residuum bond, are further divided on to three subtypes: 5b.1. - salts of organic acids, 5b.2. - hydrocarbons and related compounds, 5b.3. - natural mixtures of organic substances, including fossil resin.

Thus, we could obtain the following scheme of intersubordination of considered taxons:

1. Type: Minerals with principal metallic and metallic-covalent bond - native metals and semimetals, metallides and semimetallides.

    1.1. Subtype: Metals and metallides.

    1.2. Subtype: Semimetals and semimetallides (only of sidero- and chalcophylic cations)

    1.2.1 Quasisubtype*: Semimetals and semimetallides of Va-semimetals.

    1.2.2 Quasisubtype*: Semimetals and semimetallides of VIa-semimetals

2. Type: Minerals with principal metallic-covalent and ionic-covalent bond, rare van der Waals forces - chalcogen compounds and native VIa-nonmetals.

    2a. Quasitype*: Elementary (native) VIa-nonmetals (residuum bond).

    2b. Quasitype*: Chalcogenic compounds - compounds of d-schrink-analogues - S and Se (metallic-covalent and ionic-covalent bond, rare van der Waals forces) - simple chalcogen compounds (isodesmical) → complex chalcogen compounds → chalcosalts (anisodesmical).

    2b.1. Subtype: Chalcogenic compounds of sidero- and chalcophylic cations.

    2b.1a. Quasisubtype*: Sulfides and sulfosalts of sidero- and chalcophylic cations.

    2b.1b. Quasisubtype*: Selenides and selenosalts of sidero- and chalcophylic cations.

    2b.2. Subtype: Chalcogen compounds of lithophylic cations.

3. Type: Minerals with principal ionic-covalent and covalent-ionic bond - nonmetallides of light (typical, cenosymmetrical) IVa-element (O) - oxygen compounds.

    3.1. Subtype: Oxides and hydroxides (isodesmical).

    3.1a. Quasisubtype*: Oxides and hydroxides of lithophylic cations with low FC - Force Characteristics.

    3.1b. Quasisubtype*: Oxides and hydroxides of lithophylic cations with middle FC.

    3.1c. Quasisubtype*: Oxides and hydroxides of chalcophylic cations (except Va- and VIa- cations) - simple and complex → arsenites, …, tellurates.

    3.1d. Quasisubtype*:Oxides and hydroxides of Va-cations (As, Sb, Bi).

    3.1e. Quasisubtype*: Oxides and hydroxides of VIa-cations (Te).

    3.1f. Quasisubtype*: Oxides and hydroxides of nonmetallic (lithophylic) cations.

    3.2. Subtype: Oxysalts (anisodesmical).

4. Type: Minerals with principal covalent-ionic and ionic bond - halogen compounds.

    4.1. Subtype: Halogenides (isodesmical).

    4.2. Subtype: Halogensalts (anisodesmical)(with hexacyanoferrates and hexatiocyanates, rhodonides).

    4.2a. Quasisubtype*: Halogensalts with d-anionformers.

    4.2b. Quasisubtype*: Halogensalts with p-anionformers.

5. Type: Carbon, its compounds (without carbonates) and related substances.

    5a.Quasitype*: Inorganic compounds of carbon (without carbonates) and related substances.

5a.1. Subtype: Native minerals.

5a.2. Subtype: Minerals with principal covalent and metallic-covalent bond - carbides and related compounds - silicides, nitrides and phosphides.

5b. Quasitype*: Organic compounds.

5b.1. Subtype: Salts of organic acids.

5b.2. Subtype: Hydrocarbons and related compounds.

5b.3. Subtype: Mixtures of organic compounds, including fossil resines.

After development of scheme of highest taxons of mineralogical systematic, we could begin to separate classes and to base their sequence, their subdivision on quasi and subclasses.

Classes and their sequence

As it have been already known, the classes of compounds in chemistry, since it have been suggested at the beginning of last century by J.J. Berzelius, are selected by anions. The same approach is preserved in mineralogy. However, there two questions which are considerable from the systematic's point of view:

1) which part of compound's formulae in which basis should be considered as anion;

2) which one should be the sequence of consideration of separate classes which place in it should occupy the compounds with several anions - aresenido-sulfides, sulfido-halogenides, silicato-carbonates, silicato-fluorides, arsenato-sulfates and so on.

For compounds with simple anions like S^{2-}, O^{2-}, F^-, Cl^-, even with polynuclear one like $[S_2]^{2-}$, $[As_4]^{4-}$, and so on, the question of selection of anion is pretty obvious and it could be ignored. It is not more clear also for compounds with complex anions like $[SiO_4]^{4-}$, $[SO_4]^{2-}$, $[CO_3]^{2-}$, including polymeric $[Si_6O_{18}]^{12-}$, $[Si_2O_5]^{2-}$ and so on, in which anionformers have the lowest of all possible values of coordination number, i.e. present their acid properties in maximal extent. However, situation for amorphous anionformers is getting even more difficult, when it is needed to make boundary between binary (complex) compounds and salts with complex anions, because anionformers could have different CN depending on acid-basic properties.

It was shown earlier, that as cations-anionformers (complexformers) could be considered:

1. Cations - complexformers of lithophylic elements with low-middle FC, including:

a). Cations of noncenosymmetrical d'-complexformers with low-middle FC -$^{(12)}Zr^{4+}$, $^{(6)}Zr^{4+}$, $^{(6)}Ta^{5+}$, $^{(6)}Nb^{5+}$, $^{(6)}W^{6+}$, $^{(4)}Ta^{5+}$, $^{(6)}Mo^{6+}$, $^{(6)}Nb^{5+}$, $^{(4)}W^{6+}$, $^{(4)}Mo^{6+}$.

b). Cations of cenosymmetrical d-complexformers with middle FC - $^{(6)}Ti^{4+}$, $^{(6)}V^{4+}$, $^{(4)}Ti^{4+}$, $^{(6)}V^{5+}$, $^{(4)}V^{4+}$, $^{(4)}V^{5+}$, $^{(4)}Cr^{6+}$, $^{(4)}Mn^{7+}$.

2. Chalcophylic cations-complexformers (mainly cations of chalcophylic elements with middle FC) - $^{(6)}Sb^{3+}$, $^{(6)}Bi^{3+}$, $^{(6)}As^{3+}$, $^{(6)}Te^{4+}$, $^{(4)}As^{3+}$, $^{(4)}Te^{4+}$, $^{(3+E)}Sb^{3+}$, $^{(3+E)}Bi^{3+}$, $^{(3+E)}As^{3+}$, $^{(6)}Sn^{4+}$, $^{(6)}Bi^{5+}$, $^{(6)}Sb^{5+}$, $^{(6)}Ge^{4+}$, $^{(6)}Te^{6+}$, $^{(4)}Sb^{5+}$,

$^{(6)}As^{5+}$, $^{(4)}Ge^{4+}$, $^{(4)}As^{5+}$, $^{(4)}Se^{6+}$; cations $^{(6)}Sb^{3+}$, $^{(6)}Bi^{3+}$, $^{(6)}As^{3+}$, $^{(6)}Te^{4+}$, $^{(4)}As^{3+}$, $^{(4)}Te^{4+}$, $^{(3+E)}Sb^{3+}$, $^{(3+E)}Bi^{3+}$, $^{(3+E)}As^{3+}$, $^{(6)}Sn^{4+}$, are coming forward similar to cations of chalcophylic elements with low FC (group 4.a).

3. p-Anionformers (cations of light and heavy anionformers) - $^{(4)}Si^{4+}$, $^{(4)}B^{3+}$, $^{(4)}P^{5+}$, $^{(3)}B^{3+}$, $^{(4)}S^{6+}$, $^{(4)}Cl^{7+}$, $^{(3)}C^{4+}$, $^{(3)}N^{5+}$, $^{(4)}S^{4+}$, $^{(3)}I^{5+}$, $^{(3)}Br^{5+}$.

Every of these three groups is responding to minerals of definite paragenesis, planning its specific sequence of mineral changing depending on changing of acid-basis conditions, oxidizing potential. Thus, cations - complexformers of lithophylic elements with low-middle FC are included in composition of oxygen compounds - complex oxides and oxysalts, placed predominantly in igneous rocks, in connected with them pegmatite and hydrothermal veins, in products of their metamorphism and weathering. The transformation from complex oxides to oxysalts is responding to increasing of acidity of cation-complexformer, decreasing of its CN from 6 through 5 up to 4. By this way we could design narrow genetically connected sequences of minerals like: (6)-vanadates → (5)-vanadates → (4)-vanadates, in which first their members are refered to typical complex oxides, and the last ones - are typical oxysalts, including of rather strong acids, to which we could refere (4)-vanadates and all the more (4)-arsenates. (4)-chromates and (4)-permanganates which are oxyslats of very strong acids, are ending these saubstances. Many of considered cations-complexformers are included in composition of genetically coneccted with each other minerals, refered to common paragenesis, for example, titanates and tantaloniobates of alkaline and rare-metallic granitic pegnatites, but $^{(6)}Zr^{4+}$, $^{(6)}Ti^{4+}$, $^{(5)}Ti^{4+}$, $^{(6)}Ta^{5+}$, $^{(6)}Nb^{5+}$ and so on cations are typical for mixed anionic radicals like zircono-, titano-, niobo-, tantalosilicatous ones which are especially characteric for paragenesises of agpaitic rocks and their derivatives.

The cations of the second group are typical for paragenesises of two types. One of them is characterized by primary minerals in which as ligands coming S and Se. Mentioned cations-complexformers are included in composition of complex chalcogenides and chalcosalts of sulfide deposits of magmatogenic, volcanogenic and metamorphogenic genesis, different hydrothermal veins, genetically connected often with acid rocks. Another paragenesis which is typical for them, is forming in zone of oxidation of mentioned earlier primary paragenetic associations of minerals and for which both complex oxides and oxysalts up to salts of relatively strong oxyacids like arsenic acid, are characteric.

The anionformers of third group are typical for mineral paragenesises of magmatic and igneous rocks, theirs pegmatites (silicates, some phosphates, carbonates, borates), skarns (silicates, borates, carbonates), hydrothermal veins (especially carbonates and some sulfates), volcanic exhalations (sulfates, halogenides, halogensalts), crusts of weathering (silicates), marine sediments (halogenides, sulfates, carbonates, phosphates, borates), evaporites (borates, halogenides), zones of oxidation (carbonates, sulfates, nitrates and so on).

Thus, every of mentioned groups of cations-anionformers are included in composition of minerals of its own paragenesis, as if it is forming its own genetic branch.

Earlier [26], [27] it was shown, that:

a). The quantitative measure of acid-basic properties of elements (cations) is their force characteristic (FC) and the value of atomic number (Z): decreasing of FC with constant Z is responding to strengthening of basic properties of cations, and vice-versa increasing of (FC), - acidic ones. Typical cations have coordinate number (CN) by one order or by two times less, then amphoteric cations, and amphoteric cations on their own turn - by one order lower, then typical anionformers.

b). The values of FC and Z of cations allow to clarify regular changing of structures of compounds in sequences, responding to transformation from simple compounds with coordinate structure to salts, and to predict with more probability the structures of unknown compounds either in total form, or in some of their details (for example, conditions of transformations from cubical "perovskites" to tetrahonal, orthorombic and minoclinic), regularities of polymorphous transformations.

c). With constant in definite limits physical-chemical parameters, first of all temperature and pressure, straightening of acid properties of amphoteric cations is clarifying in increasing of their valency, and with permanent valency (with permanent in definite limits oxidizing potential) - in decreasing of theirs CN. On the contrary, increasing of their basic properties are responding to decreasing of valency, decreasing of theirs CN.

d). In compounds, containing only amphoteric cations, last ones play typical cation role; their basic properties are increasing by straightening of acid properties of ligands, including complex anions too.

e). For cations with basic properties low valences (lower or equal to 2), high CN (usually higher then 6) are typical. For cations with acid properties (anionformers) - high valency (equal or higher then 4) and low CN (usually lower then 6); amphoteric cations in this relations placed intermediate position.

f). The structures of simple compounds, containing cations with basic properties, are coordinate (PbS, MgO, $NaCl$, CaF_2 and so on); containing cations with acid properties - are molecular or quasimolecular (As_4S_4, As_2S_3, B_2O_3, CO_2, SO_3, B_2F_4 and so on). In agreement with it the first ones differ by their high melting temperatures, low vapor resiliency; and second ones - by low melting temperatures, high vapor resiliency.

g). Amphotherity from crystallochemical positions is clarifying initially in changing of coupling motive of cation polyhedres with amphoteric cations from typical cationic, when cation polyhedres are connecting by edges or even verges in compounds, not containing cations with very low FC (as it is, for example, in ilmenite) up to motives in which these polyhedres are connected with each other only by bridge ligands. Further decreasing of FC bring to change of CN by cation from $CN > 6$ to $CN = 6$, defined by difference in FC of elements - partners in compound, their average or summary atomic numbers, formation of complex anion. Visually this could be seen in sequences of compounds, containing, along with basic cations, cations with lower CN which are playing more distinct role of anionformers up to formation of complex anions, as it is, for example, in sequences: simple oxides $\rightarrow$ binary oxides $\rightarrow$ oxysalts like $2AO_2$ ($2TiO_2$) $\rightarrow$ ABO_4 ($FeWO_4$) $\rightarrow$ $M[TO_4]$ ($Ca[WO_4]$).

All this show us, that the boundary "complex compounds $\rightarrow$ salts" for compounds of different types is coming differently, defining first of all by fundamental properties of atoms, forming them, - their FC and Z.

At the same time, it is not correct to refer to compounds of definite type the substances only by their brutto-formulae which is not consider their structure. Thus by now, practically all authors refer oxygen compounds, containing Te^{6+}, to salts of telluric acid or to tellurates, considering them as analogues of sulfates in narrow connection with last ones. By the first view this is absolutely natural, especially, if base on primitive-straightforward ideas about analogies of properties of elements in limits of separate subgroups of Periodic system, for example, expressing in association of O, S, Se, Te and Po in "chalcogenes", as it is offered by "Nomenclature rules of IUPACK in chemistry" [52], and what is done by , for example G.B. Bokiy and N.A. Golubkova later [7]. They are not considered differences between ceno- and noncenosymmetrical elements [62], [20], [21], nonmetallic and semimetallic elements, elements shell- and shrink-analogues, their differences from nonshell- and nonshrink analogues [21], [31].

The similarity of brutto-formulaes of sulfur, selenic and so-called telluric acids - H_2SO_4, H_2SeO_4 and H_2TeO_4 respectively is pushing on this way. However, if first two compounds are fully respond to structural ideas about acids, because they contain separated tetrahedral complex radicals $[SO_4]^{2-}$ or $[SeO_4]^{2-}$ with CN S and Se, equal to 4, what gives us opportunity to write their structural formulaes as $H_2[SO_4]$ and $H_2[SeO_4]$, this could not bee said about "telluric acid". The investigation of this compound was not established in its structures the anionic groups $[TeO_4]^{2-}$ with CN of Te = 4. Instead of it, it was found that cations Te^{6+} have CN = 6, i.e. respond to CN of amphoteric or weak acid anionformers. The structure of this compound is to be consisting of chains of TeO_4 $(OH)_2$ - octahedrons, in two opposite edges of which OH-ions are placed, connected with each other by common O atoms of equatorial edges of octahedrons [66]. It is easy to see, that by separating of repetition period of such structure we get structural formulae in form of $Te(OH)_2O_2$. Thus, this compound is hydroxido-oxide of Te^{6+} with very weak acid properties, clearly differing it from sulfur and selenic acids. That is why it should be considered not among acids and their salts in subtype 3.2. - oxysalts, but rather in subtype 3.1. - oxides and hydroxides at the end of it as compound with intermediate properties from oxides and hydroxides to acids and salts. The existence of so-called telluric acid with formulae $Te(OH)_6$ or H_6TeO_6, for which the octahedral coordination of Te, different polymeric groups, established in it salts [66], close by this and other properties to complex oxides, is characteric, is not changing this conclusion.

In advantage to what was said clear differences in properties of SO_3, SeO_3, H_2SO_4, H_2SeO_4 from one side, and TeO_3, $Te(OH)_6$ or H_6TeO_6 - from another are evident.

Thus, SO_3 in normal conditions - gas (b.p. 44,8°C), lower 16,8°C - SO_3 - transparent icy mass; energically reacts with water, forming sulfur acid $H_2[SO_4]$ which has pKa = 1,94;

SeO_3 - solid glassy or asbestic substance (m.p. 118,5°C; b.p. 185°C); energically reacts with water, forming selenic acid $H_2[SeO_4]$ with pKa = 1,92;

TeO_3 - solid respectively inert substance, on which are not act neither cold water, nor diluted bases (m.p. - higher then 400°C); obtained by dehydrotation of H_6TeO_6 in oxygen atmosphere with presence of concentrated $H_2[SO_4]$; on the first stage of dehydrotation of H_6TeO_6 results $TeO_2(OH)_2$; the compound H_6TeO_6 has properties of weak acid with pKa = 7,61.

All this considered to be natural, if take into account, that S and Se are connected by shrink-analogue, what is narrowing their properties, as far as Te differ from them, not being theirs shrink analogue. This responds also to systematic of cations which was given earlier, in accordance with which the typical in minerals for Te cation $^{(6)}Te^{6+}$ is refered to chalcophylic cations with middle FC, clearly differing in this aspect from $^{(4)}S^{6+}$ and $^{(4)}Se^{6+}$, for which CN = 6 has not been established in minerals. Shrink analogue of S and Se and not shrink-analogue with them of Te clarifying in elementary substance too, among which S and Se are nonmetals, as far as Te - semimetal[15]. By what

---

[15] Misunderstanding of different stages of analogues between elements of one group [21], [31] often brings to absolutely absurd associations of elements in limits of subgroups. Unfortunately, this could be found even in reference books and manuals, standing several editions. As example to what was said we could note the subdivision of VIa-elements by N.S. Achmetov [1] on O, after that S and, at last "elements of subgroup of Se" - Se, Te and Po, as far as in fact they should be divided on: 1) nonmetallic cenosymmetrical elements - O, 2) nonmetallic d-shrink analogues - S and Se, 3) semimetallic f-shrink

was said it is not amazing, that so-called salts of telluric acid and tellurates should be considered as complex oxides, containing $^{(6)}Te^{6+}$, what is confirmed by results of investigation of their crystalline structures. For example, we could mention mineral yafsoanite. The authors, who have discovered it [37], have opinion, that it is "refered to group (correct - to class - A.G.) of tellurates and it is complex salt of telluric acid" (p. 120) with formulae $(Zn_{1,38} \ Ca_{1,36} \ Pb_{0,26})_{\Sigma 3} \ TeO_6$. The investigation of its structure [36] shows, that the structural formulae of yafsoanite has form of $^{(8)}Ca_3 \ ^{(6)}Te_2 \ [^{(4)}ZnO_4]_3$ and it should be considered as complex oxide - zincate with structure, analogical to those of garnet in which the role Si is played by Zn, and Te is coming as typical cation - Al. In this relation yafsoanite is analogical to isostructural synthetic complex oxides, called incorrectly garnets, and more correct - garnatites - with common formulae $^{(8)}R^{3+}{}_3 \ ^{(6)}M^{3+}{}_2 \ [^{(4)}X^{3+}O_4]_3$, where $R^{3+}=$ Y, Ln; M^{3+}, $X^{3+}=$ Fe, Al, Ga.

Analogical reasons force to consider the substances of Te^{4+}, often called tellurites, among complex oxides to which by the same purposes we refer also arsenites, antimonites and antimonates.

To complex oxides, considered at their end, we refer also compounds with d-oxoradicals, even with tetrahedral coordination of anionformer - vanadates, molybdates, tungstenates, because respective "acids" and "anhydrides" are not only solid substances with octahedral coordination of V^{5+} , Mo^{6+}, W^{6+}, but also have very wear acid properties, more weak, then those of silicium acid. Such position of considering substances is imagine to be even more well-grounded, because it reverberates the natural connections of them with complex oxides, in which CN of electronegative cations (including V^{5+}, Mo^{6+} , W^{6+}) is sequentially changing from 6 through 5 to 4. By this the considerable chemical differences in properties of d- and p-elements are clarifying, in particular as anionformers, forcing us to separate vanadates from class of phosphates and arsenates, and molybdates and tungstenates from class of sulfates and selenates in which they have been usually considered since last century by such common feature as group-analogue, and formal similarity of oxoradicals.

Among d-elements in this relation position only cenosymmetrical Cr and Mn in their highest oxidizing extent, oxygen compounds of which refered to strong anhydrides (acids), forming typical salts with heterodesmical bonds - chromates and permanganates. This allows to consider, for example, chromates exactly after sulfates.

The sequence of classes in most numerous by number of representatives subtype of oxysalts should be also considered in detail, because there is no common or somehow regular way, what is illustrated by respective parts of systematic by different authors:

E.S.Dana, [13]
VI.Salts of oxygen acids
VI.Oxygen salts (oxysalts)
   1.Carbonates
   2.Silicates
   3.Niobates, tantalates
   4.Phosphates, arsenates, vanadates, antimonates, antimonites, arsenites

---

analogues - Te and Po. The same could be said about all other groups, on which we could unstop, however special attention attracts accepted by him division of VIIIb-elements. N.S. Achmetov divide them strongly by verticals, selecting 3 groups : 1) Fe, Ru, Os; 2) Co, Rh, Ir; 3) Ni, Pd, Pt. Along with it, it has not been token into account, that Fe, Co and Ni are cenosymmetricals. This bring to their high oxidizing, low normal electrode potentials, rareness in nature in native form. Other VIIIb elements - noncenosymmetrical f-shrink-analogues and they are not accidentally usually considered in common as platinoides. High normal electrode potentials, usualness of their native form in nature - are the common features for them [28].

Phosphates and so on with sulfate-anions and so on
Nitrates
5.Borates
Uranates
6.Sulfates, chromates
Tellurates; also tellurites, selenites
7.Tungstenates, molybdates

A.G. Betekhtin, [6]
VI.Oxygen salts (oxysalts)

1 class.Iodates	6 class.Molybdates and tungstenates
2 class.Nitrates	7 class.Phosphates, arsenates and vanadates
3 class.Carbonates	8 class.Arsenites
4 class.Sulfates, selenates, tellurates	9 class.Borates
5 class.Chromates	10 class.Silicates

A.S Povarennykh, [55]

Class III.Silicates, borosilicates, alumosilicates, beryllosilicates titanosilicates, zirconosilicates and uranosilicates	Class VII.Phosphates
	Class VIII.Tellurites and selenites
	Class IX.Tungstenates and molybdates
	Class X.Chromates and selenates
Class IV.Borates	Class XI.Sulfates
Class V.Vanadates	Class XII.Carbonates
Class VI.Arsenates	Class XIII.Iodates
	Class XIV.Nitrates

I.Kostov, [40]

Class V.Silicates	Class X.Chromates
Class VI.Borates	Class XI.Carbonates
Class VII.Phosphates, arsenates and vanadates	Class XII.Nitrates and iodates
Class VIII.Tungstenates and molybdates	A.Nitrates
Class IX.Sulfates	B.Iodates

     A.Sulfates
     B.Selenates, selenites, tellurates and tellurites

H.Strunz, [63]
    V.Nitrates, carbonates, borates
        Va.Nitrates
        Vb.Carbonates
        Vc.Borates
    VI.Sulfates (chromates, molybdates, tungstenates)
        A.- D.Sulfates
        E.Chromates
        F.Molybdates and tungstenates
    VII.Phosphates, arsenates, vanadates
    VIII.Silicates

A.A.Godovikov, [22]
Subtype II.Oxysalts   Class 4.Borates
  Class 1.Silicates   Quasiclass 1.(4)-Borates
  Class 2.Phosphates (arsenates)   Quasiclass 2.(4)-(3)-Borates
  Class 3.Sulfates   Quasiclass 3.(3)-Borates
                                             Class 5. Carbonates
                                             Class 6. Nitrates

A.R.Hoelzel,[34]
5.Carbonates, nitrates, borates   7.Phosphates, arsenates, vanadates
6.Sulfates. chromates, molybdates, tungstenates   8.Silicates

A.M.Clark, [12]
9.Borates
10.Borates with other anions
11.Carbonates
12.Carbonates with other anions
13.Nitrates
14.Silicates, not containing Al
15.Silicates of Al
16.Silicates, containing Al and other metals
17.Silicates, containing also other anions
18.Niobates and tantalates
19.Phosphates
20.Arsenates (also arsenates with phosphate-ion, but without other anions)
21.Vanadates (and vanadates with arsenat- or phosphate-ions)
22.Phosphates, arsenates or vanadates with other cations
23.Arsenites
24.Antimonates and antimonites
25.Sulfates
26.Sulfates with halogenid-ions
27.Sulfites, chromates, molybdates and tungstenates
28.Selenites, selenates, tellurites and tellurates
29.Iodates
30.Tioceanates

It is easy to see big discord, presented in position of separated classes by different authors. The situation is aggravating also by, that, usually, the explanations of accepted sequence are not given in publications.

In accepted by us sequence of substances change, responding to decreasing in them of extent of metallicity of chemical bond by increasing of its covalent extent, and after that ionicity, up to appearing of ionic compounds; oxysalts begin to appear at the end of class of oxides and hydroxides. In this case right after complex oxides and hydroxides, ending with minerals of silica family, it is natural to place the class of silicates - oxysalts of the most weak oxyacids, beginning of it with alumosilicates, structures of many of which are derivatives of structures of different polymorphs of silica. The connection between properties and formation conditions of minerals of silica family and silicates is so significant, that some mineralogists placed silica family into one class with silicates, excluding it from oxides [40], [43].

The silicates are close to complex oxides not only because they are salts of very weak silicium acids, but also by such properties as low solubility, high temperatures of melting point, and many of them also because they contain in their structure polymeric radicals, what makes them close to many of titanates, tantalo-niobates and so on minerals.

Oxides and silicates of lithophylic elements, usually, are forming in endogenious, often in hightemperature, processes; for oxides and silicates of chalcophylic elements it is more usual to form in oxidation zone conditions.

Further sequence of placement of classes it is more convenient to subdue to changing sequence of salts, responding to decreasing of straight of acids, forming these salts. Such sequence replies in total for increasing of ionicity extent of chemical bond between cations and acid radicals, decreasing of polymeric radicals role in salts, increasing of their solubility, changing of formation conditions from hypogenic to more and more lowtemperature ones, up to surface ones, regular changing of many physical properties of respective minerals.

This sequence, however, is disturbed in class of borates, including minerals, in which CN of B in oxyradicals could be 4; 3; 4 and 3. In dependence from it have been suggested [22] to divide borates onto 3 quasiclasses: 1) (4)-borates, 2) (3)-borates, 3) (4)-(3)-borates, considering them in systematic in mentioned sequence. First of them are structurally close to silicates; radical groups of (4)-borates could polycondensate with silicium-oxygen radicals, forming common heteronuclear polymeric radicals of borosilicates. (4)-Borates are characteric for hypogenic conditions; high pressure and increased basicity are favorable for their formation. (3)-Borates are mainly hypogenic minerals too, as far as formation of (4)-(3)-borates is typical along with evaporation from marine and lacustrine water with formation of evaporates, in some crusts of weathering, rarely in lowtemperature hydrotermal veins.

In light of what was said we have accepted the following sequence of consideration of separate classes of oxysalts with lithophylic p-anionformers (in brackets the values of pKa and formulaes of acids, for which they are defined are given [44], for polybasic acids acid index is given of first stage of dissociation):

Class 1.Silicates	$(9,9; H_4 [SiO_4])$
Class 2.Borates	
Quasiclass a).(4)-Borates	
Quasiclass b).(3)-Borates	$(9,15; H_3 [BO_3])$
Quasiclass c).(4)-(3)-Borates	
Class 3.Carbonates	$(3,25; H_2 [CO_3])$
Class 4.Phosphates	$(2,15; H_3 [PO_4])$
Class 5.Sulfates	$(1,94; H_2 [SO_4])$
Class 6.Nitrates	$(-1,64; H[NO_3]$; from [9] p. 98)
Class 6a.Iodates	$(0,77; H [IO_3])$
Class 6b.Rodanates	$(-1; H[CNS])$

Referring of minerals to definite class of oxysalts is disputable, when in its composition there are several different anionic radicals, for example, $[SiO_4]^{4-}$ and $[PO_4]^{3-}$; $[CO_3]^{2-}$ and $[SO_4]^{2-}$; $[CO_3]^{2-}$ and F^-, i.e. when mineral is mixed salt [23]. Along with to mixed salts are not refered:

a) Inclusion compounds which have structure with big cavities, in which addition ions are placed, sometimes with cations (oxides of Mn with tunnel structures like hollandite, minerals of cancrinite-vishnevite seria, zeolites);

b) Minerals with hybrid structures in which separated layers are neutral or weakly charged packages, containing in their composition "heterogeneous" anion (tundrite, lomonosovite and so on minerals).

By now it have been suggested to refer such minerals to definite class by the most strong anion [22], [23], However, modeling of developed systematic in exposition of A.E. Fersman Mineralogical museum RAS showed, that from the paragenetic point of view it is more convenient to do vice versa, referring minerals - mixed salt to this or that class by anion of the most weak acid [25].

Analogical approach is used by us for many other minerals with several anions which are not refered to oxysalts, such as arsenido-tellurides, arsenido-sulfides, sulfido-halogenides and so on.

In subtype of halogenides two classes are separated: 1. Fluorides and 2. Chlorides, bromides, iodides. Here fluorides are separated in independent class as derivatives of cenosymmetrical F, considerably differing by its properties and genesis, paragenesis from all other halogenides. The rest of halogenides is associated in one class somewhat conditionally. In it chlorides and bromides, as derivatives of d-shrink-analogues - Cl and Br - have mach more similarities with each other, then with iodides. The last ones are refered to this class because they have few representatives among minerals. In systematic of inorganic crystalline substances they should be separated in independent class.

In subtype of halogensalts the separation of classes is executed on bases which were explained earlier [24], although they have different sequence. It, as in oxysalts case, respond to increasing of straight of respective halogenacids too.

The sequence of highest taxons in developed structural-chemical systematic of minerals

1. Type: Minerals with principal metallic and metallic-covalent bond - native metals and semimetals, metallides and semimetallides.

1.1. Subtype: Metals and metallides.

1.1.1. Class: Metals and metallides of sidero- and chalcophylic elements.

1.1.2. Class: Metals and metallides of lithophylic elements.

1.2. Subtype: Semimetals and semimetallides (of sidero and chalcophylic cations only).

1.2.1. Quasisubtype*: Semimetals and semimetallides of Va-semimetals.

1.2.1a. Class: Native Va-semimetals.

1.2.1b. Class: Va-Semimetallides - arsenides, antimonides, bismutides.

1.2.2. Quasisubtype*: Semimetals and semimetallides of VIa-semimetals.

1.2.2a. Class: Native VIa-semimetals.

1.2.2b. Class: VIa-semimetallides - tellurides.

2. Type: Minerals with principal metallic-covalent and ionic-covalent bond, rare van der Waals forces - native VIa-nonmetals, chalcogen compounds: - chalcogenides (isodesmical) → chalcosalts (anisodesmical).

2a. Quasitype*: Native VIa-nonmetals (van der Waals forces).

2b. Quasitype*: Chalcogenic compounds (metallic-covalent and ionic-covalent bond, rare van der Waals forces) - simple (isodesmical) → complex → chalcosalts (anisodesmical).

2b.1. Subtype: Chalcogenic compounds of sidero- and chalcophylic cations.

2b.1a. Quasisubtype*: Sulfides and sulfosalts of sidero- and chalcophylic cations.

2b.1a.1. Class: Sulfides of sidero- and chalcophylic cations.

2b.1a.2. Class: Sulfosalts of sidero and chalcophylic cations.

2b.1b. Quasisubtype*: Selenides and selenosalts of sidero- and chalcophylic cations.

2b.1b.1. Class: Selenides of sidero- and chalcophylic cations.

2b.1b.2. Class: Selenosalts of sidero- and chalcophylic cations.

2b.2. Subtype: Chalcophylic compounds of lithophylic cations.

2b.2.1. Class: Sulfides (and selenides) of lithophylic cations.

2b.2.2. Class: Sulfosalts of lithophylic cations.

3 Type: Minerals with principal ionic-covalent and covalent-ionic bond - nonmetallides of light (typical, cenosymmetrical) VIa-element (O) - oxygen compounds: Oxides and hydroxides (isodesmical → anisodesmical) → oxosalts (anisodesmical).

3.1. Subtype: Oxides and hydroxides (isodesmical).

3.1a. Quasisubtype*: Oxides and hydroxides of lithophylic cations with low FC - Force Characteristics.

3.1a.1. Class: Oxides and hydroxides of s-, d_s- and p_s-cations.

3.1a.2. Class: Oxides and hydroxides of f-cations with low FC - of 4-valent f-cations.

3.1a.3. Class: Oxides and hydroxides of f-cations with middle FC - of 6-valent f-cations (U^{6+}) → uranyl $(UO_2)^{2+}$ compounds- uranyl acids, uranates and their derivatives (uranium micas and related minerals).

3.1b. Quasisubtype*: Oxides and hydroxides of lithophylic cations with middle FC.

3.1b.1. Overclass*: Oxides of Zr .

3.1b.1a. Class: Simple oxides of Zr.

3.1b.1b. Class: Complex oxides of Zr → titanates of zirconium → zirconotitanates.

3.1b.2. Overclass*: Oxides of Sn^{4+} and Ti^{4+}.

3.1b.2a. Class: Simple oxides and hydroxides of Sn^{4+} and Ti^{4+}.

3.1b.2b. Class: Complex oxides of Ti^{4+} (Sn^{4+}) → titanates (stannates), ((6)-titanates, (6)-stannates only).

3.1b.3. Overclass*: Oxides and hydroxides of Nb^{5+} and Ta^{5+}.

3.1b.3a. Class: Simple oxides and hydroxides of Nb^{5+} and Ta^{5+}.

3.1b.3b. Class: Complex oxides of Nb^{5+} and Ta^{5+} ((6)-tantaloniobates → (4)-tantaloniobates).

3.1b.4. Overclass*: Oxides and hydroxides of Mo and W .

3.1b.4a. Class: Simple oxides and hydroxides of Mo and W .

3.1b.4b. Class: Complex oxides and hydroxides of Mo and W ((6)-molybdates and tungstenates → (4)-molybdates and tungstenates).

3.1b.5. Overclass*: Oxides and hydroxides of Mn^{4+}.

3.1b.5a. Class: Simple oxides and hydroxides of Mn^{4+}.

3.1b.5b. Class: Complex oxides and hydroxides of Mn^{4+}.

3.1b.6. Overclass*: Oxides and hydroxides of V^{4+}.

3.1b.6a. Class: Simple oxides and hydroxides of V^{4+}.

3.1b.6b. Class: Complex oxides and hydroxides of V^{4+} (vanadites).

3.1b.7. Overclass*: Oxides and hydroxides of V^{5+}.

3.1b.7a. Class: Simple oxides and hydroxides of V^{5+}.

3.1b.7b. Class: Complex oxides and hydroxides of V^{5+} ((6)-vanadates → (5)-vanadates → (4)-vanadates).

3.1b.7b.1. Quasiclass: (6)-Vanadates.

3.1b.7b.2. Quasiclass: (5)-Vanadates.

3.1b.7b.3. Quasiclass: (4)-Vanadates.

3.1c. Quasisubtype*: Oxides and hydroxides of chalcophylic cations (except Va- and VIa-cations).

3.1c.1. Overclass*: Oxides and hydroxides of Ib-cations.

3.1c.2. Overclass*: Oxides and hydroxides of IIb-cations.

3.1c.3. Overclass*: Oxides and hydroxides of IIIa-cations.

3.1c.4. Overclass*: Oxides and hydroxides of IVa-cations.

3.1d. Quasisubtype*: Oxides and hydroxides of Va-cations.

3.1d.1. Overclass*: Oxides and hydroxides of As^{3+}, Sb^{3+}, Bi^{3+}.

3.1d.1a. Class: Simple oxides and hydroxides of As^{3+}, Sb^{3+}, Bi^{3+}.

3.1d.1b. Class: Complex oxides and hydroxides of As^{3+}, Sb^{3+}, Bi^{3+} →
(6)-Arsenites, antimonites, bismutites.

3.1d.2. Overclass : Oxides and hydroxides of As^{5+}, Sb^{5+}, Bi^{5+} (all are complex) →
arsenates, antimonates and bismutates (only (6)-arsenates, (6)-antimonates and (6)-bismutates).

3.1e. Quasisubtype*: Oxides and hydroxides of VIa-cations (Te).

3.1e.1. Overclass*: Oxides and hydroxides of Te^{4+}.

3.1e.1a. Class: Simple oxides and hydroxides of Te^{4+}.

3.1e.1b. Class: Complex oxides and hydroxides of Te^{4+} → tellurites.

3.1e.2. Overclass: Oxides and hydroxides of Te^{6+} (all are complex) → tellurates (all (6)-tellurates).

3.1f. Quasisubtype*: Oxides and hydroxides of nonmetals (lithophylic) elements.

3.1f.1. Class: Oxides and hydroxides of Si and Ge (silicic and germanium anhydrides, silicic and germanium acids).

3.1f.2. Class: Oxides and hydroxides of B (boric anhydride and boric acids).

3.1f.3. Class: Oxides and hydroxides of Se (selenic anhydride).

3.2. Subtype: Oxosalts (anisodesmical).

3.2.1. Class: Silicates.

3.2.1a. Class: Germanates (zone of oxidization of Tsumeb and Franse).

3.2.2. Class: Borates.

3.2.2.1. Quasiclass: (4)-Borates.

3.2.2.2. Quasiclass: (3)-Borates.

3.2.2.3. Quasiclass: (4)-(3)-Borates.

3.2.3. Class: Carbonates.

3.2.4. Class: Phosphates

3.2.4.1. Quasiclass: Orthophosphates

3.2.4.2. Quasiclass: Pyrophosphates

3.2.4.3. Quasiclass: Triphosphates

3.2.4a.Class: Arsenates.

3.2.4a.1. Quasiclass: (6)-Arsenates.

3.2.4a.2. Quasiclass: (4)-Arsenates (orthoarsenates).
*3.2.4b. Class: Arsenites
3.2.5. Class: Sulfates.
3.2.6. Class: Sulfites.
3.2.6a. Class: Selenites.
3.2.7. Class: Chromates.
3.2.8. Class: Nitrates.
3.2.8a.Class: Iodates.
*3.2.8b.Class: Iodites
3.2.8c.Class: Rhodanates (tiocyanates).

4. Type: Minerals with principal covalent-ionic and ionic bond - halogen compounds.

4.1. Subtype: Halogenides (isodesmical).
4.1.1. Class: Fluorides.
4.1.2. Class: Chlorides, bromides.
4.1.2a. Class: Iodides.

4.2. Subtype: halogensalts (anisodesmical) (with hexacyanoferrates and hexatiocyanates, rhodonides).
4.2a. Quasisubtype*: Halogensalts with d-anionformers.
4.2a.1. Class: Cloroferrites and clorocuprites (of s-cations and NH^+_4 only).
4.2a.2. Class: Hexachlorferrates and hexachlormanganates (of s-cations only).
4.2b. Quasisubtype*: Halogensalts with p-anionformers.
4.2b.1. Class: Fluoraluminates (of s-cations only).
4.2b.2. Class: Fluorborates (of s-cations only).
4.2b.3. Class: Fluorsilicates (of s-cations and NH^+_4 only).
4.2b.4. Class: Chloraluminates (of s-cations only).

5. Type: Carbon, its compounds (without carbonates) and related substances.
5a. Quasitype*: Inorganic compounds (without carbonates) and related substances.
5a.1. Subtype: Native minerals.
5a.2. Subtype: Minerals with principal covalent or metallic-covalent bond - carbides and related compounds - silicides, nitrides and phosphides.
5a.2.1. Class: Carbides.
5a.2.1a. Class: Silicides.
5a.2.2. Class: Nitrides.
5a.2.2a. Class: Phosphides.
5b. Quasitype*: Organic carbon compounds (mineral with principal van der Waals forces bond).
5b.1. Subtype: Salts of organic acids.
5b.1.1. Class: Salts of benzopolycarbonic acids (C_6H_{6-n} $(COOH)_n$; n = 6).
5b.1.2. Class: Salts of citric acids (citrates).
5b.1.3. Class: Salts of acetic acids (acetates).
5b.1.4. Class: Salts of oxalic acids (oxalates).
*5b.1.5. Class: Salts of formic acids (formates).
5b.2. Subtype: Hydrocarbons and related compounds.
5b.2.1. Class: Hydrocarbons cyclic (by decreasing of x = H : C).
5b.2.2. Class: Hydrocarbons oxygenbearing (by increasing of O : C).

5b.2.3. Class: Nitrogenbearing organic compounds.

5b.3. Subtype: Mixtures of organic substances, including amber and related substances..

So, the principles of taxon separation in developed structural-chemical systematic of minerals are presented in table #1.

Table #1.General enumeration of the taxons of structural-chemical classification of minerals

Taxon	Feature	Examples
1	2	3
Type	It is principle type of chemical bond (but not a single type of chemical bond)	The five types are uniting all minerals species: 1.Type: Minerals with principal metallic and metallic-covalent bond - native metals and semimetals, metallides and semimetallides. 2.Type: Minerals with principal metallic-covalent and ionic-covalent bond, rare van der Waals forces - chalcogen compounds and native VIa nonmetals. 3.Type: Minerals with principal ionic-covalent and covalent-ionic bond - nonmetallides of light (typical, noncenosymmetrical) VIa element (O) - oxygen compounds. 4.Type: Minerals with principle covalent-ionic and ionic bond - halogen compounds. 5.Type: Carbon, its compounds (without carbonates) and related substances.
Quasi-type*	Type of chemical bond (this taxon is divided when more higher taxon unites the minerals with three or more types of chemical bond)	There are two quasitypes at the second type of minerals with principal metallic-covalent and ionic-covalent bond, rare van der Waals forces - chalcogen compound and native VIa nonmetals: 2a. Native VIa nonmetals (van der Waals forces); 2b. Chalcogenic compounds (metallic-covalent and ionic-covalent bond rare van der Waals forces) - simple (isodesmical) -> complex -> chalcosalts (anisodesmical)
Subtype	1.Type of chemical bond, (only single type of chemical bond)	There are two subtypes at the 1. taxon of minerals with principal metallic and metallic-covalent bond - native metals and semimetals, metallides and semimetallides: 1.1 Metals and metallides; 1.2 Semimetals and semimetallides.
	2. Type of cation (siderophylic, chalcophylic or lithophylic)	There are two subtypes at the 2b. quasitype of the second type: 2b.1. Chalcogenic compounds of sidero- and chalcophylic cations (metallic-covalent bond); 2b.2. Chalcogenic compounds of lithophylic cations (ionic-covalent bond).
	3. Belonging of	There are two subtypes at the 3 type of minerals with

	mineral to isodesmical and anisodesmical compounds	principal ionic-covalent and covalent-ionic bond - nonmetallides of light (typical, noncenosymmetrical) VIa-element (O) - oxygen compounds: 3.1. Oxides and hydroxides (isodesmical); 3.2. Oxysalts (anisodesmical) There are two subtypes at the 4 type of minerals with principle covalent-ionic and ionic bond - halogen compounds: 4.1. Halogenides (isodesmical); 4.2. Halogenosalts (anisodesmical) (with hexacyanoferrates and hexatiocyanates, rhodonides).
Quasi-subtype*	1.Anion 2. Type of cation and FC of cation	There are two quasitypes at the 2b.1. subtype chalcogen compounds of sidero- and chalcophylic cations (metallic-covalent bond): 2b.1a. Sulfides and sulfosalts of sidero- and chalcophylic cations; 2b.1b. Selenides and selenosalts of sidero- and chalcophylic cations. There are six consequently changing quasitypes at the 3.1 Subtype oxides and hydroxides (isodesmical), that are corresponding for transferal from the cations with low FC to the cations with high FC, from lithophylic cations to chalcophylic and to nonmetallic cations of the elements with mostly high FC: 3.1a - Oxides and hydroxides of lithophylic cations with low FC; 3.1b - Oxides and hydroxides of lithophylic cations with middle FC; 3.1c - Oxides and hydroxides of chalcophylic cations (without Va- and VIa- cations); 3.1d - Oxides and hydroxides Va- cations (As, Sb, Bi); 3.1e -Oxides and hydroxides VIa- cation (Te); 3.1f - Oxides and hydroxides of nonmetals (lithophylic) elements;
Over-class*	Cation	There are seven overclasses at the 3.1b. taxon - oxides and hydroxides of lithophylic cations with middle FC: 3.1b.1. Oxides Zr; 3.1b.2. Oxides Ti (Ti^{4+}); 3.1b.3. Oxides and hydroxides Nb^{5+} и Ta^{5+}; 3.1b.4. Oxides and hydroxides Mo и W; 3.1b.5. Oxides and hydroxides Mn^{4+}; 3.1b.6. Oxides and hydroxides V^{4+}; 3.1b.7. Oxides and hydroxides V^{5+}.

Class	Type of anion (simple, complex) or compound (simple, complex)	There are two classes at the 2b.1a. quasitype- sulfides and sulfosalts of sidero- and chalcophylic cations:
		2b.1a.1. Class: Sulfides of sidero- and chalcophylic cations; 2b.1a.2. Class: Sulfosalts of sidero- and chalcophylic cations;
	2. Anionforming, when minerals are anisodesmical compounds	There are two classes at the 3b.1b.1. overclass - oxides Zr: 3.1b.1a. Class: Simple oxides of Zr; 3.1b.1b. Class: Complex oxides of Zr -> titanates of Zr -> zirconotitanates.
		There are eleven classes at the 3.2. subtype - Oxosalts (anisodesmical): 3.2.1. Class: Silicates; 3.2.2. Class: Borates; 3.2.3. Class: Carbonates; 3.2.4. Class: Phosphates; 3.2.4a. Class: Arsenates; 3.2.5. Class: Sulfates; 3.2.6. Class: Sulfites; 3.2.6a. Class: Selenites; 3.2.7. Class: Nitrates; 3.2.7a. Class: Iodates; 3.2.76. Class: Rhodonates (tiocyanates).
Quasi class	Coordination number of the anionforming	There are three quasiclasses at the 3.1b.7b. class - complex oxides and hydroxides of V^{5+} ((6)-vanadates -> (5)-vanadates -> (4)-vanadates): 3.1b.7b.1. Qusiclass: (6)-vanadates; 3.1b.7b.2 Qusiclass: (5)-vanadates; 3.1b.7b.3 Qusiclass: (4)-vanadates;
		There are tree quasiclasses at the borates class: 1) (4)-borates; 2) (3)-borates; 3) (4)-(3)-borates: 3.2.2.1. Quasiclass: (4)-Borates; 3.2.2.2. Quasiclass: (3)-Borates; 3.2.2.3. Quasiclass: (4)-(3)-Borates;
Subclass	The size of FC	There are three subclasses at the class of silicates: 1) silicates with low FC; 2) silicates with middle FC; 3) silicates of chalcophylic elements.
Family	The minerals of one family have similar of equal compound, single	The family of zeolites unite the subfamilies: thomsonite, scolecite-natrolite, garronite, wairakite, gmelinite, stilbite, stellerite, mordenite. The micas family unite dioctahedral and trioctahedral micas

	genesis or paragenesis	and all polytypes.
Subfamily	Similar or equal compound and same type of structure	There are five subfamilies at the chalcopyrite family: talnakhite, actually chalcopyrite, germanite, briartite, morozeviczite. There are three subfamilies at the stannite family: stannoidite, actually stannite, rodostannite.
Series (genus)	Uninterrupted solid solutions between two or greater number of the extreme members	The forsterite genus and garnet genus among of the middle tetrasilicates
Group	The same type of the compound or structure	The dolomite group include dolomite, ankerite, kutnohorite, benstonite, eitelite. All its minerals have one type structure, but they have not the uninterrupted solid solutions between ones.
Mineral species	There is an individual chemical compound, extreme member of the solid solutions, middle member of the uninterrupted solid solutions.	a) There are three mineral species at the genus monticellite: monticellite, glaucochroite, kirnschsteinite. b) There are five mineral species at the forsterite genus: forsterite, fayalite, tephroite, liebenbergite, laihunite.

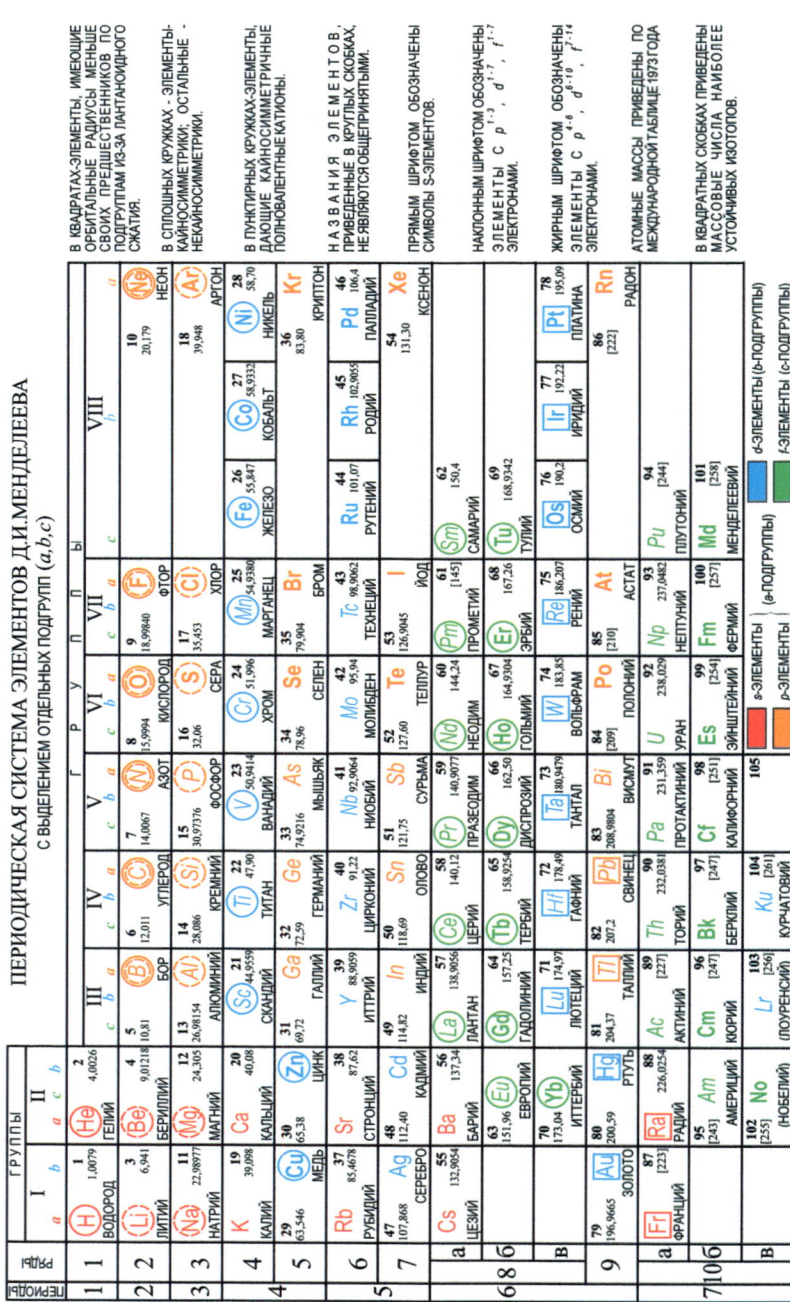

Fig. 4 Periodical table of elements D.I. Mendeleev's with little Sub-Groups (a,b,c)

Structural-Chemical Systematic of Minerals

1.TYPE: MINERALS WITH PRINCIPAL METALLIC END METALLIC-COVALENT BOND – NATIVE METALS AND SEMIMETALS, METALLIDES AND SEMIMETALLIDES

1.1. SUBTIPE: METALS AND METALLIDES

1.1.1. *Class*: Metals and metallides of sidero- and chalcophylic elements

1.1.1.1. <u>Minerals of heavy noncenosymmetrical d-elements (with 5 – 10 d-electrons)</u>

1.1.1.1.1. Minerals of **VIII**b-elements (platinoides Pn)

1.1.1.1.1.1. Native metals

Platinum group
Platinum	Pt
Iridium	Ir
Palladium	Pd
Rhodium	(Rh,Pt)

Osmium group
Osmium	Os
Ruthenium	Ru
*Rutheniridosmin	(Ir,Os,Ru)

1.1.1.1.1.2. Metallides

*Iridrhodruthenium

1.1.1.1.1.2.1. Only Pn

$Ru_{6,4}Rh_{1,7}Ir_{1,2}Pt_{0,7}Os_{0,04}$

1.1.1.1.1.2.2. Ferreeds

Isoferroplatinum group
Isoferroplatinum	Pt_3Fe
Unnamed 023	$Pt_2(Ir,Os)Fe$
*Chengdeite	$(Ir,Pt,Os,Ru)_3(Fe,Ni)$
*Unnamed	Pt_2Fe

Tetraferroplatinum group
Unnamed 021	RhFe
Tetraferroplatinum	$Pt_{1+x}Fe$ $(0 \le x \le 1)$
Unnamed 022	Pt(Fe,Ni,Cu)
Ferronickelplatinum	Pt_2FeNi
Tulameenite	Pt_4Fe_2CuNi or Pt_2FeCu
Unnamed 305	Pt_3Cu_2Fe

Group **Pn$_2$(Fn,Cu)$_3$**
Unnamed 308	$(Os,Ir,Ru,Pt)_2Fe_3$
Unnamed 306	$(Ru,Os,Ir,Pt)_2(Fe,Ni,Cr)_3$
Unnamed 454	$Pt_2(Fe,Bi)_3$

© Springer Nature Switzerland AG 2020
A. A. Godovikov and S. N. Nenasheva, *Structural-Chemical Systematics of Minerals*, https://doi.org/10.1007/978-3-319-72877-3_2

Group **Pn(Fn,Cu)$_3$**
Unnamed 304   Pt(Ni,Cu,Fe)$_3$
Unnamed 307   (Pt,Ir,Os)(Fe,Ni,Cr,Cu)$_3$
Unnamed 302   PtCu$_2$Fe
                              1.1.1.1.1.2.3. Pn and Ag
Unnamed 498   PdAg
                              1.1.1.1.1.2.4. Pn, Ag and Tl
Unnamed 455   2,(6)   (Pd,Ag)$_8$Tl$_3$
Unnamed 456   1,75   Pd$_6$AgTl$_4$
                              1.1.1.1.1.2.5. Pn and Tl
Unnamed 453   2,5   (Pd,Sn,Cu,Fe)$_5$(Tl,S)$_2$
                              1.1.1.1.1.2.6. Pn and Au
Unnamed 443   PdAu$_3$
                              1.1.1.1.1.2.7. Mercureeds (amalgams)
Potarite   PdHg
                              1.1.1.1.1.2.8. Cuprides
                              1.1.1.1.1.2.8.1. Proper cuprides
Unnamed 514   Pt$_3$Cu
Unnamed 515   (Pd,Pt,Au)$_2$Cu
*Skaergaardite   PdCu
Hongshiite   PtCu
*Unnamed   PtCu$_{1-x}$Sb$_x$
Unnamed 433   Pt$_2$Cu$_3$
Unnamed 303   PtCu$_3$
*Nielsenite   PdCu$_3$
                              *1.1.1.1.1.2.8.2. Cuprido-zincides
*Bortnikovite   Pd$_4$Cu$_3$Zn
                              1.1.1.1.1.2.8.3. Cuprido-stanides

Cabriite family
Unnamed 019   Pd$_6$Cu$_2$(Sn,Sb)
Cabriite   Pd$_2$CuSn
Unnamed 389   Rh$_2$CuSn
Stannopalladinite   (Pd,Cu)$_3$Sn$_2$

                              1.1.1.1.1.2.9. Stanides
                              1.1.1.1.1.2.9.1. Proper stanides
Unnamed 411   (Pd,Pt)$_7$(Sn,Pb)$_2$
Zvyagintsevite family
 Zvyagintsevite series
 Rustenburgite   Pt$_3$Sn
 Atokite   Pd$_3$Sn
 Zvyagintsevite   Pd$_3$(Pb,Sn)
 Unnamed   Pd$_3$Pb
 Maslenytskovite series
 Maslenytskovite-(Pt)   Pt$_{3-x}$Pd$_x$Sn$_{1-y}$□$_y$ $(0{,}6 \leq x \leq 1{,}5)$; $(0 \leq y \leq 0{,}1)$
 Maslenytskovite-(Pd)   Pd$_{3-x}$Pt$_x$Sn$_{1-y}$□$_y$ $(0{,}6 \leq x \leq 1{,}5)$; $(0 \leq y \leq 0{,}1)$
 Taimyrite   (Pd,Cu,Pt)$_3$Sn
*Tatyanaite   (Pt,Pd,Cu)$_9$Cu$_3$Sn$_4$

Paolovite family

Palarstanide	$Pd_5(Sn,As)_2$
Unnamed 020	$(Pt,Pd)_5(Sn,Sb)_2$
Paolovite	Pd_2Sn

  Niggliite group

Niggliite	$PtSn$
Unnamed 436	$(Ni,Pt)Sn$

1.1.1.1.1.2.9.2. Stanides-arsenides

Unnamed 233	Pd_6Sn_2As

1.1.1.1.1.2.10. Plumbides
1.1.1.1.1.2.10.1. Proper plumbides

Plumbopalladinite	Pd_3Pb_2
*Norilskite	$(Pd,Ag)_{2-x}Pb\ (0.08 <= x <= 0.11)$

1.1.1.1.1.2.10.2. Plumbido-sulfides

*Shandite	$Pb_2Ni_3S_2$
Rhodplumsite	$Rh_3Pb_2S_2$
*Laflammeite	$Pd_3Pb_2S_2$

1.1.1.1.2. Minerals of Ib-elements
1.1.1.1.2.1. Minerals of Ag
1.1.1.1.2.1.1. Native metals
Silver family

Silver-*3C*	Ag
Silver -*2H*	
Silver -*4H*	
1.1.1.1.2.1.2. Metallides	1.1.1.1.2.1.2.1. Ag, Au and Cu
Unnamed 005	$AuAgCu$
	1.1.1.1.2.1.2.2. Mercureeds (amalgams)
*Amalgam	(Ag,Hg)
	1.1.1.1.2.1.2.2.1. Simple

Moschellandsbergite family

*Eugenite	$Ag_{11}Hg_2$
Luanheite	Ag_3Hg
Moschellandsbergite	Ag_2Hg_3
Paraschachnerite	$Ag_{1,2}Hg_{0,8}$
Schachnerite	$Ag_{1,1}Hg_{0,9}$
1.1.1.1.2.1.2.2.2. Complex	1.1.1.1.2.1.2.2.2.1. Ag, Au and Hg
Unnamed 016	$Au_5Ag_{10}Hg$
Unnamed 015	$Au_{1.6}Ag_{7.4}Hg$

1.1.1.1.2.2. Minerals of Au
1.1.1.1.2.2.1. Native metals
  Gold series

Gold	Au
Electrum	(Au,Ag)
1.1.1.1.2.2.2. Metallides	1.1.1.1.2.2.2.1. Mercureeds (amalgams)
Unnamed 3	$(Au,Ag)_3Hg$
Weishanite	$(Au,Ag)_{1.2}Hg_{0.8}$

*Amalgam $(Au,Ag)_2Hg$
*Amalgam $\alpha\text{-}AuAgHg$
*Amalgam $(Au,Ag)Hg$
                                                     1.1.1.1.2.2.2.2. Cuprides
Auricupride family
Unnamed 464 Au_3Cu
Auricupride $AuCu_3$
Tetra-auricupride $AuCu$
                                                     *1.1.1.1.2.2.2.3. Stannides
*Yuanjiangite $AuSn$
*Unnamed $AuSn_2$
*Nisnite Ni_3Sn
                                                     1.1.1.1.2.2.2.4. Plumbides
*Novodneprite $AuPb_3$
Anyuiite $AuPb_2$
*Hunchunite Au_2Pb

1.1.1.1.3. Minerals of **II**b-elements
1.1.1.1.3.1. Native metals
Mercury Hg
1.1.1.1.3.2. Metallides   1.1.1.1.3.2.1. Cuprides
Kolymite Cu_7Hg_6
*Belendorffite Cu_7Hg_6 trig.,pseudo-cub.
                                                     1.1.1.1.3.2.2. Plumbides
Leadamalgam (altmarkite) $Hg_{0.3}Pb_{0.7}$

1.1.1.2. Minerals of cenosymmetrical d-elements
*1.1.1.2.1. Minerals of **VII**b-elements
*1.1.1.2.1.1. Native metals
1.1.1.2.2. Minerals of **VIII**b-elements (Fn)
1.1.1.2.2.1. Native metals
Iron family
Iron $\alpha\text{-}Fe$
*Hexaferrum $(Fe_{0.65}Ir_{0.14}Os_{0.08}Ru_{0.08}Rh_{0.03}Ni_{0.01}Cu_{0.01})_{\Sigma 1.00}$
   Nickel series
   Nickel Ni
   Taenite $\gamma\text{-}(Ni,Fe)$
   Kamacite $\alpha\text{-}(Fe,Ni)$
1.1.1.2.2.2. Metallides   1.1.1.2.2.2.1. Only Fn
Awaruite   up $FeNi_2$ to $FeNi_3$
   Tetrataenite group
   Tetrataenite $FeNi$
   Wairauite $FeCo$
Unnamed 014 Fe_2Co
                                                     1.1.1.2.2.2.2. Chromides
   Ferchromide series
   Ferchromide $Cr_{1.5}Fe_{0.2}$
   Chromferide $Cr_{0.2}Fe_{1.5}$
                                                     1.1.1.2.2.2.3. Stanides
Unnamed 288 Ni_3Sn_2

1.1.1.2.2.2.4. Plumbides
1.1.1.2.2.2.4.1. Plumbido-sulfides
Shandite $Ni_3Pb_2S_2$
*1.1.1.2.2.2.5.Tantalido-niobides
*Jedwabite $Fe_7(Ta,Nb)_3$

1.1.1.2.3. Minerals of **I**b-elements (Cu)
1.1.1.2.3.1. Native metals
Copper   Cu
1.1.1.2.3.2. Metallides   1.1.1.2.3.2.1. Zincides
Brass family
Unnamed 009 $Cu_{4.45}Zn$
Unnamed 008 Cu_3Zn
*Unnamed $(Cu, Au,Ag)_4Zn$
*Unnamed $Cu_{1.81}Zn_{1.2}Fe_{0.07}$
Brass $\beta\text{-}CuZn$
Zhanghengite   CuZn
Danbaite $CuZn_2$
                                 1.1.1.2.3.2.2. Stanides
Bronze family
Unnamed 013 (Cu,Ni,Sn)
Unnamed 012 Cu_3Sn
Bronze-n Cu_6Sn_5
Unnamed 007 $Cu(Sn,Sb)$
*Unnamed solid solution $Cu_3AuHg_{0.4}Sn_{0.7}\text{-}Cu_3Au_{1.8}HgSn$
                                 1.1.1.2.3.2.3. Aluminides
Cupalite $(Cu,Zn)Al$
Khatyrkite $(Cu,Zn)Al_2$

1.1.1.2.4. Minerals of **II**b-elements(Zn and Cd)
1.1.1.2.4.1. Native metals
 Zinc group
 Zinc   Zn
 Cadmium   Cd
1.1.1.2.4.2.Metallides   1.1.1.2.4.2.1. Aluminides
Unnamed 011 Zn_2Al
Unnamed 010 $ZnAl_2$

*1.1.1.2.5. Minerals of **VI**b-elements
*1.1.1.2.5.1. Native metals
*Titanium   Ti

1.1.1.3. <u>Minerals of noncenosymmetrical p-elements</u>
1.1.1.3.1. Minerals of **III**a-elements
1.1.1.3.1.1. Native metals
Indium   In

1.1.1.3.2. Minerals of **IV**a-elements
1.1.1.3.2.1. Native metals
Tin   Sn

Lead   Pb

1.1.2. *Class*: Metalls and metallides of lithophylic elements
1.1.2.1. Minerals of light *d*-elements_ (with 1 – 4 d-electrons)
1.1.2.1.1. Minerals of noncenosymmetrical *d*-elements
1.1.2.1.1.1. Minerals of V*b*-elements
1.1.2.1.1.1.1. Native metals
*1.1.2.1.1.2. Minerals of **VI**b-elements
*1.1.2.1.1.2.1. Native metals
*Hexamolybdenum   (Mo,Ru,Fe)
*Tungsten   W

1.1.2.1.2. Minerals of cenosymmetrical *d*-elements
1.1.2.1.2.1.Minerals of **VI***b*-elements
1.1.2.1.2.1.1. Native metals
Chromium   Cr

1.1.2.2. Minerals of light *p*-elements
1.1.2.2.1. Minerals of light cenosymmetrical *p*-elements
1.1.2.2.1.1. Minerals of **III**a-elements
1.1.2.2.1.1.1. Native metals
Aluminium   Al
*1.1.2.2.1.1.2. Metallides
*Unnamed $Al_{0.98}(Si,Cu,Ag)_{0.02}$
*Unnamed $Al_{0.72}(Si,Cu,Ag,Mn,Fe)_{0.28}$
*Unnamed $Al_{0.7}(Si,Mn,Fe)_{0.3}$
*Icosahedrite $Al_{63}Cu_{24}Fe_{13}$

1.2. Subtype: Semimetals and semimetallides (ohly of sidero- and chalcophylic cations)

1.2.1.Quasisubtype*: Semimetals and semimetallides of V*a*-elements

1.2.1a. *Class*: Native V*a*-semimetalls
 Arsenic family
   Arsenic group
   Arsenic   As
   *Pararsenolamprite   As
   Stibarsen (allemontite)   SbAs
   Antimony   Sb
   Bismuth   Bi
   Arsenolamprite   As
   Paradocrasite $Sb_2(Sb,As)_2$

1.2.1b. *Class*: Va-Semimetals- arsenides, antimonides, bismuthides
1.2.1b.1. Minerals of heavy *d*-elements (with 5 – 10 *d*-electrons)
1.2.1b.1.1. Minerals of heavy noncenosymmetrical *d*-elements (with 5 – 10 *d*-electrons)
1.2.1b.1.1.1. Minerals of **VIII***b*-cations (Pn^{n+})
1.2.1b.1.1.1.1. Arsenides

1.2.1b.1.1.1.1.1. Polyanionic
1.2.1b.1.1.1.1.1.1. Pn^{2+}
1.2.1b.1.1.1.1.1.1.1. Proper polyarsenides with As : Pn = 2 - diarsenides (simple)
Sperrylite family 0,5
Sperrylite $Pt[As_2]$
Iridarsenite $Ir[As_2]$
  Omeiite group
  Omeiite $(Os,Ru)[As_2]$
  Anduoite $(Ru,Os)[As_2]$
  Unnamed 442 $(Fe,Pt)[(As,S)_2]$
1.2.1b.1.1.1.1.1.2. Pn^{3+}
1.2.1b.1.1.1.1.1.2.1. Polyarsenido-sulfides with (As+S) : Pn = 2 - diarsenido-sulfides
(simple)
  Hollingworthite series
  Platarsite 0,5 $(Pt,Ph,Ru)[AsS]$
  Hollingworthite $(Rh,Pt,Pd)[AsS]$
  Osarsite group
  Osarsite 0,5 $(Os,Ru)[AsS]$
  Irarsite 0,5 $(Ir,Ru,Rh,Pt)[AsS]$
  Ruarsite 0,5 $Ru[AsS]$

1.2.1b.1.1.1.1.1.3. $Pn^{>3+}$ (?)
1.2.1b.1.1.1.1.1.3.1. Polyarsenides → subarsenides
1.2.1b. 1.1.1.1.1.3.1.1. Proper polyarsenides (simple)
Unnamed 400   0,2 $OsAs_5$
Unnamed 175   0,6 Pd_3As_5
Unnamed 176   0,(6) Pd_2As_3

1.2.1b.1.1.1.1.1.3.2. Polyarsenido-sulfides
1.2.1b.1.1.1.1.1.3.2.1.Simple
Unnamed 463   0,1(6) $PtAs_2S_4$
Unnamed 441   0,25 $PtAs_2S_2$
Unnamed 174   0,25 $Pt_2As_5S_3$
1.2.1b.1.1.1.1.1.3.2.2. Complex
Unnamed 163   0,1(6) $PdCu(As,S)_6$
Unnamed 312   0,(3) $Pd_2Cu_2As_5S_7$
1.2.1b.1.1.1.1.2. Subarsenides
1.2.1b.1.1.1.1.2.1. Proper subarsenides
1.2.1b.1.1.1.1.2.1.1. Simple
Unnamed 461   3,(3) $Pd_{10}As_3$
  Series **(Pd,Ni)$_3$As** 3
  Unnamed 410 $(Pd,Ni)_3As$
  Unnamed 414 $(Pd,Pt,Pb)_3(As,Sb)$
  *Unnamed $(Pd,Pt)_3(Sb,Sn,As)$
  *Vincentite $(Pd,Pt,)_3(As,Sb,Te)$
Unnamed 388 $(Ru,Os,Fe,Rh,Ir,Ni)_3As$
Stillwaterite family
Stillwaterite 2,(6) Pd_8As_3
Arsenopalladinite $Pd_8(As,Sb)_3$
Unnamed 224   2,5 Pd_5As_2

Unnamed 231 2,5	$(Pd,Ni)_5As_2$
Palladoarsenide family 2	
Palladoarsenide	Pd_2As
*Rhodarsenide	$(Rh,Pd)_2As$
*Palladodymite	$(Pd,Rh)_2As$
Palladobismutharsenide	$Pd_{10}(BiAs_4)_{\Sigma 5}$
Unnamed	$Pd_8(BiAs_3)_{\Sigma 4}$
*Unnamed	$Pd_3(Sb,As)$
*Polkanovite 1,7	$Rh_{12}As_7$
1.2.1b.1.1.1.1.2.1.2. Complex	1.2.1b.1.1.1.1.2.1.2.1. Only Pn
Unnamed 403 2	$OsRuAs$
	1.2.1b.1.1.1.1.2.1.2.2. Pn and Fn
Unnamed 232 2,(6)	$Pd_2Ni_6As_3$
*Unnamed 2,5	$Pd_5(As,Te,Sn)_2$
*Unnamed 2,5	$(Pd,Pt)_5(Sn,As,Sb)_2$
Unnamed 223 2,(3)	$Pd_3Ni_4As_3$
Majakite group	
Majakite	$PdNiAs$
Zaccariniite = Unnamed 387	$RhNiAs$
Unnamed 381 1,7(3)	$Pd_{1.6}NiAs_{1.5}$
*Menshikovite	$Pd_3Ni_2As_3$
	1.2.1b.1.1.1.1.2.1.2.3. Pn and Cu
Unnamed 434 8,(3)	$(Pt,Pd)_{17}Cu_8As_3$
Unnamed 221 3,5	$(Pd,Cu)_7(As,Sb)_2$
	1.2.1b.1.1.1.1.2.1.2.4. Pn and Hg
Atheneite 2	$Pd_2(As_{0.75}Hg_{0.25})$
	1.2.1b.1.1.1.1.2.1.2.5. Pn and Pb
Borishanskiite 2	$Pd_{1+x}(As,Pb)_2$ (x=0-0.2)
	1.2.1b.1.1.1.1.2.1.2.6. Pn and Sn
Group **Pd$_6$SnAs** 7	
Unnamed 310	Pd_6SnAs
Unnamed 311	Pd_6SnSb

1.2.1b.1.1.1.2.2.2. Subantimonido-arsenides (simple)

Mertieite family	
Mertieite-I 2,2	$Pd_{5+x}(Sb,As)_{2-x}$ (x=0.1-0.2)
Isomertieite 2,2	$Pd_{11}Sb_2As_2$
Mertieite-II 2,(6)	$Pd_8(Sb,As)_3$

1.2.1b.1.1.1.1.2.3. Subarsenido-sulfides
1.2.1b.1.1.1.1.1.2.3.1. Simple

Unnamed 229 2,(6)	Pd_8As_2S

1.2.1b.1.1.1.1.2.3.2. Complex

Daomanite 0,(6)	$CuPtAsS_2$

1.2.1b.1.1.1.1.3. Monoanionic
1.2.1b.1.1.1.1.3.1. Pn^{2+} and Pn^{3+} (complex)
1.2.1b.1.1.1.1.3.1.1. Proper arsenides

Unnamed 225 1,(3)	$Pd_4As_3 \rightarrow Pd^{2+}_3Pd^{3+}As_3$

1.2.1b.1.1.1.1.3.1.2. Arsenido- sulfides
Unnamed 401   1 $Os^{2+}Rh^{3+}AsS$

1.2.1b.1.1.1.1.3.2. Pn^{3+}
1.2.1b.1.1.1.1.3.2.1. Proper arsenides (simple)
  Ruthenarsenite group   1
  Cherepanovite   RhAs
  Ruthenarsenite   (Ru,Ni)As
  Unnamed 309   Pd(As,Te)

1.2.1b.1.1.1.1.3.2.2. Arsenido- sulfides (simple)
Unnamed 177   0,(8) $Pd^{3+}_8As_6S_3$
*1.2.16.1.1.1.1.3.2.3. Arsenido -tellurides
*Törnroosite $Pd_{11}As_2Te_2$
1.2.1b.1.1.1.2. Antimonides
1.2.1b.1.1.1.2.1. Polyantimonides
1.2.1b.1.1.1.2.1.1. Pn^{2+}
1.2.1b.1.1.1.2.1.1.1. Proper Polyantimonides (simple)
  Geversite group (?)
  Geversite $Pt[Sb_2]$
  Unnamed 161   0,5 $Pd[Sb_2]$ (?)

1.2.1b.1.1.1.2.1.1.2. Polybismuthido-anthimonido-tellurides (simple)
1.2.1b.1.1.1.2.1.2. Pn^{3+}
1.2.1b.1.1.1.2.1.2.1. Polyantimonido -sulfides (simple)
  Tolovkite grcup (?)
  Tolovkite   Ir[SbS]
   *Changchengite   Ir[BiS]
  Unnamed 158   0,5   Rh[SbS]
1.2.1b.1.1.1.2.2. Subantimonides (simple)
Unnamed 472   4 Pd_4Sb
Unnamed 429   3 $(Pd,Cu)_3Sb$
*Naldrettite Pd_2Sb
Stibiopalladinite 2,5 Pd_5Sb_2
Unnamed 234   2 $(Pd,Pt,Ni)_2(Sb,Sn)$
  Genkinite series (?)
  Genkinite 1,(3) $(Pt,Pd,Rh,Ni)_4Sb_3$
  Unnamed 493 1,(3) $(Pd,Pt)_4Sb_3$
   *Ungavaite Pd_4Sb_3

1.2.1b.1.1.1.2.3. Monoanionic
1.2.1b.1.1.1.2.3.1. Pn^{2+}
1.2.1b.1.1.1.2.3.1.1. Proper antimonides (simple)
Unnamed 445 1,5 $(Pt,Pd)_3Sb_2 \rightarrow (Pt,Pd)^{2+}_3Sb_2$
1.2.1b.1.1.1.2.3.2. Minerals Pn^{3+}
1.2.1b.1.1.1.2.3.2.1. Proper antimonides
1.2.1b.1.1.1.2.3.2.1.1. Simple
  Sudburyite group
  Unnamed 142   1   RhSb
  Sudburyite   1   PdSb

Unnamed 471   1   Pd(Sb,Te,Bi)
Stumpflite   1   PtSb
1.2.1b.1.1.1.2.3.2.1.2. Simple→ Complex
Unnamed 179 0,(6) $Pd_2(Sb,Te)_3$

1.2.1b.1.1.1.3. Bismuthides
1.2.1b.1.1.1.3.1. Polybismuthides
1.2.1b.1.1.1.3.1.1. Pn^{2+}
1.2.1b.1.1.1.3.1.1.1. Polybismuthides with Bi : Pn = 2 - dibismuthides (simple)
  Froodite group
  Insizwaite   0,5 $Pt[(Bi,Sb)_2]$
  Froodite $Pd[Bi_2]$

1.2.1b.1.1.1.3.1.2. $Pn^{>3+}$ (?)
1.2.1b.1.1.1.3.1.2.1. Polybismuthides → subbismuthides (?) with Bi : Pn = 0,(3) (simple)
Unnamed 230 0,(3) $PdBi_3$
1.2.1b.1.1.1.3.2. Subbismuthides (complex)
Unnamed 018 6 Pd_3Pb_3Bi
Unnamed 228 4 $(Pd,Rh,Pt)_3Pb(Bi,Te)$

1.2.1b.1.1.1.3.3. Monoanionic
1.2.1b.1.1.1.3.3.1. Pn^{3+}
1.2.1b.1.1.1.3.3.1.1. Proper bismuthides (simple)
  Series **(Pt,Pd)(Bi,Sb)**
  Unnamed 446 1 $(Pt,Pd)(Bi,Sb)$
  Unnamed 143 1 $(Pt,Pd)(Bi,Sb)$
Polarite $Pd(Bi,Pb)$

1.2.1b.1.1.1.3.3.1.2. Pn^{2+} and Pn^{4+}
1.2.1b.1.1.1.3.3.1.2.1. Proper bismuthides (complex)
Urvantsevite $Pd(Bi,Pb)_2$

1.2.1b.1.1.2. Minerals of **Ib**-cations
1.2.1b.1.2.1. Minerals of Ag
1.2.1b.1.1.2.1.1. Arsenides
1.2.1b.1.1.2.1.1.1. Polyarsenido (?)-sulfides (simple)
Dervillite 0,(6) Ag_2AsS_2
1.2.1b.1.1.2.1.1.2. Monoanionic (complex)
Kutinaite $(Ag_6Cu_{14})_{\Sigma 20}As_7 \to (Ag_6Cu_{13})^+_{\Sigma 19}Cu^{2+}As_7$

1.2.1b.1.1.2.1.2. Antimonides
1.2.1b.1.1.2.1.2.1. Polyantimonides (?) (simple)
Allargentum $Ag_{1-x}Sb$ (0,09 < x < 0,16)
1.2.1b.1.1.2.1.2.2. Monoanionic (simple)
Dyscrasite Ag_3Sb
1.2.1b.1.1.2.2. Minerals of Au
1.2.1b.1.1.2.2.1. Antimonides
1.2.1b.1.1.2.2.1.1. Polyantimonides (?) (simple – Au^{2+})
Aurostibite   0,5 $Au[Sb_2]$

1.2.1b.1.1.2.2.1.2. Monoanionic
1.2.1b.1.1.2.2.1.2.1. Au^+
1.2.1b.1.1.2.2.1.2.1.1. Antimonido-sulfides (complex)
Criddleite $TlAu_3Ag_2Sb^{3+}_{10}$

1.2.1b.1.1.2.2.2. Bismutides
1.2.1b.1.1.2.2.2.1. Subbismutides (simple)
Maldonite Au_2Bi

1.2.1b.1.1.3. Minerals of **II**b-cations (Hg)
1.2.1b.1.1.3.1. Antimonides
1.2.1b.1.1.3.1.1. Monoanionic
1.2.1b.1.2.3.1.1.1. Antimonido-sulfides (complex)
Tvalchrelidzeite Hg_3SbAsS_3

1.2.1b.1.2. <u>Minerals of cenosymmetrical d-cations</u>
1.2.1b.1.2.1. Minerals of **VIII**b-cations (Fn^{n+})
1.2.1b.1.2.1.1. Arsenides
1.2.1b.1.2.1.1.1. Polyanionic
1.2.1b.1.2.1.1.1.1. Fn^{2+} (seldom Fn^{4+} ?)
1.2.1b.1.2.1.1.1.1.1. Proper arsenides with As : Fn = 2 – diarsenides (simple)
Löllingite family
Krutovite $Ni[As_2]$
Pararammelsbergite $Ni[As_2]$
  Löllingite group (compare with marcasite (group); ullmannite (series))
  Rammelsbergite $Ni[As_2]$
  Safflorite $Co[As_2]$
  Löllingite $Fe[As_2]$
Clinosafflorite $Co[As_2]$

1.2.1b.1.2.1.1.1.2. Minerals of Fn^{3+}
1.2.1b.1.2.1.1.1.2.1. Proper arsenides with As : Fn = 3 – Triarsenides (simple)
  Skutterudite series
  Nickelskutterudite (chloanthite) $(Ni,Co,Fe)As_3$
  Skutterudite $CoAs_3$
  *Ferroskutterudite $(Fe,Co)As_3$
1.2.1b.1.2.1.1.1.2.2. Arsenido-sulfides with (As+S) : Fn = 2 – diarsenido-sulfides (simple)
  Gersdorffite series (?) (compare with pyrite (group))
  Gersdorffite $Ni[AsS]$
  Unnamed 450 $(Ni,Fe,Co)[AsS]$
Arsenopyrite family (compare with marcasite (group); ullmannite (series))
Cobaltite $Co[AsS]$
Glaucodote $(Co,Fe)[AsS]$
Alloclasite $(Co,Fe)[AsS]$
Arsenopyrite $Fe[AsS]$
Seinäjokite $(Fe,Ni)[(Sb,As)_2]$
*Unnamed $(Fe,Ni)SbAs$
*Oenite $CoSbAs$

1.2.1b.1.2.1.1.1.1.2.3. Arsenido-selenides with (As+Se) : Fn = 2 – diarsenido-selenides
Unnamed 479   Co[AsSe](?)
*Jolliffeite   NiAsSe

1.2.1b.1.2.1.1.1.2. Subarsenides
1.2.1b.1.2.1.1.1.2.1. Proper arsenides (simple)
Maucherite family
Dienerite = nickelskutterudite $(Ni,Co,Fe)As_3$
Orcelite   2,5 $Ni_{5-x}As_2$
Unnamed 385 2,(3) $(Ni,Fe)_7(As,Sb)_3$
Maucherite   1,375 $Ni_{11}As_8$

1.2.1b.1.2.1.1.1.2.2. Subarsenido-antimonido-sulfides (simple)
Vozhminite $(Ni,Co)_4(As,Sb)S_2$

1.2.1b.1.2.1.1.1.3. Monoanionic
1.2.1b.1.2.1.1.1.3.1. Fn^{2+}
1.2.1b.1.2.1.1.1.3.1.1. Proper arsenides (simple)
Oregonite $FeNi_2As_2$

1.2.1b.1.2.1.1.1.3.2. Fn^{3+}
1.2.1b.1.2.1.1.1.3.2.1. Proper arsenides (simple)
Nickeline family
  Nickeline series
  Nickeline   NiAs
  Langisite   CoAs
  Breithauptite   NiSb
   Modderite group
   Modderite   orth.   CoAs
   Westerveldite   orth.   (Fe,Ni)As

1.2.1b.1.2.1.1.1.3.3. Fn^{3+} and Fn^{4+}
1.2.1b.1.2.1.1.1.3.3.1. Arsenido-sulfides (complex)
Unnamed 157 $(Co,Ni)_2AsS_2 \rightarrow Fn^{3+}Fn^{4+}AsS_2$
1.2.1b.1.2.1.1.1.3.4. Fn^{4+}
1.2.1b.1.2.1.1.1.3.4.1. Proper arsenides (simple)
Unnamed 162 $(Ni,Pd)_3As_4$

1.2.1b.1.2.1.2. Antimonides
1.2.1b.1.2.1.2.1. Polyanionic
1.2.1b.1.2.1.2.1.1. Fn^{2+}
1.2.1b.1.2.1.2.1.1.1. Proper polyantimonides with Sb : Fn = 2 - diantimonides (simple)
  Nisbite group (?) (compare with pyrite (series))
  Nisbite $Ni[Sb_2]$
  Unnamed 159 $Co[Sb_2]$
*1.2.1б.1.2.1.2.1.1.2. Proper polyantimonides with Sb : Fn = 3 – триантимониды
*Kieftite $CoSb_3$

1.2.1b.1.2.1.2.1.2. Fn^{3+}

1.2.1b.1.2.1.2.1.2.1. Polyantimonido-sulfides with (Sb+S) : Fn = 2 - diantimonides-sulfides (simple)
Ullmannite family (compare with marcasite (droup.); arsenopyrite (series))
Ullmannite   Ni[SbS]
Willyamite   (Co,Ni)[SbS]
Costibite   Co[SbS]
Paracostibite   Co[SbS]
Gudmundite   Fe[SbS]

1.2.1b.1.2.1.2.2. Subantimonides (simple)
Unnamed 384 3 Ni_3Sb
Unnamed 383 2 $(Ni,Cu)_2Sb$

1.2.1b.1.2.1.2.3. Monoanionic
1.2.1b.1.2.1.2.3.1. Fn^{2+}
1.2.1b.1.2.1.2.3.1.1. Bismuthido-sulfides (complex – Ni^{2+} and Pb^{2+})
Parkerite (an. 3) $Ni_3(Bi,Pb)_2S_2$

1.2.1b.1.2.1.2.3.2. Fn^{2+} and Fn^{3+}
1.2.1b.1.2.1.2.3.2.1. Antimonido-sulfides → bismuthido-tellurido-sulfides →
Bismuthido-sulfides (complex – Ni^{2+} and Ni^{3+}))
Hauchecornite family ($Ni^{2+}_5Ni^{3+}_4X_2S_8$, X = As^{3-}, Sb^{3-}, Bi^{3-}, Te^{2-})
Arsenohauchecornite $Ni^{2+}_{10}Ni^{3+}_8Bi_3AsS_{16}$
  Hauchecornite group
  Tučekite $Ni^{2+}_5Ni^{3+}_4Sb_2S_8$
  Hauchecornite $Ni^{2+}_5Ni^{3+}_4Bi(Sb,Bi)S_8$
  Bismutohauchecornite $Ni^{2+}_5Ni^{3+}_4Bi_2S_8$
  Tellurohauchecornite $Ni^{2+}_6Ni^{3+}_3BiTeS_8$

1.2.1b.1.2.1.2.3.3. Fn^{3+} and Fn^{4+}
1.2.1b.1.2.1.2.3.3.1. Bismuthido-sulfides (complex- Ni^{3+} and Ni^{4+})
Parkerite (an. 1 and 2) $Ni^{3+}_2Ni^{4+}Bi_2S_2$

1.2.1b.1.2.2. Minerals of I*b*-cations
1.2.1b.1.2.2.1. Arsenides
1.2.1b.1.2.2.1.1. Polyanionic (only Cu^{2+})
1.2.1b.1.2.2.1.1.1. Poper arsenides (simple)
Paxite $Cu[As_2]$

1.2.1b.1.2.2.1.1.2. Arsenido-sulfides (simple)
Lautite $Cu[AsS]^\infty$

1.2.1b.1.2.2.1.2. Subarsenides (simple)
Algodonite $Cu_{17}As_3$

1.2.1b.1.2.2.1.3. Monoanionic
1.2.1b.1.2.2.1.3.1. Cu^+
1.2.1b.1.2.2.1.3.1.1. Poper arsenides (simple)
Domeykite family
Domeykite α-Cu_3As

Metadomeykite β-Cu_3As

1.2.1b.1.2.2.1.3.2. Cu^+ and Cu^{2+}
1.2.1b.1.2.2.1.3.2.1. Poper arsenides (complex)
Koutekite family
Koutekite $Cu_5As_2 \rightarrow Cu^+_4Cu^{2+}As_2$
Novakite $(Cu,Ag)_{21}As_{10} \rightarrow (Cu,Ag)^+_{12}Cu^{2+}_9As_{10}$
Unnamed 220 $Cu_2As \rightarrow Cu^+Cu^{2+}As$

1.2.1b.1.2.2.2. Antimonides
1.2.1b.1.2.2.2.1. Subantimonides (simple)
1.2.1b.1.2.2.2.2. Monoanionic
1.2.1b.1.2.2.2.2.1. Cu^+ and Cu^{2+}
1.2.1b.1.2.2.2.2.1.1. Proper antimonides (complex)
Cuprostibite $Cu_2Sb \rightarrow Cu^+Cu^{2+}(Sb,Tl)$
*Zlatogorite $CuNiSb_2$
*Sorosite $Cu(Sn,Sb)$

1.2.1b.1.2.2.2.2.1.2. Antiminido-sulfides (complex)
                                       1.2.1b.1.2.2.2.2.1.2.1.1. Cu^+, Cu^{2+} and Tl^+
Rohaite $TlCu_5SbS_2 \rightarrow Tl^+Cu^+_4Cu^{2+}SbS_2$
                                       1.2.1b.1.2.2.2.2.1.2.1.2. Cu^+, Cu^{2+},Tl^+ andFe^{2+}
Chalcothallite   3 $Tl_2(Cu,Fe)_6SbS_4 \rightarrow Tl^+_6(Cu,Ag)^+_9(Cu,Fe)^{2+}_9Sb_3S_{12}$

1.2.1b.2. Minerals of noncenosymmetrical *p*-cations
1.2.1b.2.1. Minerals of **IV***a*-cations
1.2.1b.2.1.1. Minerals of Sn^{2+} (?)
1.2.1b.2.1.1.1. Antimonides
1.2.1b.2.1.1.1.1. Polyanionic(?) antimonido-arsenides (simple)
Unnamed 181   0,25 $Sn[(Sb,As)_2]_2$

1.2.1b.2.1.1.1.1.Моноанионные (сложные - Sn^{2+} and Sn^{4+})
Stistaite $2SnSb \rightarrow Sn^{2+}Sn^{4+}Sb_2$

1.2.2. Quasisubtype*: Semimetals and semimetallidess of **V***a*-semimetals

1.2.2a. **Class:** Native **VI**a-semimetals
Tellurium   Te

1.2.2b. *Class*: **VI**a-semimetalls -tellurides
1.2.2b.1. Minerals of heavy *d*-cations (with 5 – 10 *d*-electrons)
1.2.2b.1.1. Minerals of noncenosymmetrical *d*-cations
1.2.2b.1.1.1. Minerals of **VIII***b*-cations (Pn^{n+})
1.2.2b.1.1.1.1. Polytellurides (?)
1.2.2b.1.1.1.1.1. Proper polytellurides (simple)
*Unnamed (Pd теллурид) Pd_2Te
Unnamed 180   0,(3) $PdTe_3$
*Merenskyite $Pd_{1-x}Pt_xBi_yTe_{2-y}$ ($0 \leq x \leq 0,5$); ($0 \leq y \leq 0,(6)$) или $PdTe_2$
*Gaotaiite   0,375 Ir_3Te_8

*Shuangfengite   0,5 $(Ir,Pt)Te_2$
*1.2.26.1.1.1.1.1.1. Complex
*Pašavaite $Pd_3Pb_2Te_2$

1.2.2b.1.1.1.1.2. Polytellurido-bismuthides (simple)
  Michenerite series (?) (Pn^{3+})
  Maslovite   0,5 $Pt[BiTe]$
  Michenerite $Pd[BiTe]$

1.2.2b.1.1.1.2. Subtellurides
1.2.2b.1.1.1.2.1. Proper subtellurides
1.2.2b.1.1.1.2.1.1. Simple
Unnamed 459   8 $(Pd,Au)_8(Te,As)$
Unnamed 343   8 $(Pd,Au)_8(Te,As)$
 Keithconnite family
  Keithconnite group (?)   3
  Unnamed 516 $(Rh,Pd)_3Te$
  Keithconnite $Pd_{20}Te_7$
Unnamed 115   2,(6) Pd_8Te_3
  Group. (?)   2.5
  Unnamed 117 $Pd_5(Te,Bi,Sb)_2$
  Unnamed 116 $(Pd,Cu,Sn)_5(Te,S)_2$
Telluropalladinite   2,25 Pd_9Te_4
*Unnamed   2 Pd_2Te
Unnamed 118   1,5 $(Pd,Ni)_3(Te,Sb,Bi)_2$
*Oulankaite   1,5 $(Pd,Pt)_5(Cu,Fe,Ag)_4SnTe_2S_2$
*Oulankaite-Ag $(Pd,Pt)_{5+x}(Ag,Cu,Fe)_{4-x}SnTe_2S_2$
*Unnamed $(Pd,Ni)_2Te_2Sb$

1.2.2b.1.1.1.2.1.2. Complex
1.2.2b.1.1.1.2.1.2.1. Pn and Ag
  Sopcheite group   1,75
  Unnamed 440 Pd_6AgTe_4
  Sopcheite $Pd_3Ag_4Te_4$
*Lukkulaisvaaraite $Pd_{14}Ag_2Te_9$
                              1.2.2b.1.1.1.2.1.2.2. Pn and Ag, Pb(Bi)
Telargpalite $PdAg_3Te$
*Telargpalite-Bi $(Pd,Ag)_3(Bi_{0,51}Te_{0,43}Pb_{0,02})_{0.96}$
                              1.2.2b.1.1.1.2.1.2.3. Pn and Hg
Temagamite Pd_3HgTe_3
                              1.2.2b.1.1.1.2.1.2.4. Pn and Bi
Unnamed 438   0,(3) $PdBiTe_2$

1.2.2b.1.1.1.2.2. Subtellurido-arsenides (simple)
Unnamed 341   4 $Pd_8(As,Te)_2$
Unnamed 460   4 Pd_8AsTe
Unnamed 342   3,(3) $Pd_{10}(As,Te)_3$
Unnamed 524   3 $(Pt,Pd)_3(Te,As)$
Unnamed 476   3 $Pd_3(Te,As)$
Unnamed 477   2,45 $(Rh,Pd)_{4.9}(As,Te)_2$

Unnamed 119   2 $Pd_2(Te,As)$
*Unnamed $Pd_{11}Te_2As_2$

1.2.2b.1.1.1.2.3. Subtellurido-sulfides (complex)
Unnamed 116   2,5 $(Pd,Cu,Sn)_5(Te,S)_2$
Vasilite   2,3 $(Pd,Cu)_{16}(S,Te)_7$
1.2.2b.1.1.1.3. Monoanionic
1.2.2b.1.1.1.3.1. Pn^{2+} and Pn^{3+} (complex)
1.2.2b.1.1.1.3.1.1. Proper tellurides
Unnamed 439   0,75 $(Pt,Pd,Ni)_3Te_4 \rightarrow (Pt,Pd,Ni)^{2+}(Pt,Pd,Ni)^{3+}_2Te_4$
1.2.2b.1.1.1.3.1.2. Tellurido-antimonides
1.2.2b.1.1.1.3.1.3. Tellurido-bismuthides (antimonides).
Unnamed 226   0,875 $Pd_7(Bi,Te)_8 \rightarrow Pd^{2+}_5Pd^{3+}_2(Bi,Te)_8$
Unnamed 173   0,8(3) $Pd_5(Bi,Sb)_2Te_4 \rightarrow Pd^{2+}Pd^{3+}_4(Bi,Sb)_2Te_4$
1.2.2b.1.1.1.3.2. Pn^{3+} (simple)
  Kotulskite series
  Kotulskite $Pd(Te,Bi)_{2-x}$ (x~0.4)
  Sobolevskite $PdBi$
Unnamed 444   0,(8) $Pd^{3+}_8Bi_6Te_3$
1.2.2b.1.1.1.3.3. Pn^{3+} and Pn^{4+} (complex)
1.2.2b.1.1.1.3.3.1. Proper tellurides
Unnamed 171   0,625 $(Ni,Pd)_5(Te,Bi)_8 \rightarrow (Ni,Pd)^{3+}_4(Ni,Pd)^{4+}(Te,Bi)_8$
1.2.2b.1.1.1.3.3.2. Tellurido-antimonides
Borovskite   0,6 $Pd_3SbTe_4 \rightarrow Pd^{3+}Pd^{4+}_2SbTe_4$
1.2.2b.1.1.1.3.4. Pn^{4+} (simple)
Moncheite $PtTe_2$
Merenskyite $PdTe_2$
(Kotulskite)   see. Sobolevskite series
*Mayingite $IrBiTe$
*Telluromayingite $Ir(Te,Bi)_2$

1.2.2b.1.1.2. Tellurides of Ib-cations
1.2.2b.1.1.2.1.Tellurides of Ag^+
1.2.2b.1.1.2.1.1. Polyanionic (?)
1.2.2b.1.1.2.1.1.1. Simple
Unnamed 155   0,25 $(Au,Ag)(Te,Pb)_4$
1.2.2b.1.1.2.1.1.2. Complex (Ag^+, Cu^+ nad Cu^{2+})
Cameronite   0,8 $AgCu_7Te_{10} \rightarrow Ag^+Cu^+_5Cu^{2+}_2[Te_2]_5$

1.2.2b.1.1.2.1.2. Subtellurides
1.2.2b.1.1.2.1.2.1. Simple
Unnamed 109   3,5 Ag_7Te_2
  Empressite family
  Stützite   1,(6) $Ag_{5-x}Te_3$
  Unnamed 382   1,5 Ag_3Te_2
  Empressite   1 $AgTe$
1.2.2b.1.1.2.1.2.2. Complex (Ag и Bi)
Unnamed 185   2 Ag_3BiTe_2
Unnamed 398   1,57 $Ag_8Bi_3Te_7$
Unnamed 237   0,75 $AgBi_2Te_4$

1.2.2b.1.1.2.1.3. Monoanionic
1.2.2b.1.1.2.1.3.1. Proper tellurides
1.2.2b.1.1.2.1.3.1.1. Simple
Hessite 2 Ag_2Te
1.2.2b.1.1.2.1.3.1.2. Complex   1.2.2b.1.1.2.1.3.1.2.1. Ag^+ and Au^+
Petzite family 2
Muthmannite $(Ag,Au)\,Te$
Petzite $AuAg_3Te_2$
           1.2.2b.1.1.2.1.3.1.2.2. Ag^+ and Au^{3+}
Sylvanite $AuAgTe_4$
           1.2.2b.1.1.2.1.3.1.2.3. Ag^+, Cu^+ and Cu^{2+}
Henryite $Ag_3Cu_4Te_4 \rightarrow Ag^+{}_3Cu^+{}_3Cu^{2+}Te_4$

1.2.2b.1.1.2.1.3.1.2.4. Ag^+ and $Sb^{3+} \rightarrow$ telluroantiminites
Unnamed 186 1 $Ag^+Sb^{3+}Te_2$
*Mazzettiite $Ag_3Hg^{2+}PbSb^{3+}Te_5$

1.2.2b.1.1.2.1.3.1.2.5. Ag^+ and $Bi^{3+} \rightarrow$ tellubismuthites Ag
Volynskite $Ag^+Bi^{3+}Te_2$

1.2.2b.1.1.2.1.3.2. Tellurido-sulfides
1.2.2b.1.1.2.1.3.2.1. Simple
Cervelleite family (?)   2
Unnamed 110 $(Ag,Cu,Bi)^+{}_6Te_2S$
Cervelleite $Ag^+{}_4TeS$
*Unnamed Ag_2Cu_2TeS
1.2.2b.1.1.2.1.3.2.2. Complex
           1.2.2b.1.1.2.1.3.2.2.1. Ag^+ and Cu^+
Unnamed 111 2 $Ag^+{}_5Cu^+TeS_2$
           1.2.2b.1.1.2.1.3.2.2.2. Ag^+ and Fe^{2+}
Unnamed 317 1,8 $Ag^+{}_{10}Fe^{2+}Te_2S_4$
           1.2.2b.1.1.2.1.3.2.2.3. Ag^+ and Pb^{2+}
Unnamed 397 1,28 Ag_4PbTe_2S
           1.2.2b.1.1.2.1.3.2.2.4. Ag^+ and 3-valence Va-
           cations $\rightarrow$ tellurosulfoantimonites of Ag^+

Benleonardite family (?)
Unnamed 316 $Ag_9SbTe_2S_4$
Benleonardite $Ag_{15}Cu(Sb,As)_2S_7Te_4$
           1.2.2b.1.1.2.1.3.2.2.5. Ag^+, Cu^{2+} and 3-valence Va-
         cations $\rightarrow$ tellurosulfobismuthites of Ag^+ and Cu^{2+}
Unnamed 154 0,8 $Ag_3Cu^{2+}Bi^{3+}Te_2S_2$

*1.2.26.1.1.2.1.3.3. Tellurido-selenides
*Kurilite Ag_8Te_3Se

1.2.2b.1.1.2.2. Tellurides of Au
1.2.2b.1.1.2.2.1. Polytellurides (?) (simple $\rightarrow$ complex)
Calaverite family   0,5   (x = 2)
Calaverite $AuTe_2$
Krennerite $(Au,Ag)Te_2$

Kostovite $AuCuTe_4$

1.2.2b.1.1.2.2.2. Subtellurides (complex)
Bezsmertnovite 11 Au_8Cu_2PbTe or, more precisely $Au_{8-x}Ag_xCu_{2-y}Fe_yPb_{1-z}Te_{1+z}$
                            $(0 \leq x \leq 0,7); (0,16 \leq y \leq 0,36); (0 \leq z \leq 0,28)$
  Bogdanovite group (?) 8
  Unnamed 131 $Au_5Cu_3(Te,Pb)$
  Bogdanovite $(Au,Te,Pb)_3(Cu,Fe)$
  Bilibinskite group (?) 3
  Unnamed 130 $Au_5Cu(Te,Pb)_2$
  Bilibinskite $Au_3Cu_2Pb \cdot nTeO_2$
  Group. (?) 2
  Unnamed 128 $Au(Fe,Cu)(Te,Pb)$
Unnamed 129 $Au_3(Fe,Cu)(Te,Pb)_2$
*Honeaite Au_3TlTe_2

1.2.2b.1.1.2.2.3. Monoanionic
1.2.2b.1.1.2.2.3.1. Tellurides of Au^+
1.2.2b.1.1.2.2.3.1.1. Tellurido-sulfides (complex of Au^+, Pb^{2+}, Sb^3, Te^{4+}) → telluro-
sulfoantimonites of Au^+ and Pb^{2+} to tellurosulfoantimonites of Ag^+, Pb^{2+} and Te^{4+}
*Museumite $[Pb_2(Pb,Sb)_2S_8][(Te,Au)_2]$
  Nagyagite series
  Nagyágite $[Pb(Pb,Sb)S_2][(Au,Te)]$
*Buckhornite $[Pb_2BiS_3][AuTe_2]$

1.2.2b.1.1.2.2.3.2. Tellurides of Au^{3+} (?)
1.2.2b.1.1.2.2.3.2.1. Proper tellurides (simple)
Montbrayite 1,5 $Au^{3+}_2Te_3$

1.2.2b.1.1.2.2.3.2.2. Tellurido-arsenides (complex of - Au^{3+}, Ag^+ и Pb^{2+})
Unnamed 313 1 $Au^{3+}_3Ag^+Pb^{2+}As_2Te_3$
1.2.2b.1.1.2.2.3.2.3. Tellurido-arsenides (complex of - Au^{3+}, Pb^{2+} и Bi^{3+})
Unnamed 474 0,5 $Au^{3+}Pb^{2+}_2Bi^{3+}Te_2S_3$

1.2.2b.1.1.3. Tellurides of **II**b-cations (only Hg^{2+})
1.2.2b.1.1.3.1. Monoanionic (simple)
Coloradoite $HgTe$

1.2.2b.1.2. <u>Minerals of cenosymmetrical *d*-cations</u>
1.2.2b.1.2.1. Minerals of **VIII**b-cations (Fn^{n+})
1.2.2b.1.2.1.1. Polyanionic
1.2.2b.1.2.1.1.1. Fn^{2+}
1.2.2b.1.2.1.1.1.1. Proper polytellurides with Te : Fn = 2 ditellurides (simple)
  Frohbergite series
  Mattagamite $Co[Te_2]$
  Frohbergite $Fe[Te_2]$
1.2.2b.1.2.1.1.1.2. Polytellurido-selenides with (Te + Se) : Fn = 2 ditellurido-selenides
(simple)
Kitkaite $Ni[TeSe]$
1.2.2b.1.2.1.2. Monoanionic

*Imgreite   NiTe

1.2.2b.1.2.1.2.1. Fn^{3+} and Fn^{4+}
1.2.2b.1.2.1.2.1.1. Tellurido-antiminides (complex)
Vavřinite 0,(6) $Ni_2SbTe_2 \rightarrow Ni^{3+}Ni^{4+}SbTe_2$
1.2.2b.1.2.1.2.2. Fn^{4+}
1.2.2b.1.2.1.2.2.1. Proper tellurides (simple)
Melonite $NiTe_2$

1.2.2b.1.2.2. Tellurides of Ib-cations
1.2.2b.1.2.2.1. Monoanionic
1.2.2b.1.2.2.1.1. Cu^+ и Cu^{2+} (complex)
Ricardite family
Rickardite $Cu_{3-x}Te_2$
Weissite $Cu_{2-x}Te$
1.2.2b.1.2.2.1.2. Cu^{2+} (simple)
Vulcanite $CuTe$

1.2.2b.2. Minerals of p-cations
1.2.2b.2.1. Tellurides of IVa-cations (Pb^{n+})
1.2.2b.2.1.1. Subtellurido-sulfides (complex)
*Saddlebackite $Pb_2Bi_2Te_2S_3$
Unnamed 208 1,(3) $PbBi_3TeS_2$
1.2.2b.2.1.2. Monoanionic
1.2.2b.2.1.2.1. Pb^{2+}
1.2.2b.2.1.2.1.1. Proper tellurides
1.2.2b.2.1.2.1.1.1. Simple
Altaite $PbTe$
1.2.2b.2.1.2.1.1.2. Complex (Pb^{2+} and Bi^{3+})
  Rucklidgeite group   (x=1(3))
  Rucklidgeite $(Pb,Ag,Bi)Bi_2Te_4$
  Aleksite $PbBi_2(Te_2S_2)_{\Sigma 4}$
Kochkarite 1,4 $PbBi_4Te_7$
1.2.2b.2.1.2.1.2. Tellurido-sulfides (simple)
Unnamed 236 1 Pb_2TeS
1.2.2b.2.1.2.1.3. Tellurido-chlorides (complex of - Pb^{2+} and Te^{4+})
Radhakrishnaite $PbTe_3(Cl,S)_2$
1.2.2b.2.1.2.2. Pb^{2+} and Pb^{4+}
1.2.2b.2.1.2.2.1. Proper tellurides (complex)
Unnamed 238 $Pb_2Te_3 \rightarrow Pb^{2+}Pb^{4+}Te_3$
1.2.2b.2.1.2.3. Pb^{4+}
1.2.2b.2.1.2.3.1. Proper tellurides (simple)
Unnamed 239 0,5 $PbTe_2$
1.2.2b.2.1.2.3.2. Tellurido-chlorides (simple)
Kolarite $PbTeCl_2$

1.2.2b.2.2. Tellurides of Va-cations
1.2.2b.2.2.1. Subtellurides
1.2.2b.2.2.1.1. Proper tellurides (simple)
Hedleyite 2,(3) Bi_7Te_3

Unnamed 396   2,25 Bi_9Te_4
Unnamed 125   1,5 Bi_3Te_2
Pilsenite   1,(3) Bi_4Te_3
Tsumoite   1   BiTe
Unnamed 412   0,75 Bi_3Te_4

1.2.2b.2.2.1.2. Subtellurido-sulfides (simple)
Unnamed 121   3 $Bi_{15}TeS_4$
Unnamed 122   2,25 $Bi_9Te_2S_2$
   Series (?)
   "Mineral K"   1,5 $Bi_9(Te_2S)_2$
   "Mineral L"   1,5 Bi_3TeS
    Unnamed 418   1,5 Bi_6TeS_3
"Mineral P"   1,5 $Bi_{15}(TeS_4)_2$
   Joseite series   1,(3)
   Joseite-B Bi_4Te_2S
   Joseite-A Bi_4TeS_2
   *****Baksanite** series
   *Baksanite $Bi_6(Te_2S_3)$
   Ingodite series   1
   Sulphotsumoite Bi_3Te_2S
   Ingodite Bi_2TeS
   Unnamed 147 Bi_2TeS
   "Mineral M" $(Bi,Pb)_2TeS$
   Unnamed 153 Bi_4TeSe_3
   Unnamed 148 $Bi(S,Te)$
   Series (?)   0,75
   Unnamed 126 $Bi_3(Te,Se)_3S$
   Unnamed 123 $Bi_3Te_2S_2$

1.2.2b.2.2.2. Monoanionic
1.2.2b.2.2.2.1. Sb^{3+} and Bi^{3+}
1.2.2b.2.2.2.1.1. Proper tellurides (simple)
   Tellurobismuthite group   0,(6)
   Telluroantimony Sb_2Te_2Te
   Tellurobismuthite Bi_2Te_2Te
*Unnamed Bi_2Te
*1.2.2b.2.2.2.1.2. Complex
*Unnamed $Sb(Ni,Fe,Pd)_2Te_2$
*Unnamed $(Sb,Bi)Pd(Ni,Fe)Te_2$

1.2.2b.2.2.2.1.2.1. Tellurido-sulfides (simple)
   Tetradymite family   0,(6)
   Tetradymite Bi_2Te_2S
   Kawazulite Bi_2Te_2Se
   *Vihorlatite $Bi_{24}Te_4Se_{17}$
*Unnamed Bi_4Te_2Se
*Unnamed $Bi_6(Te,Se)_3$

1.2.2b.2.2.2.2. Bi^{3+} and Bi^{5+} (?) (polytellurides (?) (simple)
Unnamed 152   0,6 $Bi^{3+}Bi^{5+}_3Te_5$ (?)
Unnamed 151   0,4 $Bi^{5+}_2Te_5$ (?)

2. TYPE: MINERALS WITH PRINCIPAL METALLIC-COVALENT AND IONIC- COVALENT BOND , RARE VAN DER WAALS FORSES (NATIVE VIA-NONMETALS) – CHALCOGENIC COMPOUNDS: CHALCOGENIDES (ISODESMICAL) →CHALCOSALTS (ANISODESMICAL)

2a. Quasitype*: Native VIa-nonmetals (van der Waals forses)
Sulfur family

α-Sulfur	S
β-Sulfur	S
Rosickyite	S
*Sulfurite amorphous	S_8
Selenium	Se

2b. Quasitype*: Chalcogenic compounds (metallic-covalent and ionic-covalent bond , rare van der Waals forses)-simple (isodesmical) → complex → chalcosalts (anisodesmical).
2b.1. Subtype: Chalcogenic compounds of sidero- and chalcophylic cations
2b.1a. Quasisubtype*: Sulfides and sulfosalts of sidero- and chalcophylic cations
2b.1a.1. *Class:* Sulfides of sidero- and chalcophylic cations
2b.1a.1.1.Minerals of heavy d-cations (with 5 – 10 d-electrons) and their crystallochemical analogues.
2b.1a.1.1.1. Minerals of cenosymmetrical d-cations
2b.1a.1.1.1.1. Minerals of **VII**b – **VIII**b-cations (Mn, Fe, Co, Ni) and their crystallochemical analogues (Cr^{3+}, V^{3+}, V^{4+}, Ti^{3+}) – cations of wide iron family – Fn'^{n+}
2b.1a.1.1.1.1.1. Polyanionic
2b.1a.1.1.1.1.1.1. Minerals of M^{2+}
2b.1a.1.1.1.1.1.1.1. Simple→ complex

Pyrite group	
Vaesite	$Ni[S_2]$
Cattierite	$Co[S_2]$
Pyrite	$Fe[S_2]$
Villamaninite	$Cu(Ni,Co,Fe)[S_2]_2$
Fukuchilite	$(Cu,Fe)[S_2]$
Hauerite	$Mn[S_2]$
Marcasite group (compare with arsenopyrite (series.))	
Marcasite	$Fe[S_2]$

2b.1a.1.1.1.1.2. Subsulfides
2b.1a.1.1.1.1.2.1. Simple

Heazlewoodite 1,5	Ni_3S_2

2b.1a.1.1.1.1.2.2. Complex
2b.1a.1.1.1.1.2.2.1. Only Fn'

Pentlandite family	
Pentlandite group 1,125	
Pentlandite	$(Co,Ni,Fe)_{<1}Fe_4Ni_4S_8$
Cobaltpentlandite	Co_9S_8
Mackinawite	$(Fe,Ni)_9S_8$
Godlevskite	$(Ni,Fe)_9S_8$
Smythite	$(Fe,Ni)_9S_{11}$

2b.1a.1.1.1.1.2.2.2. Fn', Pt, Cu and Pb

Kharaelakhite $(Cu,Pt,Pb,Fe,Ni)_9S_8$

*Sugakiite $Cu(Fe,Ni)_8S_8$

*Tarkianite $(Cu,Fe)(Re,Mo)_4S_8$

2b.1a.1.1.1.1.2.2.3. Fn', Cu, Cd and Pb

Shadlunite series

Manganese-shadlunite $(Mn,Pb,Cd)Cu_4Fe_4S_8$

Shadlunite $(Pb,Cd)Cu_4Fe_4S_8$

2b.1a.1.1.1.1.2.2.4. Fn' and Ag

Argentopentlandite $Ag(Fe,Ni)_8S_8$

2b.1a.1.1.1.1.2.2.5. Fn' and Hg

Donharrisite $Ni_8Hg_3S_9$

*2б.1a.1.1.1.1.2.2.6. Fn' и Nb

*Edgarite $FeNb_3S_6$

2b.1a.1.1.1.1.3. Monoanionic

2b.1a.1.1.1.1.3.1. M^{2+} (simple)

Troilite family 1

Troilite group

Millerite NiS

Jaipurite γ-CoS

Troilite FeS

Unnamed 139 $(Fe,Ni,Ir)S$

Unnamed 386 $(Ni,Fe,Ir,Cu,Rh,Pt)S$

Alabandite MnS

Unnamed 138 β-MnS

*Rambergite γ-MnS

Unnamed 141 CrS

2b.1a.1.1.1.1.3.2. Fe^{2+} and Fe^{3+} (complex)

2b.1a.1.1.1.1.3.2.1. Only Fn

2b.1a.1.1.1.1.3.2.1.1. $Fe^{2+} \gg Fe^{3+}$

Pyrrhotite family

Hexapyrrhotite series (?) $Fn_{1-x}S$ (x = от 0 до 0,17)

Pyrrhotite-$1C$ $Fe_{1-x}S$

Unnamed 331 $(Fe,Ni)_{1-x}S$

Unnamed 140 $(Fe,Ag)_xS$

Polysomatic series of **clino(mono)pyrrhotites** $nFeS\ Fe_2S_3$ or $Fe^{2+}_nFe^{3+}_2S_{3+n}$

Pyrrhotite-$6C$ (n = 9) $Fe_{11}S_{12} \rightarrow Fe^{2+}_9Fe^{3+}_2S_{12}$

Pyrrhotite-$11C$ (n = 8) $Fe_{10}S_{11} \rightarrow Fe^{2+}_8Fe^{3+}_2S_{11}$

Pyrrhotite-$5C$ (n = 7) $Fe_9S_{10} \rightarrow Fe^{2+}_7Fe^{3+}_2S_{10}$

Pyrrhotite-$4C$ (n = 5) $Fe_7S_8 \rightarrow Fe^{2+}_5Fe^{3+}_2S_8$

2b.1a.1.1.1.1.3.2.1.2. M^{2+}: M^{3+}= 1,5

Smythite $Fe_5S_6 \rightarrow Fe^{2+}_3Fe^{3+}_2S_6$

*Murchisite Cr_5S_6

2b.1a.1.1.1.1.3.2.1.3. M^{2+}: M^{3+} = 0.5;

Sulfospinelides family Fn; M^{2+} = Fn^{2+}, Cu^{2+}, Zn^{2+}; M^{3+} = Fn'^{3+}; In^{3+}(compare with sulfospinelides of Pn (series.); selenospinelides (series.); oxospinelides (series.))

Linnaeite series - only of Fn^{2+} и Fn^{3+}

Polydymite	$Ni^{2+}Ni^{3+}_2S_4$
Siegenite	$Co^{2+}(Ni,Co)^{3+}_2S_4$
Nickel Linnaeite = polydymite	$Ni^{2+}Ni^{3+}_2S_4$
Linnaeite	$Co_3S_4 \rightarrow Co^{2+}Co^{3+}_2S_4$
Violarite	$Fe^{2+}Ni^{3+}_2S_4$
Greigite	$Fe_3S_4 \rightarrow Fe^{2+}Fe^{3+}_2S_4$

Carrollite series - $M^{2+} = Cu^{2+}$; $M^{3+} = Fn^{3+}$

Fletcherite	$Cu(Ni,Co)_2S_4$
Carrollite	$Cu(Co,Ni)_2S_4$

Daubreelite group - $M^{2+} = Fe^{2+}$; $M^{3+} = Cr^{3+}$

Daubreelite	$Fe^{2+}Cr^{3+}_2S_4$

Brezinaite group - $M^{2+} = Fe^{2+}, Cr^{2+}$; $M^{3+} = Cr^{3+}, Ti^{3+}$

Brezinaite	$Cr_3S_4 \rightarrow Cr^{2+}Cr^{3+}_2S_4$
Heideite	$(Fe,Cr)^{2+}_{1+x}(Ti^{3+},Fe^{2+})_2S_4$

Kalininite group - $M^{2+} = Cu^{2+}, Zn^{2+}$; $M^{3+} = Cr^{3+}, Sb^{3+}$

Florensovite	$Cu(Cr_{1,5}Sb_{0,5})_{\Sigma 2}S_4$
Kalininite	$ZnCr_2S_4$
*Kuprokalininite	$CuCr_2S_4$
Indite	$FeIn_2S_4 \rightarrow Fe^{2+}_{1-3x}Fe^{3+}_{2x}In_2S_4$ $(0 \leq x \leq 0,33)$
*Cadmoindite	$CdIn_2S_4$
*Jichengite	$Cu^+_2Ir^{3+}_6(Ni,Fe)^{2+}_{10}S_{20}$

2b.1a.1.1.1.1.3.2.1.4. $2M^{3+} \rightarrow M^+M^{2+}M^{3+}$

Cubanite family - $M^+ = Cu^+, Ag^+$; $M^{2+} = Fe^{2+}$; $M^{3+} = Fe^{3+}$

Isocubanite	$CuFe_2S_3 \rightarrow Cu^+Fe^{2+}Fe^{3+}S_3$
Cubanite	$CuFe_2S_3 \rightarrow Cu^+Fe^{2+}Fe^{3+}S_3$
Argentopyrite	$AgFe_2S_3 \rightarrow AgFe^{2+}Fe^{3+}S_3$
Sternbergite	$AgFe_2S_3 \rightarrow AgFe^{2+}Fe^{3+}S_3$
*Unnamed	$(Cu,Ag,Fe)_6S_4 \rightarrow Cu_3Ag_2FeS_4$

2b.1a.1.1.1.2. Minerals of Ib-cations
2b.1a.1.1.1.2.1. Polyanionic (simple)

Unnamed	$Cu[S_2]$

2b.1a.1.1.1.2.2. Mono-polyanionic (complex)

Covellite	$3CuS \rightarrow Cu^+_2S\,Cu^{2+}[S_2]$

2b.1a.1.1.1.2.3. Monoanionic
2b.1a.1.1.1.2.3.1. $M^+ \rightarrow M^+$ and M^{2+} (simple$\rightarrow$ complex)
2b.1a.1.1.1.2.3.1.1. Only Cu

Polysomatic (?) series of **chalcocite** $mCu_2S\,nCuS$ или $Cu^+_{2m}Cu^{2+}_nS_{m+n}$

Chalcocite	(m=1 ; n= 0)	Cu_2S
Tetrachalcocite	(m=24 ; n= 1)	$Cu_{49}S_{25} \rightarrow Cu^+_{48}Cu^{2+}S_{25}$
Djurleite	(m=15 ; n= 1)	$Cu_{31}S_{16} \rightarrow Cu^+_{30}Cu^{2+}S_{16}$
Digenite	(m= 4 ; n= 1)	$Cu_9S_5 \rightarrow Cu^+_8Cu^{2+}S_5$
Roxbyite	(m= 4 ; n= 1)	$Cu_9S_5 \rightarrow Cu^+_8Cu^{2+}S_5$
Anilite	(m= 3 ; n= 1)	$Cu_7S_4 \rightarrow Cu^+_6Cu^{2+}S_4$
Geerite	(m= 3 ; n= 2)	$Cu_8S_5 \rightarrow Cu^+_6Cu^{2+}_2S_5$
Spionkopite	(m=11 ; n=17)	$Cu_{39}S_{28} \rightarrow Cu^+_{22}Cu^{2+}_{17}S_{28}$
Yarrowite	(m= 1 ; n= 7)	$Cu_9S_8 \rightarrow Cu^+_2Cu^{2+}_7S_8$

2b.1a.1.1.1.2.3.1.2. Cu^+ and Hg^{2+}

Gortdrumite $\qquad$ $Cu^+_6Hg^{2+}_2S_5$

2b.1a.1.1.1.2.3.1.3. Cu^+ and Pb^{2+}

Betekhtinite $\qquad$ $(Cu,Fe)_{21}Pb_2S_{15}$

2b.1a.1.1.1.2.3.2. M^+ and $2M^{2+} \rightarrow M^+M^{3+}$ (complex)

2b.1a.1.1.1.2.3.2.1. $M^+ : M^{2+} \cong 2$; $Cu : Fe \cong 5$

Bornite series
Bornite (orange) $\qquad$ $Cu_{5-x}FeS_{4+x}$
Bornite (brown- usual) $\qquad$ $Cu^+_5Fe^{3+}S_4$
Bornite (pink) $\qquad$ $Cu_{5+x}FeS_{4-x}$

2b.1a.1.1.1.2.3.2.2. $Cu : Fe = 3$

*Wilhelmramsayite $\qquad$ $Cu_3FeS_3 \cdot 2H_2O$
Idaite $\qquad$ $Cu_3FeS_4 \rightarrow Cu^+Cu^{2+}_2Fe^{3+}S_4$
*Unnamed $\qquad$ $Cu_4FeS_4 \rightarrow Cu^+_2Cu^{2+}_2Fe^{2+}S_4$
*Unnamed $\qquad$ $(Cu_{0,96}K_{0,04})(Fe_{0,6}Cu_{0,4})(S_{1,98}O_{0,02})$

2b.1a.1.1.1.2.3.2.3. $2M^{2+} \rightarrow M^+M^{3+}$
when $M^+ \cong M^{3+}$

Chalcopyrite family (x = number of additional cations or vacancies to 1 $CuFeS_2$)
*Horomanite $\qquad$ $Fe_6Ni_3S_8$
*Samaniite $\qquad$ $Cu_2Fe_5Ni_2S_8$

Talnakhite subfamily (x > 0; $Cu \cong Fn$)
Mooihoekite   (x = 1/2) $\qquad$ $Cu_9Fe_9S_{16} \rightarrow Cu^+_9Fe^{2+}_4Fe^{3+}_5S_{16}$
$(2\square 3Fn^{3+} \rightarrow Cu^+4Fn^{2+})$
Putoranite   (x = 1/4) $\qquad$ $Cu_9Fe_9S_{16} \rightarrow Cu^+_9Fe^{2+}_4Fe^{3+}_5S_{16}$
$(2\square 3Fe^{3+} \rightarrow Cu^+4Fe^{2+})$
Haycockite   (x = 1/4) $\qquad$ $Cu_8Fe_{10}S_{16} \rightarrow Cu^+_8Fe^{2+}_6Fe^{3+}_4S_{16}$
$(2\square 4Fn^{3+} \rightarrow 6Fn^{2+})$
Talnakhite   (x = 1/8) $\qquad$ $Cu_9(Fe,Ni)_8S_{16} \rightarrow Cu^+_9Fe^{2+}Fe^{3+}_7S_{16}$
$(\square Fn^{3+} \rightarrow Cu^+Fn^{2+})$
Proper chalcopyrite subfamily (x = 0; $Cu^+ = Fe^{3+}(Ga^{3+},In^{3+})$)
Isochalcopyrite $\qquad$ $CuFe_2S_3$

Chalcopyrite group
Chalcopyrite $\qquad$ $CuFeS_2$
Gallite $\qquad$ $CuGaS_2$
Roquesite $\qquad$ $CuInS_2$
Germanite subfamily ($2Fe^{3+} \rightarrow M^{2+}M^{4+}$; $M^{4+} = Ge^{4+}$)
Proper germanite subfamily(x > 0; $M^+ = Cu^+$; $M^{2+} = Cu^{2+}$, Zn^{2+})
Germanite series (x = 1/8)
Germanite $\qquad$ $Cu^+_8Cu^{2+}_5Fe_2Ge_2S_{16}$
*Maikainite $\qquad$ $Cu_{20}(Fe,Cu)_6Mo_2Ge_6S_{32} \rightarrow Cu^+_{20}(Fe,Cu)^{2+}_6Mo^{4+}_2Ge^{4+}_6S_{32}$
*Ovamboite $\qquad$ $Cu_{20}(Fe,Cu,Zn)_6W_2Ge_6S_{32} \rightarrow Cu^+_{20}(Fe,Cu,Zn)^{2+}_6W^{4+}_2Ge^{4+}_6S_{32}$
*Calvertite $\qquad$ $Cu_{10}GeS_8$ $Cu^+_8Cu^{2+}_2Ge^{4+}S_8$
Renierite $\qquad$ $Cu^+_{10}(Zn,Cu)^{2+}Fe^{3+}_4Ge^{4+}_2S_{16}$

Briartite subfamily (x = 0; $M^+ = Cu^+$; $M^{2+} = Fe^{2+}$, Zn^{2+})
Briartite $\qquad$ $Cu^+_8(Fe,Zn)^{2+}_4Ge_4S_{16}$
*Barquillite $\qquad$ $Cu^+_2(Cd,Fe)^{2+}GeS_4$
Morozeviczite subfamily (x < 0; $M^+ = Cu^+$; $M^{2+} = Pb^{2+}$, Cu^{2+}, Ge^{2+}; $M^{3+} = Fe^{3+}$, As^{3+})
Morozeviczite $\qquad$ $Pb_3Ge_{1-x}S_4$

Polkovicite (x = 0,625) $(Fe,Pb)_3(Ge,Fe)_{1-x}S_4$

Stannite family ($2Fe^{3+} \rightarrow M^{2+}M^{4+}$; $M^{4+} = Sn^{4+}$; Mo^{4+}, W^{4+})

 Stannoidite subfamily (x > 0; $M^+ = Cu+$; $M^{2+} = Fe^{2+}$, Cu^{2+}; $M^{3+} = Fe^{3+}$, V^{3+})

  Mawsonite group (x = 1/4)

  Mawsonite $Cu^+_6Fe^{3+}_2Sn^{4+}S_8$
                                    ($\square 2Fe^{3+} \rightarrow 2Cu^+Sn^{4+}$)

  Chatkalite $Cu^+_6Fe^{2+}Sn^{4+}_2S_8$
                                    ($\square 4Fe^{3+} \rightarrow 2Cu^+Fe^{2+}2Sn^{4+}$)

  Stannoidite series (x = 1/6)

  Stannoidite $Cu^+_8(Fe,Zn)^{2+}Fe^{3+}_2Sn^{4+}_2S_{12}$
                                    $\square 4Fe^{3+} \rightarrow 2Cu^+Fe^{2+}2Sn^{4+}$)

  Cuprostannoidite $Cu^+_8Cu^{2+}Fe^{3+}_2Sn^{4+}_2S_{12}$
                                    ($\square 4Fe^{3+} \rightarrow 2Cu^+Cu^{2+}2Sn^{4+}$)

 Nekrasovite series (x = 1/8)

  Nekrasovite $Cu_{26}V_2Sn_6S_{32} \rightarrow Cu_{18}Cu^{2+}_8V^{3+}_2Sn^{4+}_6S_{32}$
                      ($2\square 16Fe^{3+} \rightarrow 2Cu^+8Cu^{2+}2V^{3+}6Sn^{4+}$)

 Proper stannite subfamily (x = 0; $M^+ = Cu^+$, Ag^+; $M^{2+} = Fe^{2+}$, Zn^{2+}, Cu^{2+}, Cd^{2+}, Hg^{2+}; $M^{3+} = Fe^{3+}$, In^{3+}; $M^{4+} = Sn^{4+}$, Mo^{4+}, W^{4+})

  Sakuraiite   (Cu,Zn,Fe,In,Sn)S

  Stannite series

  Stannite Cu_2FeSnS_4
  Kuramite Cu_2CuSnS_4
  Cernyite Cu_2CdSnS_4
  Velikite Cu_2HgSnS_4
  Hocartite Ag_2FeSnS_4
  Pirquitasite Ag_2ZnSnS_4

  Kësterite series

  Ferrokësterite $Cu_2(Fe,Zn)SnS_4$
  Kësterite $Cu_2(Zn,Fe)SnS_4$
  Petrukite $Cu_2(Fe,Zn)SnS_4$

 Hemusite group ($M^{2+}=Cu^{2+}$; $M^{4+} = Sn^{4+}$, Mo^{4+}, W^{4+})

  *Catamarcaite Cu_6GeWS_8

  Hemusite $Cu_6MoSnS_8 \rightarrow Cu_4Cu^{2+}_2Mo^{4+}Sn^{4+}S_8$
                    ($4Fe^{3+} \rightarrow \square 2Mo^{4+}Sn^{4+}$)

  Kiddcreekite $Cu_6WSnS_8 \rightarrow Cu_4Cu^{2+}_2W^{4+}Sn^{4+}S_8$
                    ($4Fe^{3+} \rightarrow \square 2W^{4+}Sn^{4+}$)

 Vinciennite ($M^{2+} = Cu^{2+}$; $M^{3+} = Fe^{3+}$, As^{3+}; $M^{4+} = Sn^{4+}$)

                    $Cu_{10}Fe_4SnAsS_{16} \rightarrow Cu^+_7Cu^{2+}_3Fe^{3+}_4Sn^{4+}AsS_{16}$
                    ($4Cu^+Fe^{3+} \rightarrow 3\square Sn^{4+}As^{3+}$)

 Mohite ($M^{2+} = 0$; $M^{4+} = Sn^{4+}$) $Cu^+_4Sn^{4+}_2S_6$
                    ($3Fe^{3+} \rightarrow Cu^+2Sn^{4+}$)

Rhodostannite subfamily (x < 0, $M^+ = Cu^+$; $M^{2+} = Fe^{2+}$; $M^{4+} = Sn^{4+}$)

 Rhodostannite (x = 1/2) $Cu^+_2Fe^{2+}Sn^{4+}_3S_8$
                                    ($2Cu^+4Fe^{3+} \rightarrow 2\square Fe^{2+}3Sn^{4+}$)

 *Toyohaite $Ag^+_2Fe^{2+}Sn^{4+}_3S_8$

2b.1a.1.1.1.2.3.2.4. $M^{2+}= Cu^{2+}$, Fe^{2+}; Cu : Fe = 5

Nukundamite $Cu_{3.4}Fe_{0.6}S_4$

2b.1a.1.1.1.3. Minerals of **IIb**-cations - Zn^{2+} (and Cd^{2+})
2b.1a.1.1.1.3.1. Monoanionic (simple)
Sphalerite family
Sphalerite-*3C*   ZnS
Sphalerite-*2H* (wurtzite)
*Wurtzite-*2H, -15R, -18R, 21R, 4H, 8H* polytypes
Sphalerite-*3R* (matraite)
*Buseckite   (Fe,Zn,Mn)S
*Rudaschevskite   (Fe,Zn)S
Hawleyite-*3C*   CdS
Hawleyite -2H (greenockite)
*26.1a.1.1.1.3.2. Complex
*Unnamed $Zn_2(Fe,Cu)S_3$

2b.1a.1.1.2. <u>Minerals of heavy noncenosymmetrycal *d*-cations</u>
2b.1a.1.1.2.1. Minerals of **VIII**b-cations (platinoides-Pn^{n+})
2b.1a.1.1.2.1.1. Polyanionic (simple)
   Laurite group
   Erlichmanite $Os[S_2]$
   Laurite $Ru[S_2]$
2b.1a.1.1.2.1.2. Subcompound
2b.1a.1.1.2.1.2.1. Simple
*Miassite $Rh_{17}S_{15}$
*Kingstonite $(Rd,Ir,Pt)_3S_4$
2b.1a.1.1.2.1.2.2. Complex
Rhodplumsite $Rh_3Pb_2S_2$
2b.1a.1.1.2.1.3. Monoanionic
2b.1a.1.1.2.1.3.1. Pn^{2+}
2b.1a.1.1.2.1.3.1.1. Simple
Cooperite family
 Cooperite   PtS
Vysotskite   (Pd,Ni,Pt)S
*Unnamed $(Pt,Pd)_3S_2$
2b.1a.1.1.2.1.3.1.2. Complex
Braggite   (Pt,Pd,Ni)S
*Unnamed   PtSnS

2b.1a.1.1.2.1.3.2. M^{2+} and M^{3+} (complex)
Sulfospinelides family Pn^{n+} (compare with sulfospinelides Fn^{n+} (series))
                        2b.1a.1.1.2.1.3.2.1. Cu^{2+}, Fe^{2+} and Pn^{3+}
*Lisiguangite $CuPtBiS_3$
*Malyshevite $PdBiCuS_3$
   Malanite group
   Malanite $(Cu,Fe)^{2+}Pt^{3+}(Ir,Co,Pd)^{3+}S_4$
   Cuproiridsite $Cu^{2+}Ir^{3+}{}_2S_4$
   Cuprorhodsite $(Cu,Fe)^{2+}Rh^{3+}{}_2S_4$
   *Ferrorhodsite $(Fe,Cu)(Rh,Ir,Pt)_2S_4$
                        2b.1a.1.1.2.1.3.2.2. Cu^{2+}, Pb^{2+} and Pn^{3+}
Inaglyite subfamily
Xingzhongite $PbCuFe^{3+}{}_{0,67}\bullet{}_{0,33}(Ir,Rh,Pt)_2S_4$

Inaglyite group

Inaglyite $Pb^{2+}Cu^{2+}_3(Ir,Pt)^{3+}_8S_{16}$

Konderite $Pb^{2+}Cu^{2+}_3(Rh,Pt,Ir)^{3+}_8S_{16}$

*Konderite-Fe $(Fe,Pb)^{2+}Cu^{2+}_3(Rh,Ir,Pd,Pt)^{3+}_8S_{16}$

2b.1a.1.1.2.1.3.3. Pn^{3+} (simple)

Bowieite group

Kashinite $(Ir,Rh)_2S_3$

Bowieite $(Rh,IrPt)_2S_3$

2b.1a.1.1.2.1.3.4. Pn^{4+} и Pn^{6+} (complex)

Beta-iridisite $4Ir_{0.75}S_2 \rightarrow Ir^{4+}Ir^{6+}_2S_8$ (?)

2b.1a.1.1.2.2. Minerals of Ib-cations and Tl^+ (CN ≤ 4)

2b.1a.1.1.2.2.1. Minerals of Tl^+ (CN ≤ 4)

2b.1a.1.1.2.2.1.1. Monoanionic

2b.1a.1.1.2.2.1.1.1. Simple

Carlinite Tl_2S

2b.1a.1.1.2.2.1.1.2. Complex

Raguinite $TlFe^{3+}S_2$

*Unnamed $Tl_2(Cu,Fe)_6S_5$

2b.1a.1.1.2.2.2. Minerals of Ag^+ and Au^+

2b.1a.1.1.2.2.2.1. Monoanionic

2b.1a.1.1.2.2.2.1.1. Proper sulfides   2b.1a.1.1.2.2.2.1.1.1. Simple

*Unnamed   AuS

Acanthite family

Acanthite Ag_2S

Argentite Ag_2S

2b.1a.1.1.2.2.2.1.1.2. Complex   2b.1a.1.1.2.2.2.1.1.2.1. Ag^+ and Fe^{3+}, In^{3+}

*Lenaite $AgFeS_2$

*Laforêtite $AgInS_2$

                                             2b.1a.1.1.2.2.2.1.1.2.2. Ag^+ and Cu^+

Stromeyerite family

Jalpaite Ag_3CuS_2

Mckinstryite $Ag_5Cu_3S_4$

Stromeyerite $AgCuS$

                                             2b.1a.1.1.2.2.2.1.1.2.2. Ag^+ and Au^+

Uytenbogaardtite family

Uytenbogaardtite $AuAg_3S_2$

Petrovskaite $AuAg(S,Se)$

                                             2b.1a.1.1.2.2.2.1.1.2.3. $Ag^+(Cu^+)$ and Hg^{2+}

Balkanite family

Danielsite $(Cu,Ag)_{14}Hg^{2+}S_8$

Balkanite group

Balkanite $Ag_5Cu_9Hg^{2+}S_8$

Imiterite Ag_2HgS_2

                                             2b.1a.1.1.2.2.2.1.1.2.4. $Ag^+(Cu^+)$ and Pb^{2+}

Furutobeite $AgCu_5PbS_4$

                                             2b.1a.1.1.2.2.2.1.1.2.5. Ag and $Sn^{4+}(Ge^{4+})$

Canfieldite group

Canfieldite Ag_8SnS_6

Argyrodite	Ag_8GeS_6
*Putzite	$(Cu_{4.7}Ag_{3.3})_8GeS_6$
*Calvertite	$Cu_{10}GeS_8 \rightarrow Cu^+_8Cu^{2+}_2Ge^{4+}S_8$
*Alburnite	$Ag_8GeTe_2S_4$

2b.1a.1.1.2.2.2.2. Sulfido-halogenides (complex)

*Iltisite	$AgHgS(Cl,Br)$
Perroudite	$Ag^+_4Hg^{2+}_5S_5(I,Br)_2Cl_2$
*Capgaronnite	$AgHgS(Cl,Br,I,)$

2b.1a.1.1.2.3. Minerals of **II**b-cations – Hg^{2+}
2b.1a.1.1.2.3.1. Monoanionic
2b.1a.1.1.2.3.1.1. Proper sulfides (simple)
Cinnabar family

Cinnabar	HgS
Metacinnabar	HgS
Hypercinnabar	HgS
Polhemusite	$(Zn,Hg)S$

2b.1a.1.1.2.3.1.2. Sulfido-halogenides (simple)
Corderoite family

*Kenshuite	$Hg_3S_2Cl_2$
Corderoite	$Hg_3S_2Cl_2$
Lavrentievite	$Hg_3S_2(Cl,Br)_2$
*Radtkeite	Hg_3S_2ClI
Arzakite	$Hg_3^{2+}S_2(Br,Cl)_2$
Grechishchevite	$Hg_3S_2(Br,Cl,I)_2$

2b.1a.1.2. Minerals of noncenosymmetrical *p*-cations
2b.1a.1.2.1. Minerals of **IV**a-cations (all monoanionic)
2b.1a.1.2.1.1. Minerals of Sn
2b.1a.1.2.1.1.1. Minerals of Sn^{2+}
2b.1a.1.2.1.1.1.1. Simple

Herzenbergite	SnS

2b.1a.1.2.1.1.1.2. Complex

*Unnamed	$SnGeS_3$
Teallite	$PbSnS_2$
*Suredaite	$PbSnS_3$

2b.1a.1.2.1.1.2. Minerals of Sn^{2+} and Sn^{4+} (complex)

Ottemannite	$Sn^{2+}Sn^{4+}S_3$

2b.1a.1.2.1.1.3. Minerals of Sn^{4+} (simple)

Berndtite-*2T*	SnS_2
Berndtite-*4H*	

2b.1a.1.2.1.2. Minerals of Pb
2b.1a.1.2.1.2.1. Minerals of Pb^{2+} (simple)
Galena-clausthalite series

Galena	PbS	
Clausthalite	$PbSe$	see selenides

2b.1a.1.2.2. Minerals of Va-cations
2b.1a.1.2.2.1. Subsulfides (simple and complex)
Realgar family

Duranusite	As_4S
Pääkkönenite	Sb_2AsS_2
Dimorphite	As_4S_3
Realgar	$4AsS \rightarrow As_4S_4$
Pararealgar	AsS
Alacranite	As_8S_9
Uzonite	As_4S_5

2b.1a.1.2.2.2. Monoanionic (only $As^{3+}, Sb^{3+}, Bi^{3+}$)
2b.1a.1.2.2.2.1. Proper sulfides (simple and complex)
Orpiment family
  Orpiment group

Orpiment	$[As_2S_3]^{\infty 2}$
*Anorpiment	$[As_2S_3]$
Getchellite	$[SbAsS_3]^{\infty 2}$

  Stibnite group

Stibnite	Sb_2S_3
Bismuthinite	Bi_2S_3
Metastibnite	Sb_2S_3
Wakabayashillite	$(As,Sb)_6As_4S_{14}$

2b.1a.1.2.2.2.2. Sulfido-oxides (simple)

Kermesite	Sb_2S_2O

*2б.1a.1.2.2.2.3. Sulfido-halogenidesы

*Demecheleite-(Br)	BiSBr
*Demecheleite-(Cl)	BiSCl

2b.1a.2. *Class:* Sulfosalts of sidero- and chalcophylic cations

2b.1a.2.1. Sulfosalts of heavy d-cations
2b.1a.2.1.1. Minerals of cenosymmetrical d-cations
2b.1a.2.1.1.1. Minerals of **VII**b- and **VIII**b-cations - the family of Fn'
2b.1a.2.1.1.1.1. Sulfoantimonites and sulfobismuthites
2b.1a.2.1.1.1.1.1. Fn'$^{2+}$
2b.1a.2.1.1.1.1.1.1. Simple
  Berthierite group

Berthierite	$FeSb_2S_4$
*Clerite	$MnSb_2S_4$
Garavellite	$FeBiSbS_4$
*Graţianite	$MnBi_2S_4$

2b.1a.2.1.1.1.1.1.2. Complex
2b.1a.2.1.1.1.1.1.2.1. Fn'$^{2+}$ and Cu^+
  Lapieite group

Lapieite	$Cu^+Ni^{2+}SbS_3$
Mückeite	$Cu^+Ni^{2+}BiS_3$
	2b.1a.2.1.1.1.1.1.2.2. Fn'$^{2+}$ and Ag^+

Samsonite $Ag_4Mn^{2+}Sb_2S_6$

2b.1a.2.1.1.1.1.2. Fn^{3+} (complex - Fn^{3+}, Cu^+, and Pb^{2+})
Miharaite $Cu^+_4Pb^{2+}Fe^{3+}BiS_6$

2b.1a.2.1.1.2. Minerals of IIb-cations
2b.1a.2.1.1.2.1. Sulfoarsenites
2b.1a.2.1.1.2.1.1. Cu^+ (simple)
Sinnerite $Cu_6As_4S_9$
*Watanabeite $Cu_4(As,Sb)_2S_5$

2b.1a.2.1.1.2.1.2. Cu^+ and Cu^{2+} (all complex)
Enargite family ($M^+= Cu^+$; $M^{2+}= Cu^{2+}$; M^+: $M^{2+}< 1$; $Y^{3+}= As^{3+}$, V^{3+})
Colusite subfamily ($M^+ : M^{2+} \cong 0,86$)
Colusite $Cu_{13}VAs_3S_{16} \rightarrow Cu^+_6Cu^{2+}_7VAs_3S_{16}$
*Stibiocolusite $Cu_{26}V_2(Sb,Sn,As)_6S_{32}$
*Germanocolusite $Cu_{26}V_2(Ge,As)_6S_{32}$
Proper enargite subfamily ($M^+ : M^{2+} = 0,5$)
Enargite $Cu_3AsS_4 \rightarrow Cu^+Cu^{2+}_2AsS_4$
*Unnamed Cu_3AsS_4
Luzonite series
Luzonite $Cu_3AsS_4 \rightarrow Cu^+Cu^{2+}_2AsS_4$
Famatinite $Cu_3SbS_4 \rightarrow Cu^+Cu^{2+}_2SbS_4$
Sulvanite series (?)
Arsenosulvanite $Cu_3(As,V)S_4 \rightarrow Cu^+Cu^{2+}_2(As,V)S_4$
Sulvanite $Cu_3VS_4 \rightarrow Cu^+Cu^{2+}_2VS_4$

2b.1a.2.1.1.2.2. Sulfoarsenito-sulfoantimonites
2b.1a.2.1.1.2.2.1. M^+ and M^{2+} (complex)
Fahlores family ($M^+ = Cu^+$, Ag^+; $M^{2+} = Cu^{2+}$, Fe^{2+}, Zn^{2+}, Cd^{2+}, Hg^{2+})
Proper fahlores subfamily ($M^+ : M^{2+} = 5$)
Fahlores series - tennantite - tetrahedrite (compare with giraudite (group))
Ferrotennantite $Cu^+_{10}Fe^{2+}_2As_4S_{13}$
Coppite $Cu_{10}Fe_2Sb_4S_{13}$
Freibergite $Ag_6Cu_4Fe_2Sb_4S_{13}$
Tennantite (cuprotennantite) $Cu_{10}Cu_2As_4S_{13}$
Tetrahedrite (cuprotetrahedrite) $Cu_{10}Cu_2Sb_4S_{13}$
Miedziankite $Cu_{10}Zn_2As_4S_{13}$
Argentotennantite $(Ag,Cu)_{10}(Zn,Fe)_2(As,Sb)_4S_{13}$
Sandbergerite $Cu_{10}Zn_2Sb_4S_{13}$
Zincsandbergerite $Ag_6Cu_4Zn_2Sb_4S_{13}$
Cadmian tetrahedrite $Cu_{10}Cd_2Sb_4S_{13}$
Mercurian tennantite $Cu_{10}Hg_2As_4S_{13}$
Schwazite (mercurian tetrahedrite) $Cu_{10}Hg_2Sb_4S_{13}$
Goldfieldite series
Goldfieldite $Cu_{10}Te_4S_{13}$
Nowackiite subfamily ($M^+: M^{2+} = 2$)
Nowackiite $Cu_6Zn_3As_4S_{12}$
Aktashite $Cu_6Hg_3As_4S_{12}$
Gruzdevite $Cu_6Hg_3Sb_4S_{12}$

2b.1a.2.1.1.2.3. Sulfoantimonites
2b.1a.2.1.1.2.3.1. Cu^+ (simple)

Skinnerite	Cu_3SbS_3
*Unnamed	$(Cu,Zn)_3(Sb,As)S_3$
Chalcostibite	$CuSbS_2$

2b.1a.2.1.1.2.4. Sulfobismuthites
2b.1a.2.1.1.2.4.1. Cu^+
2b.1a.2.1.1.2.4.1.1. Simple
Wittichenite family

Wittichenite	Cu_3BiS_3
Emplectite	$CuBiS_2$
Cuprobismutite	$Cu_8AgBi_{13}S_{24}$
Hodrušhite	$Cu_4Bi_6S_{11}$

2b.1a.2.1.1.2.4.1.2. Complex (Cu^+, Ag^+, Pb^{2+})

Larosite	$(Cu,Ag)_{21}PbBiS_{13}$
*Kupčikite	$Cu_{3,4}Fe_{0,6}Bi_5S_{10}$
*Pizgrischite	$Cu_{17}PbBi_{17}S_{35}$

2b.1a.2.1.2. <u>Sulfosalts of noncenosymmetrical *d*-cations and *p_d*-cations</u>
2b.1a.2.1.2.1. Minerals of **1***b*-cations and Tl^+
2b.1a.2.1.2.1.1. Tl^+
2b.1a.2.1.2.1.1.1. Sulfoarsenites
2b.1a.2.1.2.1.1.1.1. Proper sulfoarsenites
2b.1a.2.1.2.1.1.1.1.1. Simple

Ellisite	Tl_3AsS_3
Lorandite	$TlAsS_2$
*Fangite	Tl_3AsS_4
*Bernardite	$Tl(As,Sb)_5S_8$
*Gillulyite	$Tl_2As_{7.5}Sb_{0.3}S_{13}$

2b.1a.2.1.2.1.1.1.1.2. Complex
2b.1a.2.1.2.1.1.1.1.2.1. Tl^+ $(Cu,Ag)^+$ and Hg^{2+}

Routhierite	$Tl(Cu,Ag)(Hg,Zn)_2(As,Sb)_2S_6$
*Arsiccioite	$TlAgHg_2(As,Sb)_2S_6$
*Sb-routhierite	$TlCuHg_2(Sb,As)_2S_6$
*Stalderite	$TlCu(Zn,Fe,Hg)_2As_2S_6$
*Gabrielite	$Tl_2AgCu_2As_3S_7$
*Erniggliite	$Tl_2SnAs_2S_6$
*Sicherite	$TlAg_2(As_2Sb)_{\Sigma3}S_6$
*Raberite	$Tl_5Ag_4As_6SbS_{15}$

2b.1a.2.1.2.1.1.1.1.2.2. Tl^+ and Hg^{2+}

Christite	$TlHgAsS_3$
Simonite	$TlHgAs_3S_6$

2b.1a.2.1.2.1.1.1.1.2.3. Tl^+, (Cu^+,Ag^+) and Pb^{2+}

Wallisite group

Wallisite	$TlCuPbAs_2S_5$
Hatchite	$TlAgPbAs_2S_5$
*Dalnegroite	$Tl_{5-x}Pb_{2x}(As,Sb)_{21-x}S_{34}$
*Unnamed	$(Tl,Ag)_2Pb_6(As,Sb)_{16}S_{31}$

2b.1a.2.1.2.1.1.1.1.2.4. Tl^+ and Pb^{2+}

Hutchinsonit	$TlPbAs_5S_9$
*Edenharterite	$TlPbAs_3S_6$
*Jentschite	$TlPbAs_2SbS_6$
*Boscardinite	$TlPb_4(Sb_7As_2)_{\Sigma 9}S_{18}$

2b.1a.2.1.2.1.1.1.2. Sulfoarsenito-sulfoarsenates (complex)

Imhofite	$Tl_{5.8}As_{15.4}S_{26}$

2b.1a.2.1.2.1.1.2. Sulfoantimonites
2b.1a.2.1.2.1.1.2.1. Simple

Weissbergite	$TlSbS_2$
*Jankovicite	$Tl_5Sb^{3+}_9(As,Sb)^{3+}_4S_{22}$

Pierrotite family

Pierrotite	$Tl_2(Sb_6As_4)_{\Sigma 10}S_{16}$
Parapierrotite	$Tl_2(Sb_9As)_{\Sigma 10}S_{16}$
*Protochabournéite	$Tl_{5-x}Pb_{2x}(Sb,As)_{21-x}S_{34}$ (x~1.2-1.5)
Chabourneite	$Tl_{10}(Sb_{22,5}As_{19,5})_{42}S_{68}$

2b.1a.2.1.2.1.1.2.2. Complex   2b.1a.2.1.2.1.1.2.2.1. Tl^+ and Hg^+

Vaughanite	$Tl^+Hg^+Sb^{3+}_4S_7$

2b.1a.2.1.2.1.1.2.2.2. Tl^+ and Hg^{2+}

Vrbaite	$Tl_4Hg_3(Sb_2As_8)_{\Sigma 10}S_{20}$

2b.1a.2.1.2.1.2. Minerals of Ag
2b.1a.2.1.2.1.2.1. Sulfoarsenites
2b.1a.2.1.2.1.2.1.1.Simple

Pearceite series

Pearceite	$(Ag,Cu)_{16}As_2S_{11}$
*Pearceite Tac	$(Ag,Cu)_{16}As_2S_{11}$
*Cupropearceite	$(Ag_9Cu_7)_{16}As_2S_{11}$
*Antimonpearceite	$(Ag,Cu)_{16}(Sb,As)_2S_{11}$
Polybasite	$(Ag,Cu)_{16}Sb_2S_{11}$
*Arsenpolybasite	$(Ag,Cu)_{16}(As,Sb)_2S_{11}$
*Selenopolybasite	$Ag_{15}CuSb_2S_9Se_2$

Proustite family
Proustite series

Proustite	Ag_3AsS_3
Pyrargyrite	Ag_3SbS_3

Xanthoconite group

Xanthoconite	Ag_3AsS_3
Pyrostilpnite	Ag_3SbS_3

Smithite family

Smithite	$AgAsS_2$
Trechmannite	$AgAsS_2$

2b.1a.2.1.2.1.2.1.2. Complex (Ag^+ and Hg^{2+})

Laffittite	$AgHgAsS_3$
*Fettelite	$[Ag_6As_2S_7][Ag_{10}HgAs_2S_8]$
*Debattistiite	$Ag_9Hg_{0.5}As_6S_{12}Te_2$

*26.1a.2.1.2.1.2.1.3. Complex (Ag^+, Cd^{2+}, Pb)

*Quadratite	$Ag(Cd,Pb)(As,Sb)S_3$
*Manganoquadratite	$AgMnAsS_3$

2b.1a.2.1.2.1.2.2. Sulfoarsenites (simple)
Billingsleyite $Ag_7(As,Sb)S_6$
*2b.1a.2.1.2.1.2.2.1. Sulfoarsenito-halogenides
*Mutnovskite $Pb_2AsS_3(I,Cl,Br)$

2b.1a.2.1.2.1.2.3. Sulfoantimonites
2b.1a.2.1.2.1.2.3.1. Simple
Stephanite family (compare with selenostephanite)
Stephanite Ag_5SbS_4
 Miargyrite subfamily
 Miargyrite $AgSbS_2$
 *Cubargyrite $AgSbS_2$
 *Baumstarkite $AgSbS_2$
 Aramayoite $Ag(Sb,Bi)S_2$
 *Ferdowsiite $Ag_8(Sb_5As_3)_{\sum 8}S_{16}$

2b.1a.2.1.2.1.2.3.2. Complex (Ag^+ and Pb^{2+})
Brongniardite = diaphorite $Ag_3Pb_2Sb_3S_8$
Roshchinite $Ag_{19}Pb_{10}Sb_{51}S_{96}$
Diaphorite $Ag_3Pb_2Sb_3S_8$
*Tubulite $Ag_2Pb_{22}Sb_{20}S_{53}$
*Unnamed $Ag_3Pb_6(Sb,Bi)_{11}S_{24}$

2b.1a.2.1.2.1.2.4. Sulfobismuthites
2b.1a.2.1.2.1.2.4.1. Simple→ complex
Matildite $AgBiS_2$
*Schapbachite $AgBiS_2$
*Unnamed cub. $AgBiS_2$
Benjaminite $Ag_3Bi_7S_{12}$
 Pavonite series
 Pavonite $AgBi_3S_5$
 Cupropavonite $AgCu_2PbBi_5S_{10}$
 *Cupromakopavonite   N = 4,5 $Ag_3Cu_8Pb_4Bi_{19}S_{38}$
 *Cupromakovickyite   N = 4 $Ag_2Cu_8Pb_4Bi_{18}S_{36}$
 *Makovickyite $Ag_{1.5}Bi_{5.5}S_9$
 *Dantopaite $Ag_5Bi_{13}S_{22}$
 *Cu-Pb-benjaminite   N = 7.86
 *Cu-Pb-mummeite   N = 8
*Unnamed $Ag_5CuPbBi_4(S,Se)_{10}$
*Borodaevite $[Ag_5(Fe,Pb)Bi_7]_{\sum 13}(Sb,Bi)_2S_{17}$

2b.1a.2.1.2.1.2.4.2. Complex
2b.1a.2.1.2.1.2.4.2.1. Ag^+ and Cu^+
Arcubisite Ag_6CuBiS_4
                          2b.1a.2.1.2.1.2.4.2.2. $Ag^+(Cu^+)$ and Pb^{2+}
Padéraite $Cu_7(Cu,Ag)_{0.33}Pb_{1.33}Bi_{11.33}S_{22}$
Mummeite $Cu_{0.58}Ag_{3.11}Pb_{1.10}Bi_{6.65}S_{13}$
 Berryite series
 Berryite-(Cu) $(Cu,Ag)_5Pb_3Bi_7S_{16}$
 Berryite-(Ag) $(Ag,Cu)_5Pb_3Bi_7S_{16}$

p-Ourayite	$Ag_{3,6}Pb_{2,8}Bi_{5,6}S_{13}$
Treasurite	$Ag_7Pb_6Bi_{15}S_{32}$
Gustavite	$AgPbBi_3S_6$
Ourayite	$Ag_3Pb_4Bi_5S_{13}$
*Terrywallaceite	$AgPb(Sb,Bi)_3S_6$

*26.1a.2.1.2.1.2.5. Sulfoarsenantimonites
*Unnamed Ag_2SbAsS_4

2b.1a.2.1.2.2. Minerals of **II**b-cations (Hg^{2+})
2b.1a.2.1.2.2.1. Sulfoarsenites (complex)
Galkhaite $(Hg_5Cu)CsAs_4S_{12}$

2b.1a.2.1.2.2.2. Sulfoantimonites (mono-polyanionic) (simple)
Livingstonite $HgSb_4S_8 \rightarrow Hg^{2+}Sb^{3+}_4S_6[S_2]$

*26.1a.2.1.2.2.3. Sulfobismuthites (complex)
*Grumiplucite $HgBi_2S_4$

2b.1a.2.2. <u>Sulfosalts of noncenosymmetrical p-cations</u>
2b.1a.2.2.1. Minerals of **IV**a-cations (only Pb^{2+})
2b.1a.2.2.1.1. Sulfoarsenites
2b.1a.2.2.1.1.1. Simple
  Jordanite series

Jordanite	$Pb_{14}(As,Sb)_6S_{23}$
Geocronite	$Pb_{14}(Sb,As)_6S_{23}$
*Marumoite	$Pb_8As_{10}S_{23}$
Gratonite	$Pb_9As_4S_{15}$
*Tsugaruite	$Pb_4As_2S_7$
Kirkiite	$Pb_{10}(As_3Bi_3)_{\Sigma6}S_{19}$
Dufrénoysite	$Pb_2As_2S_5$
Baumhauerite	$Pb_3As_4S_9$ or $Pb_{12}As_{16}S_{36}$
*Argentobaumhauerite = baumhauerite 2a	$Ag_{1.5}Pb_{22}As_{33.5}S_{72}$
Liveingite	$Pb_{18.5}As_{25}S_{56}$
Sartorite	$PbAs_2S_4$

2b.1a.2.2.1.1.2. Complex
2b.1a.2.2.1.1.2.1. Pb^{2+} and Cu^+
Seligmannite ($y = MS : M_2S = 2$) $CuPbAsS_3$
                              2b.1a.2.2.1.1.2.2. Pb^{2+} and $Ag^+(Cu^+)$
Marrite $AgPbAsS_3$
Lengenbachite ($y = \leq 6$) $(Ag,Cu)_2Pb_6As_4S_{13}$
                                    *26.1a.2.2.1.1.3. Pb^{2+},Cd
*Tazievite $Pb_{20}Cd_2(As,Bi)_{22}S_{50}Cl_{10}$
*26.1a.2.2.1.1.3. Pb^{2+}, Hg
*Daliranite $PbHgAs_2S_6$

2b.1a.2.2.1.2. Sulfoantimonites
2b.1a.2.2.1.2.1. Proper sulfoantimonites
2b.1a.2.2.1.2.1.1. Simplee
Falkmanite $Pb_{5,4}Sb_{3,6}S_{11} \sim Pb_3Sb_2S_6$

Boulangerite	$Pb_5Sb_4S_{11}$
*Moëloite	$Pb_6Sb_6S_{17}$ or $Pb_6Sb_6S_{14}(S_3)$
Semseyite	$Pb_9Sb_8S_{21}$
Madocite	$Pb_{19}(Sb,As)_{16}S_{43}$
Veenite	$Pb_2(Sb,As)_2S_5$
Sorbyite	$Pb_9Cu(Sb,As)_{11}S_{26}$
Heteromorphite	$Pb_7Sb_8S_{19}$
Launayite	$Pb_{10}Cu(Sb,As)_{13}S_{30}$
Robinsonite	$Pb_4Sb_6S_{13}$
Plagionite	$Pb_5Sb_8S_{17}$
Twinnite series (?)	
Guettardite	$Pb(Sb,As)_2S_4$
Twinnite	$Pb(SbAs)_{\Sigma2}S_4$
Rathite	$Pb_{12-x}Ag_2Tl_{x/2}As_{18+x/2}S_{40}$
Zinkenite	$Pb_9Sb_{22}S_{42}$
Fülöppite	$Pb_3Sb_8S_{15}$
Playfairite	$Pb_{16}(Sb,As)_{19}S_{44}Cl$
2b.1a.2.2.1.2.1.2. Complex	2b.1a.2.2.1.2.1.2.1. Pb^{2+} and Fn'^{2+}
Jamesonite family	
Jamesonite series (?)	
Jamesonite	$Pb_4FeSb_6S_{14}$
Benavidesite	$Pb_4(Mn,Fe)Sb_6S_{14}$
*Marrucciite	$Hg_3Pb_{16}Sb_{18}S_{46}$
	2b.1a.2.2.1.2.1.2.2. Pb^{2+} and Cu^+
Bournonite (y = 2)	$CuPbSbS_3$
Tintinaite series (y = 11)	
Tintinaite-(Sb)	$Pb_{10}Cu_2Sb_{16}S_{35}$
Tintinaite-(Bi)	$Pb_{10}Cu_2(Bi,Sb)_{16}S_{35}$
Meneghinite series (y=26-20)	
Meneghinite	$CuPb_{13}Sb_7S_{24}$
Jaskolskiite	$Cu_xPb_{2+x}(Sb,Bi)_{2-x}S_5$ (x = 0,2)
*Rouxelite	$Cu_2HgPb_{22}Sb_{28}S_{64}(O,S)_2$
*Izoklakeite	$(Cu,Fe)_2Pb_{27}(Sb,Bi)_{19}S_{57}$
*Unnamed	$Cu_5Fe_6Pb_6Bi_2S_{21}$ (?)
	2b.1a.2.2.1.2.1.2.3. Pb^{2+} and Tl^+, Ag^+
Rayite	$TlAg_3Pb_{16}Sb_{16}S_{42}$
	2b.1a.2.2.1.2.1.2.4. Pb^{2+} and $Ag^+(Cu^+)$
Andorite family (y=2)	
Andorite group	
Freieslebenite	$AgPbSbS_3$
Andorite	$AgPbSb_3S_6$
Senandorite	$AgPbSb_3S_6$
Ramdohrite (y = 4)	$Ag_3Pb_6Sb_{11}S_{24}$
Fizelyite (y = 5,6)	$Ag_5Pb_{14}Sb_{21}S_{48}$
Owyheeite family (y = 8)	
Owyheeite	$Ag_{3+x}Pb_{10-2x}Sb_{11+x}S_{28}$ ($-0,13 \leq x \leq 0,2$)
Zoubekite	$AgPb_4Sb_4S_{10}$
*Parasterryite	$Ag_4Pb_{20}(Sb_{14.5}As_{9.5})_{24}S_{58}$
Sterryite	$(Ag,Cu)_2Pb_{10}(Sb,As)_{12}S_{29}$

2b.1a.2.2.1.2.1.2.5. Pb^{2+}, $Fn^{,2+}$ and Ag^+

Uchucchacuaite $AgPb_3Mn^{2+}Sb_5S_{12}$
*Menchettite $AgPb_{2.40}Mn^{2+}_{1.60}Sb_3As_2S_{12}$
*Unnamed $AgPb_9(Sb,As)_{13}S_{29}$

2b.1a.2.2.1.2.1.2.6. Pb^{2+}, Fn^{2+}, Sn^{2+} and Sn^{4+}

Franckeite series
Franckeite $Pb_5Fe^{2+}Sn^{2+}Sn^{4+}_2Sb_2S_{14}$
Cylindrite $Pb_3Fe^{2+}Sn^{2+}Sn^{4+}_3Sb_2S_{14}$

2b.1a.2.2.1.2.2. Sulfoantimonito-halogenides (simple)
Dadsonite $Pb_{23}Sb_{25}S_{60}Cl$
Ardaite $Pb_{19}Sb_{13}S_{35}Cl_7$

*2б.1a.2.2.1.2.3. Sulfoantimonito-chlorido-oxides (simple)
*Pillaite $Pb_9Sb_{10}S_{23}ClO_{0.5}$
*Pellouxite $(Cu,Ag)_2Pb_{21}Sb_{23}S_{55}ClO$
*2б.1a.2.2.1.2.4. Sulfoantimonito-oxides (simple)
*Scainiite $Pb_{14}Sb_{30}S_{54}O_5$
*Chovanite $Pb_{15-2x}Sb_{14+2x}S_{36}O_x$

2b.1a.2.2.1.3. Sulfobismuthites
2b.1a.2.2.1.3.1. Proper sulfobismuthites
2b.1a.2.2.1.3.1.1. Simple
Aschamalmite $Pb_6Bi_2S_9$
Lillianite family
Lillianite $Pb_{3-2x}Ag_xBi_{2+x}S_6$
Xilingolite $Pb_3Bi_2S_6$
Cosalite $Pb_2Bi_2S_5$
Cannizzarite $Pb_8Bi_{10}S_{23}$
Galenobismutite $PbBi_2S_4$
*Kudriavite $(Cd,Pb)Bi_2S_4$
Sakharovaite $(Pb,Fe)_5(Bi,Sb)_6S_{14}$ (?)
*Mozgovaite $PbBi_4(S,Se)_7$
Ustarasite $Pb(Bi,Sb)_6S_{10}$
*Crerarite $(Pb,Pt)Bi_3(S,Se)_{4-x}$ ($x \sim 0.7$)
2b.1a.2.2.1.3.1.2. Complex
2b.1a.2.2.1.3.1.2.1. Pb^{2+} and Cu^+
*Pizgrischite $Cu_{17}PbBi_{17}S_{35}$
Aikinite homologous series – $Cu_{1-x}Pb_{1-x}Bi_{1+x}S_3$
 Aikinite   ($0 < x < 0,11$) $CuPbBiS_3$
 Friedrichite   ($0,13 < x < 0,20$) $Cu_5Pb_5Bi_7S_{18}$
 *Felbertalite   ($x = 0,26$) $Cu_2Pb_6Bi_8S_{19}$
 Hammarite   ($0,32 < x < 0,38$) $Cu_2Pb_2Bi_4S_9$
 *Emilite   ($x = 0,32$) $Cu_{10,7}Pb_{10,7}Bi_{21,3}S_{48}$
 Lindströmite   ($x = 0,4$) $Cu_3Pb_3Bi_7S_{15}$
 Krupkaite   ($0,41 < x < 0,48$) $CuPbBi_3S_6$
 *Paarite   ($x = 0,58$) $Cu_{1,7}Pb_{1,7}Bi_{6,3}S_{12}$
 *Zalzburgite   ($x = 0,6$) $Cu_{1,6}Pb_{1,6}Bi_{6,4}S_{12}$
 Gladite   ($0,62 < x < 0,77$) $CuPbBi_5S_9$
 *Unnamed   ($x = 0,75$) $CuPbBi_7S_{12}$
 *Pekoite   ($x = 0,83$) $CuPbBi_{11}(S,Se)_{18}$

*Unnamed	$(x = 0,92)$	$CuPbBi_{23}S_{36}$
Nuffildite		$Cu_{1.4}Pb_{2.4}Bi_{2.4}Sb_{0.2}S_7$
*Angelaite		$Cu_2AgPbBiS_4$
Neyite	$(Cu,Ag)_2Pb_7Bi_6S_{17}$ or $Ag_2Cu_6Pb_{25}Bi_{26}S_{68}$	
*Cuproneyite		$Cu_7Pb_{27}Bi_{25}S_{68}$
Kobellite		$Pb_{11}(Cu,Fe)_2(Bi,Sb)_{15}S_{35}$
Eclarite		$(Cu,Fe)Pb_9Bi_{12}S_{28}$
Giessenite		$(Cu,Fe)_2Pb_{26.4}(Bi,Sb)_{19.6}S_{57}$

2b.1a.2.2.1.3.1.2.2. Pb^{2+} and Ag^+, Au^+

Heyrovskyite		$Pb_6Bi_2S_9$
Vikingite family		
Vikingite		$Ag_5Pb_8Bi_{13}S_{30}$
Eskimoite		$Ag_7Pb_{10}Bi_{15}S_{36}$
Ourayite		$Ag_3Pb_4Bi_5S_{13}$
*Jonassonite $Au(Bi,Pb)_5S_4$		$Au(Bi,Pb)_5S_4$

2b.1a.2.2.1.3.1.2.3. Pb^{2+}, Cu^+ and Sn^{4+}

Levyclaudite		$Cu^+{}_3Pb_8Sn^{4+}{}_7(Bi,Sb)_3S_{28}$
*Coiraite		$(Pb,Sn)_{12.5}Sn^{4+}{}_5Fe^{2+}As_3S_{28}$

*2б.1a.2.2.1.3.1.2.4. Pb^{2+}, In^{3+}, Sn^{4+}

*Znamenskyite		$Pb_4In_2Bi_4S_{13}$
*Abramovite		$Pb_2InSnBiS_7$

2б.1a.2.2.1.3.2. Sulfobismuthito-halogenides
*2б.1a.2.2.1.3.2.1. Simple
*2б.1a.2.2.1.3.2.2. Complex

*2б.1a.2.2.1.3.2.2.1. Pb, Sn^{4+}

*Vurroite		$Pb_{20}Sn_2(Bi,As)_{22}S_{54}Cl_6$

*2б.1a.2.2.1.4. Sulfoselenobismuthites
*2б.1a.2.2.1.4.1. Simple

*Babkinite		$Pb_2Bi_2(S,Se)_3$

*2б.1a.2.2.1.5. Sulfoselenotelluroantimonites

*Tsnigriite		$Ag_9SbTe_3(S,Se)_3$

2b.1b. Quasisubtype*: Selenides and selenosalts of sidero- and chalcophilic cations
2b.1b.1. **Class:** Selenides sidero- and chalcophilic cations
2b.1b.1.1. Minerals of heavy d-elements and their crystallochemical analogues.
2b.1b.1.1.1. Minerals of cenosymmetrical d-cations
2b.1b.1.1.1.1. Minerals of **VIIIb**-cations - families of Fn^{n+}
2b.1b.1.1.1.1.1. Polyanionic
2b.1b.1.1.1.1.1.1. Fn^{2+} (simple)
Trogtalite family

Trogtalite group	
Penroseite	$(Ni,Co,Cu)[Se_2]$
Trogtalite	$Co[Se_2]$
Kullerudite group	
Kullerudite	$Ni[Se_2]$
Ferroselite	$Fe[Se_2]$

*Dzharkenite $Fe[Se_2]$
2b.1b.1.1.1.1.2. Monoanionic
2b.1b.1.1.1.1.2.1. Fn^{2+} (simple)
Sederoholmite family
  Sederoholmite group
  Sederoholmite β-NiSe
  Freboldite $CoSe$
Mäkinenite γ-NiSe

2b.1b.1.1.1.1.2.2. Minerals of M^{2+} and M^{3+} (complex)
Selenospinelides family ($M^{2+} : M^{3+}$ = 0,5; compare with sulfospinelides of Fn (family);
sulfospinelides of Pn (family))
  Bornhardtite family   (M^{2+} and M^{3+} only Fn)
Wilkmanite $Ni_3Se_4 \rightarrow Ni^{2+}Ni^{3+}_2Se_4$
    Bornhardtite group
  Trüstedtite Ni_3Se_4
  Bornhardtite $Co_3Se_4 \rightarrow Co^{2+}Co^{3+}_2Se_4$
  Tyrrellite series ($M^{2+} = Cu^{2+}$, $M^{3+}= Fn^{3+}$)
  Tyrrellite-(Ni) (an.1) $Cu(Ni,Co)_2Se_4$
  Tyrrellite-(Co) (an.2) $Cu(Co,Ni)_2Se_4$

2b.1b.1.1.1.2. Minerals of Ib-elements – Cu (and Tl^+ with $CN \leq 4$)
2b.1b.1.1.1.2.1. Polyanionic (simple)
  Krutaite family
  Krutaite $Cu[Se]_2$
  Bambollaite $Cu[(Se,Te)_2]$

2b.1b.1.1.1.2.2. Mono polyanionic (complex)
Klockmannite (compare with covellite) $3CuSe \rightarrow Cu_2Se\,Cu[Se_2]$

2b.1b.1.1.1.2.3. Monoanionic
2b.1b.1.1.1.2.3.1. Cu^+
2b.1b.1.1.1.2.3.1.1. Simple
  Berzelianite family
Berzelianite Cu_2Se
Bellidoite Cu_2Se

2b.1b.1.1.1.2.3.1.2. Complex   2b.1b.1.1.1.2.3.1.2.1. Cu^+ and Tl^+
Crookesite $TlCu_7Se_4$
2b.1b.1.1.1.2.3.2. Cu^+ and Cu^{2+} (complex)
                                   2b.1b.1.1.1.2.3.2.1. Cu^+, Tl^+ and Cu^{2+}
                                   and Cu^{2+} (Fe^{3+});
Sabatierite family $M^+ : M^{2+}$ = up 6 (sabatierite) to 2 (when $2M^{2+} \leftarrow M^+M^{3+}$) (bukovite)
Sabatierite $Tl^+Cu^+_5Cu^{2+}Se_4$
Bukovite $Tl^+_2Cu^+_{3+x}Fe^{3+}Se_{4-x}$
                                   2b.1b.1.1.1.2.3.2.2. $Cu^+ : Cu^{2+}$ = 2
Umangite $Cu_3Se_2 \rightarrow Cu^+_2Cu^{2+}Se_2$
                                   2b.1b.1.1.1.2.3.2.3. $Cu^+ : Cu^{2+}$ = 0,(6)
Athabascaite $Cu_5Se_4 \rightarrow Cu{+}_2Cu^{2+}_3Se_4$

2b.1b.1.1.1.2.3.2.4. $M^+ : M^{2+} = 0,2$

Geffroyite $(Cu,Fe,Ag)_9Se_8$

2b.1b.1.1.1.2.3.3. M^{2+}
2b.1b.1.1.1.2.3.3.1. Complex
2b.1b.1.1.1.2.3.3.1.1. $2M^{2+} \rightarrow Cu^+M^{3+}(Fe^{3+})$
Eskebornite (comp. chalcopyrite (subfam.)) $Cu^+Fe^{3+}Se_2$
2b.1b.1.1.1.2.3.3.1.2. $3M^{2+} \rightarrow 2M^+(Cu^+)M^{4+}(Sn^{4+})$
Selenocernyite (comp.stannite (series)) $Cu^+_2Cd^{2+}Sn^{4+}Se_4$

2b.1b.1.1.1.3. Minerals of **II**b-cations - Zn^{2+} (and Cd^{2+})
2b.1b.1.1.1.3.1. Monoanionic (simple)
Stilleite family (compare with sphalerite (family))
 Stilleite   ZnSe
 Cadmoselite   CdSe

2b.1b.1.1.2. <u>Minerals of heavy noncenosymmetrical *d*-cations</u>
2b.1b.1.1.2.1. Minerals of **VIII**b-cations (Pn^{n+})
2b.1b.1.1.2.1.1. Subselenides
2b.1b.1.1.2.1.1.1. Simple
*Sudovikovite $PtSe_2$
*Verbeekite $PdSe_2$
Palladseite $Pd_{17}Se_{15}$
*Luberoite Pt_5Se_4
2b.1b.1.1.2.1.1.2. Complex
*Jagueite $Pd_3Cu_2Se_4$
*Unnamed $Pd_3Cu_2Se_4$
Oosterboschite $(Pd,Cu)_7Se_5$
*Miessiite $Pd_{11}Te_2Se_2$
*Chrisstanleyite $Pd_3Ag_2Se_4$
*Padmaite   PdBiSe
*Jacutingaite Pt_2HgSe_3
*Tischendorfite $Pd_8Hg_3Se_9$
*Unnamed $(Pb,Cu,Hg)_{1,16}Se$

2b.1b.1.1.2.2. Minerals of **I**b-cations
2b.1b.1.1.2.2.1. Monoanionic
2b.1b.1.1.2.2.1.1. Ag^+
2b.1b.1.1.2.2.1.1.1. Proper selenides
2b.1b.1.1.2.2.1.1.1.1. Simple
Naumannite Ag_2Se

2b.1b.1.1.2.2.1.1.1.2. Complex   2b.1b.1.1.2.2.1.1.1.2.1. Cu^+ and Ag^+
Eucairite   AgCuSe
*Selenojalpaite Ag_3CuSe_2
*Unnamed $(Ag,Cu)_{14}S_6Se_3$
                                     2b.1b.1.1.2.2.1.1.1.2.2. Ag and Au^+
Fischesserite $AuAg_3Se_2$

2b.1b.1.1.2.2.1.1.1.2. Selenido-sulfides

2b.1b.1.1.2.2.1.1.1.2.1. Simple
Aguilarite Ag_4SeS
2b.1b.1.1.2.2.1.1.1.2.2. Complex
Penzhinite $Au^+Ag^+(Ag_{2,65}Cu_{0,35})^{2+}_{\Sigma3}(S_{3,31}Se_{0,69})_{\Sigma4}$

2b.1b.1.1.2.3. Minerals of IIb-cations (Hg^{2+})
2b.1b.1.1.2.3.1. Monoanionic (simple)
Tiemannite $HgSe$
*26.16.1.1.2.3.2.Complex
*Brodtkorbite Cu_2HgSe_2

2b.1b.1.2. <u>Minerals of noncenosymmetrical p-cations</u>
2b.1b.1.2.1. Minerals of IVa-cations
2b.1b.1.2.1.1. Minerals of Pb
2b.1b.1.2.1.1.1. Monoanionic (simple)
2b.1b.1.2.1.1.1.1. Pb^{2+}
Clausthalite (comp.galena – clausthalite (series)) $PbSe$
2b.1b.1.2.2. Minerals of Va-cations
2b.1b.1.2.2.1. Subselenido-sulfides (simple)
*Antimonselite Sb_2Se_3
 Laitakarite family
  Laitakarite series
   Ikunolite $Bi_4(S,Se)_3$
   Laitakarite $Bi_4(Se,S)_3 \rightarrow Bi_4Se_2S$
  Nevskite $(Bi,Pb)(Se,S)$
 Laphamite As_2Se_3
2b.1b.1.2.2.2. Monoanionic
2b.1b.1.2.2.2.1. Selenido-sulfides (at that number selenido-tellurides) (simple)
 Guanajuatite family
  Paraguanajuatite group
   Paraguanajuatite Bi_2Se_3
   Skippenite $Bi_2(Se_2Te)_{\Sigma3}$
   Guanajuatite Bi_2Se_3
  *Telluronevskite Bi_3TeSe_2
 *Vihorlatite $Bi_{24}Se_{17}Te_4$

2b.1b.2. *Class*: Selenosalts of sidero- and chalcophylic cations
2b.1b.2.1. Selenosalts of heavy d- cations
2b.1b.2.1.1. <u>Minerals of cenosymmetrical d-cations</u>
2b.1b.2.1.1.1. Minerals of Ib-cations
2b.1b.2.1.1.1.1. Cu^+
2b.1b.2.1.1.1.1.1. Selenoarsenites (simple)
Mgriite Cu_3AsSe_3

2b.1b.2.1.1.1.2. Cu^+ and Cu^{2+}(Fe^{2+}, Zn^{2+}, Hg^{2+}, Pb^{2+}) (complex)
2b.1b.2.1.1.1.2.1. Selenoarsenites
Chameanite $Cu^+_3(Cu,Fe)^{2+}As(Se,S)_4$
*Unnamed $(Cu,Co,Ni)_7As_3Se_6$

2b.1b.2.1.1.1.2.2. Selenoarsenito-selenoantimonites

Giraudite series (compare with fahlores (series))
Giraudite (Se-sandbergerite) $Cu_6[Cu_4(Fe,Zn)_2]As_4Se_{13}$
Hakite (Se-schwazite) $Cu_{10}Hg_2Sb_4(Se,S)_{13}$

2b.1b.2.1.1.1.2.3. Selenoantimonites
Permingeatite (comp. enargite (family)) $Cu_3SbSe_4 \rightarrow Cu^+Cu^{2+}_2SbSe_4$

2b.1b.2.1.1.1.2.4. Selenobismuthites and seleno-sulfobismutites
                          *2б.1б.2.1.1.1.2.4.1. Cu^+
*Eldragonite $Cu^+_6BiSe^{2-}_4(Se_2)^{2-}$
                          2b.1b.2.1.1.1.2.4.2. Cu^+, Hg^{2+} and Pb^{2+}
Petrovicite $Cu_3HgPbBiSe_5$
                          2b.1b.2.1.1.1.2.4.3. Cu^+ and Pb^{2+}
*Schlemaite $(Cu,\square)_6(Pb,Bi)Se_4$
Součekite (compare with aikinite (series)) $CuPbBi(SeS_2)_{\Sigma3}$
*Součekite-like mineral $Cu_{2,1-2,6}Ag_{0,7}Pb_{0,3}Bi_{0,2}Se_3$ (?)
Proudite $Pb_8Bi_{10}S_{23}$
Watkinsonite $PbCu_2Bi_4Se_8$
Nordströmite $CuPb_3Bi_7(S,Se)_{14}$ S : Se = 2,4
Junoite $Cu_2Pb_3Bi_8(S,Se)_{16}$ S : Se = 1,7 - 4,8
Pekoite $CuPbBi_{11}(S,Se)_{18}$ S : Se = 5,2

2b.1b.2.1.2. Selenosalts of noncenosymmetrical d-cations
*2b.1b.2.1.2.1. Minerals of **VIII**b-cations
*Kalungaite   PdAsSe
*Milotaite   PdSbSe

2b.1b.2.1.2.2. Minerals of **I**b-elements
2b.1b.2.1.2.2.1. Minerals of Ag
2b.1b.2.1.2.2.1.1. Selenoantimonites (simple)
Selenostephanite (comp. stephanite (group.)) $Ag_5Sb(Se,S)_4$
*Selenopolybasite $Ag_{15}CuSb_2S_9Se_2$

2b.1b.2.1.2.2.1.2. Selenobismutites (simple)
Bohdanowiczite $AgBiSe_2$
*Litochlebite $Ag_2PbBi_4Se_8$

2b.1b.2.2. Selenosalts of noncenosymmetrical p-cations
2b.1b.2.2.1. Minerals of **IV**a-cations (Pb^{2+})
2b.1b.2.2.1.1. Seleno-sulfobismutites (at that number telluro-selenosulfobismutit)
(simple)
Weibullite family
Weibullite $Ag_{0,3}Pb_{5,3}Bi_{8,3}(S,Se)_{18}$
Wittite $Pb_8Bi_{10}(S,Se)_{23}Se$
Poubaite (Te,Se-galenobismutite) $PbBi_2(Se,Te,S)_4$

2b.2. Subtype: Chalcogenic compounds of lithophylic cations
2b.2.1. *Class*: Sulfides (and selenides) of lithophylic cations
2b.2.1.1. Minerals of light d-elements (with $1-4$ d-electrons)
2b.2.1.1.1. Minerals of cenosymmetrical d-elements

*2б.2.1.1.1.1. Minerals of **IV**b-elements
*Wassonite   TiS

2b.2.1.1.1.1. Minerals of **V**b-elements
2b.2.1.1.1.1.1. Minerals of M^{5+} 2b.2.1.1.1.1.1.1. Simple
Patronite $V[S_2]_2$
*Colimaite K_3VS_4

2b.2.1.1.2. <u>Minerals of noncenosymmetrical d-elements</u>
2b.2.1.1.2.1.Minerals of **VI**b-elements
2b.2.1.1.2.1.1. Monoanionic
2b.2.1.1.2.1.1.1. Minerals of M^{4+} 2b.2.1.1.2.1.1.1.1. Simple
 Molybdenite family
   Molybdenite group
   Molybdenite -*2H* MoS_2
   Molybdenite -3R
   Tungstenite-*2H* WS_2
   *Tungstenite-*3R* WS_2
   Drysdallite $MoSe_2$
   Jordisite MoS_2
   *Rheniite ReS_2

*2б.2.1.1.2.2. Minerals of VIb-элементов and VIIb-elements
*Tarkianite $(Re,Mo)_4(Cu,Fe)S_8$
*Buseckite $(Fe,Zn,Mn)S$

2b.2.1.2. Sulfides of s-elements
2b.2.1.2.1. Sulfides of **I**a-cations and Tl^+ (with CN=8-12) (all monoanionic)
2b.2.1.2.1.1. Sulfides of $Tl^+(K)$
2b.2.1.2.1.1.1. Proper sulfides   2b.2.1.2.1.1.1.1.Complex
                                                    2b.2.1.2.1.1.1.1.1. $M^+(K^+,Tl^+,Cu^+)$ and
                                                    $M^{3+}(Fe^{3+})$
 Murunskite group
 Thalcusite $Tl_2Cu^+_3Fe^{3+}S_4$
 Murunskite $K_2Cu^+_3Fe^{3+}S_4$
                                                    2b.2.1.2.1.1.1.1.2. $M^+(K^+,Tl^+)$, $M^{2+}(Fe^{2+})$ and $M^{3+}(Fe^{3+})$
 Rasvumite group
 Picotpaulite $TlFe^{2+}Fe^{3+}S_3$
 Rasvumite $KFe^{2+}Fe^{3+}S_3$
                                                    2b.2.1.2.1.1.1.1.3. $M^+(K^+)$,$M^{2+}(Fe^{2+})$ and $M^{3+}(Fe^{3+})$
*Owensite $(Ba,Pb)_6(Cu^{1+},Fe,Ni)_{25}S_{27}$
Bartonite $K_6(Fe,Cu)_{20}S_{26}S$
*Chlorbartonite $K_6(Fe,Cu)_{24}S_{26}(Cl,S)$

2b.2.1.2.1.1.2. Sulfido-chlorides   2b.2.1.2.1.1.2.1. Complex
 Djerfisherite group
 Thalfenisite $Tl_6(Fe,Ni,Cu)_{25}S_{26}Cl$
 Djerfisherite $K_6(Fe,Cu,Ni)_{25}S_{26}Cl$

2b.2.1.2.1.2. Sulfides of Na

2b.2.1.2.1.2.1. Complex   2b.2.1.2.1.2.1.1. Anhydrous
                                 2b.2.1.2.1.2.1.1.1. $M^+(Na^+)$ and $M^{2+}(Fe^{2+}, Cu^{2+}, Zn^{2+})$
Chvilevaite $Na^+_2Cu^+_2(Fe,Cu,Zn)^{2+}_2S_4$
                                 2b.2.1.2.1.2.1.1.2. $M^+(Na^+)$ and $M^{3+}(Cr^{3+})$
Caswellsilverite $NaCr^{3+}S_2$
                                 2b.2.1.2.1.2.1.2. Hydrous
                                 2b.2.1.2.1.2.1.2.1. $M^+(Na^+)$ and $M^{3+}(Fe^{3+})$
Erdite $Na^+Fe^{3+}S_2 \cdot 2H_2O$
Coyoteite $Na^+Fe^{3+}_3S_5 \cdot 2H_2O$
                                 2b.2.1.2.1.2.1.2.2. $M^+(Na^+, K^+, Cu^+)$, $M^{2+}(Fe^{2+})$
                                 and $M^{3+}(Fe^{3+})$
Orickite $CuFeS_2 \cdot nH_2O$
                                 2b.2.1.2.1.2.1.2.3. $M^+(Na^+)$, $M^{3+}(Cr^{3+})$ and $M^{6+}(Cr^{6+})$
Schöllhornite $Na_{0.3}CrS_2 \cdot H_2O$
*Pautovite $CsFe_2S_3$

2b.2.1.2.2. Sulfides of IIa-cations and their crystallochemical analogues (all
monoanionic)
2b.2.1.2.2.1. Proper sulfides   2b.2.1.2.2.1.1. Simple
Niningerite family
 Oldhamite CaS
  Niningerite series
  Niningerite -(Mg) $(Mg,Fe,Mn)S$
  Niningerite -(Fe) $(Fe,Mg,Mn)S$
  *Keilite $(Fe,Mg)S$
*2б.2.1.2.2.1.2. Complex
*2б.2.1.2.2.1.2.1.Hydrates
*Cronusite $Ca_{0.2}(H_2O)_2CrS_2$

2b.2.1.2.2.2. Sulfido-oxides   2b.2.1.2.2.2.1. Complexe
Sarabauite $CaSb^{3+}_{10}S_6O_{10}$
*Apuanite $Fe^{2+}Fe^{3+}_4Sb^{3+}_4O_{12}S$

*2б.2.1.2.2.3. Sulfido-oxido-carbonates   *2б.2.1.2.2.3.1. Hydrates
*Ignicolorite $FeS_2 \cdot 0.7CaCO_3 \cdot 2.8 H_2O$

2b.2.1.2.2.3. Sulfido-hydroxides 2b.2.1.2.2.3.1. Complex
Valleriite family
 Valleriite $(Mg,Al)_3(Fe,Cu)_4(OH)_6S_4 \rightarrow$
                                  $4 (Fe,Cu)S \cdot 3 (Mg,Al)(OH)_2$
*Ferrovalleriite $2(Fe,Cu)S \cdot 1.53[(Fe,Al,Mg)(OH)_2]$
Haapalaite $(Mg,Fe^{2+})_3(Fe,Ni)^{2+}_4(OH)_6S_4 \rightarrow$
                                  $4 (Fe,Ni)S \cdot 3 (Mg,Fe^{2+})(OH)_2$
Tochilinite $(Mg,Fe)_5Fe_6(OH)_{10}S_6 \rightarrow$
                                  $6FeS \cdot 5 (Mg,Fe^{2+})(OH)_2$
*Ferrotochilinite $FeS \cdot 0.85[Fe(OH)_2]$
Yushkinite $(Mg,Al)(OH)_2 \cdot VS_2$
*Vyalsovite $FeS \cdot Ca(OH)_2 \cdot Al(OH)_3$
*Ekplexite $(Mg_{1-x}Al_x)(Nb,Mo,W)(OH)_{2+x}S_2$
*Kaskasite $Mg_{1-x}Al_x(OH)_{2+x}(Mo,Nb)S_2$

*Manganokaskasite $Mn_{1-x}Al_x(OH)_{2+x}(Mo,Nb)S_2$

2b.2.1.2.2.4. Sulfido-tiosulfates
2b.2.1.2.2.4.1. Hydrate
2b.2.1.2.2.4.1.1. Basic
Bazhenovite $Ca_8(OH)_2S_5[S_2O_3]\,20H_2O$

2b.2.2. *Class:* Sulfosalts of lithophylic cations
2b.2.2.1. Minerals of **I**a-elements and Tl^+ (with CN = 8-12)
2b.2.2.1.1. Sulfoantimonites
2b.2.2.1.1.1. Proper sulfoantimonites
2b.2.2.1.1.1.1. Simple
2b.2.2.1.1.1.1.1. Crystalline hydrate (middle)
Gerstleyite $Na_2(Sb,As)_8S_{13}\,2H_2O$
*Ambrinoite $(K,NH_4)_2(As,Sb)_8S_{13}\cdot H_2O$
2b.2.2.1.1.2. Sulfoantimonito-antimonites
2b.2.2.1.1.2.1. Hydrate (basic)
Cetineite $NaK_5Sb_{14}S_6O_{18}\cdot 6H_2O$
*Ottensite $Na_3(Sb_2O_3)_3(SbS_3)\cdot 3H_2O$

3. TYPE: MINERALS WITH PRINCIPAL IONIC-COVALENT AND COVALENT- IONIC BOND – NONMETALLIDES OF LIGTH (TYPICAL NONCENOSYMMETRICAL) VIa-ELEMENT (O) – OXIGEN COMPOUNDS: OXIDES AND HYDROXIDES (ISODESMICAL → ANISODESMICAL) → OXOCALTS (ANISODESMICAL)

3.1. SUBTIPE: OXIDES AND HYDROXIDES (ISODESMICAL)
3.1a. *QUASISUBTIPE: OXIDES AND HYDROXIDES OF LITHOPFYLLIC CATIONS WITH LOW FC*
3.1a.1. *Class:* Oxides and hydroxides of s-, d_s- and p_s-cations
3.1a.1.1. Oxides and hydroxides of s-, d_s- and p_s-cations without Li^+, Be^{2+}
3.1a.1.1.1. Monoanionic
3.1a.1.1.1.1. Proper oxides
3.1a.1.1.1.1.1. M^{2+}
3.1a.1.1.1.1.1.1. Simple

   Periclase group

Periclase	MgO
Hongquiite дискредитирован	TiO
Manganosite	MnO
Wüstite	FeO
Bunsenite	NiO
Lime	CaO

3.1a.1.1.1.1.1.2. M^+,M^{2+} and M^{3+}
3.1a.1.1.1.1.1.2.1. Complex   CN $M(A,B) = 4 – 6$
Diaoyudaoite $NaAl_{11}O_{17}$
Oxsospinelides family - AB_2O_4; $A(M^{2+}) : B(M^{3+}) = 2$
  Spinel subfamily $A = Mg^{2+}$, Fe^{2+}, Mn^{2+}, Ni^{2+}, Co^{2+}; $B = Al^{3+}$, V^{3+}, Cr^{3+}, Mn^{3+}, Fe^{3+}, Co^{3+}, Ni^{3+}, Mg^{2+}
  Magnetite series $^{(6)}A^{(6)}B|^{(4)}B'O_4|$; $A = Mg^{2+}$, Fe^{2+}, Ni^{2+}, Mn^{2+}; $B = Fe^{3+}$, Mn^{3+}; $B' = Fe^{3+}$

*Cuprospinel	$(Cu,Mg)Fe_2O_4$
Magnesioferrite	$MgFe_2O_4$
Trevorite	$NiFe_2O_4$
Magnetite	$Fe_3O_4 \rightarrow {}^{(6)}Fe^{2+(6)}Fe|^{(4)}FeO_4$
Jacobsite	$Mn^{2+}Fe_2^{3+}O_4$

Chromite series $^{(4)}A^{(6)}B_2O_4$; $A = Mg^{2+}, Fe^{2+}, Mn^{2+}, Co^{2+}, Ni^{2+}$; $B = Mg, Cr^{3+}, Al^{3+}, V^{3+}$

Magnesiochromite	$MgCr_2^{3+}O_4$
Cochromite	$CoCr_2O_4$
Chromite	$FeCr_2O_4$
Manganochromite	$Mn^{2+}Cr_2O_4$
*Xieite orth.	$FeCr_2O_4$

Spinel series $^{(4)}A^{(6)}B_2O_4$; $A = Mg^{2+}, Mn^{2+}, Fe^{2+}$; $B = Al^{3+}, Fe^{3+}$

Spinel	$MgAl_2O_4$
*Krotite	$CaAl_2O_4$
Hercynite	$Fe^{2+}Al_2O_4$
Galaxite	$Mn^{2+}Al_2O_4$
*Brunogeierite	$GeFe_2O_4 \rightarrow (Ge^{2+},Fe^{2+})Fe^{3+}_2O_4$

Coulsonite series $^{(4)}A^{(6)}B_2O_4$; $A = Mn^{2+}, Fe^{2+}$; $B = V^{3+}, Cr^{3+}$

Vuorelainenite	$Mn^{2+}V_2^{3+}O_4$
*Magnesiocoulsonite	$Mg\ V_2^{3+}O_4$
Coulsonite	FeV_2O_4
*Unnamed	$Mn_2La_2O_5$

Hausmannite group $^{(4)}A^{(6)}B_2O_4$; $A = Fe^{2+}, Mg^{2+}, Mn^{2+}$; $B = Cr^{3+}, Fe^{3+}, Mn^{3+}$

Iwakiite	$Mn^{2+}Fe_2^{3+}O_4$
Hausmannite	$MnMn_2O_4$
	CN $M^{2+} = 8$
Marokite	$^{(8)}Ca^{(6)}Mn_2O_4$

3.1a.1.1.1.1.2. M^{2+}, M^{3+}	$M^{2+} : M^{3+} = \geq 1$
Muskoxite	$Mg_7Fe_4^{3+}(OH)_{26} \cdot H_2O$
Brownmillerite family	
Brownmillerite	$Ca_2(Al,Fe)_2O_5 \cong$
	$\cong {}^{(<9)}Ca_2{}^{(6)}(Fe,Al)O^{(4)}(Al,Fe)O_4$
Srebrodolskite	$Ca_2Fe_2O_5 \rightarrow {}^{(<9)}Ca_2{}^{(6)}FeO^{(4)}FeO_4$
*Tululite	$Ca_{14}(Fe^{3+},Al)(Al,Zn,Fe^{3+},Si,P,Mn,Mg)_{15}O_{36}$
	$M^{2+} : M^{3+} = < 1$
*Aciculite	$CaFe_2O_4$
*Harmunite	$CaFe_2O_4$
Mayenite	$^{(8)}Ca_{12}{}^{(5;4)}Al_{14}O_{33}$
*Dmitryivanovite	$CaAl_2O_4$
*Grossite	$CaAl_4O_7$
*Barioferrite	$BaFe^{3+}_{12}O_{19}$

3.1a.1.1.1.1.3. M^{3+}
3.1a.1.1.1.1.3.1. Simple
Corundum family

*Deltalumite	Al_2O
Corundum group	
Corundum	Al_2O_3
*Tistarite	Ti_2O_3

Hematite	Fe_2O_3
Unnamed	$(Ru,Fe)_2O_3$
Eskolaite	Cr_2O_3
Karelianite	V_2O_3
Maghemite series	
Maghemite	$\gamma\text{-}Fe_2O_3 \rightarrow Fe_{2,67}O_4 \rightarrow {}^{(6)}Fe^{3+}{}_{0,67}\square_{0,33}{}^{(6)}Fe^{3+}|{}^{(4)}Fe^{3+}O_4|$
Titanomaghemite	$Fe(Fe,Ti)_2O_4$
Luogufengite	$\varepsilon\text{-}Fe_2O_3$
*Ittriaite-(Y)	Y_2O_3
Bixbyite	$Mn_2{}^{3+}O_3$
	3.1a.1.1.1.1.3.1. Hydrates
Akdalaite	$(Al_2O_3)_5 \cdot H_2O$

*3.1a.1.1.1.1.4. M^{2+}, M^{3+}, M^{4+}, M^{5+}

*Wernerkrauseite	$CaFe^{3+}{}_2Mn^{4+}O_6$
*Bitikleite (SnAl)	$Ca_3SnSb[AlO_4]_3$
*Dzhuluite – new name of bitikleite-(SnFe)	$Ca_3SbSn[FeO_4]_3$
*Usturite - new name of bitikleite-(ZrFe).	$Ca_3SbZr[FeO_4]_3$

3.1a.1.1.1.2. Hydroxido-oxides
3.1a.1.1.1.2.1. Proper hydroxido-oxides M^{3+} 3.1a.1.1.1.2.1.1. Simple

Diaspore family

Diaspore group

Diaspore	$\alpha\text{-}Al(OH)O$
Montroseite	$\alpha\text{-}(V,Fe)(OH)O$
Bracewellite	$\alpha\text{-}Cr(OH)O$
Groutite	$\alpha\text{-}Mn(OH)O$
Goethite	$\alpha\text{-}Fe(OH)O$
Grimaldiite	$(Cr,Al)(OH)O$
Guyanaite	$(Cr,Fe,Al)(OH)O$
Feroxyhyte	$\delta\text{-}Fe(OH)O$

Akaganeite group

Akaganeite	$\beta\text{-}Fe(OH,Cl)O$
Feitknechtite	$Mn(OH)O$

Böhmite group

Böhmite	$\gamma\text{-}Al(OH)O$
Lepidocrocite	$\gamma\text{-}Fe(OH)O$
*Tsumgallite	$Ga(OH)O$

3.1a.1.1.1.3. Hydroxides
3.1a.1.1.1.3.1. M^{2+} 3.1a.1.1.1.3.1.1. Simple

Brucite family

Brucite	$Mg(OH)_2$
Amakinite	$(Fe,Mg)(OH)_2$

Pyrochroite group

Pyrochroite	$Mn(OH)_2$
Theophrastite	$Ni(OH)_2$
Portlandite	$Ca(OH)_2$

3.1a.1.1.1.3.2. M^{3+}
Gibbsite family
Gibbsite $Al(OH)_3$
Bayerite $Al(OH)_3$
Nordstrandite $Al(OH)_3$
Doyleite $Al(OH)_3$
Söhngeite $Ga(OH)_3$
*Bernalite $Fe(OH)_3$

3.1a.1.1.1.3.3. M^{2+}, M^{3+} 3.1a.1.1.1.3.3.1. Complex
*Taschelgite $CaMgFe^{2+}Al_9O_{16}(OH)$
                                      3.1a.1.1.1.3.3.1.1. Hydrates
Meixnerite $Mg_6Al_2(OH)_{18} \cdot 4H_2O$
*Fougerite $Fe^{2+}_6Fe^{3+}_2(OH)_{18} \cdot 4H_2O$

3.1a.1.1.2. Polyanionic
3.1a.1.1.2.1. Proper oxides
3.1a.1.1.2.2. Oxido-silicates   3.1a.1.1.2.2.1. Complex
Braunite family
Braunite II $Ca^{2+}Mn^{3+}_{14}O_{20}[SiO_4]$
Braunite $Mn^{2+}Mn^{3+}_6O_8[SiO_4]$
*Gatedalite $Mn^{2+}_2Mn^{3+}_4ZrO_8[SiO_4]$
Neltnerite $CaMn^{3+}_6O_8[SiO_4]$
*Abswurmbachite $Cu^{2+}Mn^{3+}_6O_8[SiO_4]$
Dorrite $Ca_4(Mg_3Fe^{3+}_9)O_4[Si_3Al_8Fe^{3+}O_{36}]$
Sapphirine family
Sapphirine $Mg_4(Mg_3Al_9)O_4[Si_3Al_9O_{36}]$
Sapphirine-1TC, -2M, -4M polytipes $(Mg,Al,Fe^{2+})_8[(Al,Si,Fe^{3+})_6O_{18}]O_2$
Surinamite $Mg_3Al_3O[Si_3BeAlO_{15}]$
                                      *3.1a.1.1.1.2.1. Hydrates

3.1a.1.1.2.3. Oxido-halogenides
*Brearleyite $Ca_{12}Al_{14}O_{32}Cl_2$
                                      3.1a.1.1.2.3.1. Hydrates
*Kyuygenite = Chlorkyuygenite $Ca_{12}Al_{14}O_{32}[(H_2O)_4Cl_2]$
*Fluorkyuygenite $Ca_{12}Al_{14}O_{32}[(H_2O)_4F_2]$

3.1a.1.1.2.4. Hydroxido-oxides   3.1a.1.1.2.4.1. Simple
Ferrihydrite $Fe_{10}^{3+}O_{14}(OH)_2$
Manganite family
Manganite $Mn[OHO]$
Heterogenite-*3R* $Co[OHO]$
Heterogenite-2H

3.1a.1.1.2.5. Hydroxido-oxido-silicates
*Macaulayite $Fe_{24}O_{43}Si_4(OH)_2$

*3.1a.1.1.2.6. Hydroxido-oxido-carboates
*3.1a.1.1.2.6.1. Simple   *3.1a.1.1.2.6.1.1. Hydrates
*Mössbauerite $Fe^{3+}_3O_2(OH)_4(CO_3)_{0.5} \cdot 1.5H_2O$
*3.1a.1.1.2.6.2. Complex   *3.1a.1.1.2.6.2.1. Hydrates

*Trébeurdenite $Fe^{2+}_2Fe^{3+}_4O_2(OH)_{10}(CO_3) \cdot 3H_2O$

*3.1a.1.1.2.7. Hydroxido-carbonates
*Karchevskyite $Mg_{18}Al_9(OH)_{54}Sr_2[CO_3]_9(H_2O)_6(H_3O)_5$

*3.1a.1.1.2.8. Hydroxido-oxido-sulfates
*3.1a.1.1. 2.8.1. Simple   *3.1a.1.1. 2.8.1.1. Hydrates
*Schwertmannite $Fe_{16}O_{16}(OH)_{9,6}(SO_4)_{3,2}\cdot10H_2O$ or→
                $Fe_{16}O_{16}(OH)_y(SO_4)_z\cdot nH_2O$, где $2\leq z \leq 3,5$; $16-y = 2z$
3.1a.1.1.2.9. Hydroxido-oxido-halogenides 3.1a.1.1.2.9.1. Simple
Zharchikhite $Al(OH)_2F$
                                  *3.1a.1.1. 2.9.1.1. Hydrates
*Lesukite $Al_2(OH)_5Cl\,2H_2O$

3.1a.1.1.2.9.2. M^{2+}, M^{3+}
3.1a.1.1.2.9.2.1. Complex   *3.1a.1.1. 2.9.2.1. Hydrates
Iowaite $Mg_6Fe_2^{3+}(OH)_{16}Cl_2 \cdot 4H_2O$
*Droninoite $Ni_3Fe^{3+}(OH)_8Cl\cdot2H_2O$

3.1a.1.2. Oxides and hydroxides Be^{2+} (all monoanionic)
3.1a.1.2.1. Proper oxides
3.1a.1.2.1.1. Simple
Bromellite BeO
3.1a.1.2.1.2. Complex
*Rhodizite $KBe_4Al_4(B_{11}Be)O_{28}$
*Londonite $CsBe_5Al_4B_{11}O_{28}$
*Byrudite $(Be,\square)(V^{3+},Ti^{4+})_3O_6$

Taaffeite family
Taaffeite $Mg_3Al_8BeO_{16}$
*Magnesiotaaffeite-2N´2S гексаг. $Mg_3Al_8BeO_{16}$
*Magnesiotaaffeite-6N´3S тригон. $Mg_2BeAl_6O_{12}$
*Ferrotaaффеит-2N´2S $(Fe^{2+},Mg,Zn)_3Al_8BeO_{16}$
*Ferrotaaффеит-6N´3S $BeFe_2^{2+}Al_6O_{12}$
Pehrmanite synonym of *Ferrotaaффеит-6N´3S
Chrysoberyl Al_2BeO_4
*Maryinskite $Be(Cr,Al)_2O_4$

3.1a.1.2.2. Hydroxides
3.1a.1.2.2.1. Simple
Behoite family
Behoite $\beta-Be(OH)_2$
Clinobehoite $Be(OH)_2$

3.1a.1.3. **Oxides and hydroxides of Zn^{2+}, Pb^{2+}, As^{3+}, Sb^{3+} and Sb^{5+} lithophylic paragenetic association of Franclin and Sterling Hill, New Jersey, USA, Langban and Jacobsberg, Sweden.**
3.1a.1.3.1. Minerals of Zn^{2+}
3.1a.1.3.1.1. Simple
3.1a.1.3.1.1.1. Neutral

Zincite $(Zn,Mn)O$

3.1a.1.3.1.2. Complex
3.1a.1.3.1.2.1. Neutral
  Franklinite group $^{(4)}A^{(6)}B_2O_4$; $A = Zn^{2+}$, Mn^{2+}, Fe^{2+}; $B = Al^{3+}$, Fe^{3+}, Mn^{3+}
                                  (compare with   oxospinelides (family))
  Gahnite $ZnAl_2O_4$
  Franklinite $(Zn,Mn,Fe)(Fe,Mn)_2O_4$
  Hetaerolite family (?) $^{(4)}A^{(6)}B_2O_4 \rightarrow H_{3x}AB_{2-x}O_4$; $A = Zn^{2+}$; $B^{3+} = Mn^{3+}$
  Hetaerolite $^{(4)}ZnMn_2O_4$
  Hydrohetaerolite $HZnMn^{3+}_{5/3}O_4$
  *Cianciulliite $Mn(Mg,Mn)Zn_2(OH)_{10} \cdot 2\text{-}4H_2O$

3.1a.1.3.2. Minerals of Pb^{2+}
3.1a.1.3.2.1. Complex
3.1a.1.3.2.1.1. Neutral
Plumboferrite $Pb_2(Fe^{3+},Mn^{2+},Mg)_{11}O_{19}$
Magnetoplumbite $PbFe_{12}^{3+}O_{19}$
(compare with hibonite (group))
*Ferricoronadite $Pb[Mn^{4+}_6(Fe^{3+},Mn^{3+})_2]O_{16}$
*Nežilovite $PbZn_2(Mn^{4+},Ti^{4+})_2Fe^{3+}_8O_{19}$
                          3.1a.1.3.2.1.2. Oxido-hydroxides
Quenselite $PbMn^{3+}(OH)O_2$
Hematophanite $Pb_4Fe^{3+}_3(OH,Cl)O_8$

3.1a.1.3.3. Minerals of As^{3+} and Sb^{3+}
3.1a.1.3.3. Complex   3.1a.1.3.3.1. Neutral
Stenhuggarite $CaFe^{3+}Sb^{3+}As^{3+}_2O_7$
Filipstadite $(Mn,Mg)_4Fe^{3+}Sb^{5+}O_8$
(compare with oxospinelides (family.); 4

3.1a.1.3.4. Minerals of Sb^{5+}
3.1a.1.3.4.1. Proper oxides
3.1a.1.3.4.1.1. Complex   3.1a.1.3.4.1.1.1. Neutral
Monimolite $(Pb,Ca)_3Sb^{5+}_2O_8$ (?)
Melanostibite $Mn^{2+}_2Fe^{3+}Sb^{5+}O_6$
Ingersonite $Ca_3MnSb^{5+}_4O_{14}$
Swedenborgite $NaBe_4Sb^{5+}O_7 \rightarrow {}^{(12)}Na[^{(4)}Be_4O(^{(6)}Sb^{5+}O_6)]^{\infty 3}$
*Rinmanite $Zn^{2+}_2Sb^{5+}_2Mg_2Fe^{3+}_4O_{14}(OH)_2$

3.1a.1.3.4.2. Oxido-silicates
3.1a.1.3.4.2.1. Complex   3.1a.1.3.4.2.1.1. Neutral
  Katoptrite series
  Katoptrite $^{(6)}(Mn^{2+}_5Sb^{5+}_2)_{\Sigma 7}{}^{(4)}(Mn^{2+}_8Al_4Si_2)_{\Sigma 14}O_{28}|^{\infty 2}$
  Yeatmanite $^{(6)}(Mn^{2+}_5Sb^{5+}_2)_{\Sigma 7}{}^{(4)}(Mn^{2+}_2Zn_8Si_4)_{\Sigma 14}O_{28}|^{\infty 2}$
  *Örebroite $Mn_6^{2+}(Fe^{3+},Sb^{5+})_2(SiO_4)_2(O,OH)_6$

3.1a.2.*Class:* Oxides and hydroxides of *f*-cations low FC of 4-valence *f*-cations
3.1a.2.1.1. Proper oxides
3.1a.2.1.1.1. Monoanionic
3.1a.2.1.1.1.1. Neutral

Uraninite group
Uraninite UO_2
Thorianite ThO_2
Cerianite-(Ce) $(Ce,Th)O_2$

3.1a.2.1.1.2. Polyanionic oxides (peroxides)
                                        3.1a.2.1.1.2.1. Hydrates
Studtite family
Studtite $UO_4 \cdot 2H_2O \rightarrow U[O_2]_2 \cdot 2H_2O$
Metastudtite $UO_4 \cdot 2H_2O \rightarrow U[O_2]_2 \cdot 2H_2O$

3.1a.3.**Class**: Oxides and hydroxides of f-cations with middle FC – 6-valence
f-cations (U^{6+}) $\rightarrow$ compounds uranyl $(UO_2)^{2+}$ – k. uranil asids, uranates and their
derivates (uranium micas and related minerals)
3.1a.3.1. Uranil acids and uranates
3.1a.3.1.1. Uranil acids (hydrates of uranyl hydroxides)
*Paulscherrerite $UO_2(OH)_2$
Paraschoepite $UO_3 \cdot 2H_2O$
Schoepite family
Schoepite $[(UO_2)_8O_2(OH)_{12}]^{\infty 2} \cdot (H_2O)_{12}$
Metaschoepite $[(UO_2)_8O_2(OH)_{12}]^{\infty 2} \cdot (H_2O)_{10}$
3.1a.3.1.2. Uranates
3.1a.3.1.2.1. Basic
*Vorlanite $CaU^{6+}O_4$
Metacalciouranoite $[(UO_2)_2O_2(OH)_2]^{\infty 2}(Ca,Na_2,Ba)$
3.1a.3.1.2.2. Hydrates
3.1a.3.1.2.2.1. Oxides-hydroxides
Ianthinite $[(UO_2)_4O_6(OH)_4]^{\infty 2}U^{4+}_2 \cdot 9H_2O$
Vandendrisscheite family (y =7) $y = UO_2 : Me^{2+}$
Vandendriesscheite $[(UO_2)_{10}O_6(OH)_{11}]Pb_{1.5} \cdot 11H_2O$
Metavandendriesscheite $[(UO_2)_7O_2(OH)_{12}]^{\infty 2}Pb \cdot nH_2O$
Becquerelite family (y = 6) $Ba(UO_2)_6O_4(OH)_6 \cdot 8H_2O$
Vandenbrandeite $[(UO_2)_2(OH)_8]^{\infty 2}Cu_2$
Becquerelite $[(UO_2)_6O_4(OH)_6]^{\infty 2}Ca(H_2O)_4 \cdot 4H_2O$
Billietite $[(UO_2)_6O_4(OH)_6]^{\infty 2}Ba(H_2O) \cdot 7H_2O$
Compreignacite $[(UO_2)_6O_4(OH)_6]^{\infty 2}K_2 \cdot 7H_2O$
Fourmarierite family (y = 4) $PbO (UO_2)_4(OH)_{4+2x} \cdot 4H_2O$
Fourmarierite $[(UO_2)_4O_{3-2x}(OH)_{4+2x}]^{\infty 2}Pb_{1-x} \cdot 4H_2O$ or
                                                   $[(UO_2)_4O_3(OH)_4]^{\infty 2}Pb \cdot 4H_2O$
Richetite $[(UO_2)_{36}O_{36}(OH)_{24}]^{\infty 2}(Fe^{3+},Mg)_xPb^{2+}_{8.6} \cdot 41H_2O$
Agrinierite family
Agrinierite (y = 3) $[(UO_2)_3O_3(OH)_2](K_2,Ca,Sr) \cdot H_2O$
Protasit (y = 3) $[(UO_2)_3O_3(OH)_2]^{\infty 2} Ba(H_2O)_3$
Rameauite (y=1.5) $[(UO_2)_3O_3(OH)_2]^{\infty 2}_2K_2Ca \cdot 6H_2O$
Curite family
Curite (y = 2.(6)) $[(UO_2)_4O_{4+x}(OH)_{3-x}]^{\infty 2}_2Pb_{3+x}(H_2O)_2$
Sayrite (y = 2.5) $[(UO_2)_5O_6(OH)_2]^{\infty 2}Pb_2 \cdot 4H_2O$
*Spriggite (y = 2) $[(UO_2)_6O_8(OH)_2]^{\infty 2}Pb_3 \cdot 3H_2O$
Masuyite (y=1.75) $[(UO_2)_3O_3(OH)_2]^{\infty 2}Pb(H_2O)_3$

Clarkeite family (y = 2)

Wölsendorfite	$[(UO_2)_{14}O_{19}(OH)_4]^{\infty 2} Pb_7 \cdot 12H_2O$
Calciouranoite	$[(UO_2)_2O_2(OH)_2]^{\infty 2}(Ca,Ba,Pb,K,Na) \cdot 4H_2O$ [1*]
Bauranoite	$[(UO_2)_2O_2(OH)_2]^{\infty 2}Ba \cdot 4H_2O$
Clarkeite	$[(UO_2)_2O_2(OH)_2]^{\infty 2} (Na,K,Ca,Pb) \cdot nH_2O$
Uranosphaerite (y = 0.(6))	$[(UO_2)O_2(OH)]^{\infty 2}Bi$

3.1a.3.2. Uranilo-titanates
3.1a.3.2.1. Basic

Orthobrannerite	$U^{+6}U^{+4}Ti_4O_{12}(OH)_2$ [2*] or
	$[(UO_2)_2(TiO_3)_4(OH)_2]^{\infty 2}U^{4+}$
*Cleusonite	$Pb(U^{4+},U^{6+})(Ti,Fe^{2+},Fe^{3+})_{20}(O,OH)_{38}$
*Holfertite	$U^{6+}_{2-x}Ti(O_{8-4x}OH_{4x})_{\Sigma 8}[(H_2O)_3Ca_x] \rightarrow$ or
	$[(UO_2)_{2-x}TiO_{4-2x}(OH)_{4x}][(H_2O)_3Ca_x]$

3.1a.3.3a. Uranilo-molybdenic acids (hydrates)

*Umohoite	$[(UO_2)(MoO_4)(H_2O)_2]^{\infty 2} \cdot H_2O$ [3*]
Iriginite	$[(UO_2)(^{[6]}Mo_2O_7)(H_2O)_2]^{\infty 2} \cdot H_2O$

3.1a.3.3b. Uranilo-molybdates
3.1a.3.3b.1. Basic → acid

Deloryite group (x = 0.5; y = 0.25, where x=UO$_2$:MoO$_4$, y= UO$_2$:Me^{2+})

Deloryite	$[(UO_2)(MoO_4)_2(OH)_6]^{\infty 2} Cu_4$

3.1a.3.3b.1.1. Hydrates
3.1a.3.3b.1.1.1. Basic → acid

Moluranite group (x = 0.43; y = 1,5)

Moluranite	$H_4U^{4+}(UO_2)_3(MoO_4)_7 18H_2O$ [4*]

Urmolite family (x= (UO$_2$) : (MoO$_4$) = 1; y = (UO$_2$) : M^{2+}; 2 ≤ y ≤ 5)

Cousinite (y = 2)	$[(UO_2)_2(MoO_4)_2(OH)_2]^{\infty 2}Mg \cdot 5H_2O$
Calcurmolite (y = 3)	$[(UO_2)_3(MoO_4)_2(OH)_{6-x}]^{\infty 2}(Ca_{1-x}Na_x)_2 \cdot nH_2O$
* Unnamed 1 (y = 4)	$[(UO_2)_4(MoO_4)_4(OH)_2]^{\infty 2}Ca_{1-x}Na_x \cdot 10\text{-}14H_2O$
Unnamed 2 (y = 4)	$[(UO_2)_4(MoO_4)_4(OH)_2]^{\infty 2}Na_2 \cdot 12H_2O$
Natrurmolite [4**] (y = 5)	$[(UO_2)_5(MoO_4)_5(OH)_2]^{\infty 2}Na_2 \cdot 8H_2O$

Uranotungstite group (x = 2; y = 2)

Uranotungstite	$[(UO_2)_2(WO_4)(OH)_4]^{\infty 2}Fe \cdot 12H_2O$

3.1a.3.3b.1.1.2. Neutral

Tengchongite group (x = 3; y = 6)

Tengchongite	$[(UO_2)_6(MoO_4)_2O_5]^{\infty 2}Ca \cdot 12H_2O$

3.1a.3.4a. Uranylo-vanadic acids (hydrates)

Ferganite	$[(UO_2)_3(V_2O_8)]^{\infty 2} \cdot 6H_2O$
Uvanite	$[(UO_2)_2(V_6O_{17})]^{\infty 2} \cdot 15H_2O$

3.1a.3.4b. Uranyl-polyvanadates
3.1a.3.4b.1. Hydrates
3.1a.3.4b.1.1. Neutral

[1*] The structure formula is given by anlogy with bauranoite.
[2*] Metamict; the crystal structure formula is assigned from morfology and crystallochemical consideration .
[3*] Initial formula has been changed by putting out the (MoO$_4$)$^{2-}$ radical.
[4*] The name does not aprooved by the CNMMN IMA and is used after G. A. Sidorenko.

Rauvite group (x = 0.2; y = 2)

Rauvite $[(UO_2)_2(V_{10}O_{28})]^{\infty2}Ca \cdot 16H_2O$

3.1a.3.4c. Uranylo-(5)-vanadates
3.1a.3.4c.1. Hydrates
3.1a.3.4c.1.1. Basic

Vanuralite family (x = 1; y = 1,(3))

Vanuralite $[(UO_2)_2(V_2O_8)(OH)]^{\infty2}Al \cdot 11H_2O$
Metavanuralite $[(UO_2)_2(V_2O_8)(OH)]^{\infty2}Al \cdot 8H_2O$

Tyuyamunite family

Tyuyamunite $[(UO_2)_2(V_2O_8)]^{\infty2}Ca \cdot 5\text{-}8H_2O$
Metatyuyamunite $[(UO_2)_2(V_2O_8)]^{\infty2}Ca \cdot 3\text{-}5H_2O$

Carnotite family (x = 1; y = 2)

Sengierite $[(UO_2)_2(V_2O_8)]^{\infty2}Cu_2(OH)_2 \cdot 6H_2O$
Strelkinite $[(UO_2)_2(V_2O_8)]^{\infty2}Na_2 \cdot 6H_2O$
Carnotite $[(UO_2)_2(V_2O_8)]^{\infty2}K_2 \cdot 3H_2O$
Margaritasite $[(UO_2)_2(V_2O_8)]^{\infty2} (Cs, H_3O, K)_2 \cdot nH_2O$, где n=1

Curienite family (x = 1; y = 2)

Curienite $[(UO_2)_2(V_2O_8)]^{\infty2\text{-}}Pb(H_2O)_4 \cdot H_2O$
Francevillite $[(UO_2)_2(V_2O_8)]^{\infty2\text{-}}(Ba,Pb)(H_2O)_4 \cdot H_2O$

3.1a.3.4d. Uranylo-(4)-vanadates
3.1a.3.4d.1. Hydrates
3.1a.3.4d.1.1. Neutral

Fritzscheite (x=1; y=2) $[(UO_2)_2(VO_4,PO_4)_2]^{\infty2}Mn^{2+} \cdot 4H_2O$ [5*]
*Mathesiusite $[(UO_2)_4(VO_5)(SO_4)_4]K_5(H_2O)_4$

3.1a.3.5a Uranylo-telluric acids (anhydrous)

Schmitterite $[(UO_2)_2(Te_2O_6)^{\infty1}]^{\infty2}$
Cliffordite $\{(UO_2)[Te_3O_7]^{\infty3}\}^{\infty3}$

3.1a.3.5b. Uranylo-tellurites
3.1a.3.5b.1. Neutral

Moctezumite $[(UO_2)Pb]^{\infty2}(TeO_3)_2$
*Markcooperite $[(UO_2)(TeO_6)Pb_2]$

3.1a.3.6a. Uranylo-silica acids (hydrates)

*Uranosilite $[(UO_2)(Si_7O_{15})]$
Soddyite $[(UO_2)_2(H_2O)_2(SiO_4)]^{\infty3.}$

3.1a.3.6b. Uranylo-silicates
3.1a.3.6b.1. Uranylo-mono-disilicates (к = 1.2)
3.1a.3.6b.1.1. Hydrates
3.1a.3.6b.1.1.1. Basic

Magursilite group (x = 0.8; y=1)

Magursilite [4*] $[(UO_2)_4(Si_5O_{13})^{\infty2}_2]^{\infty3}Mg_4(OH)_4 \cdot 15H_2O$ [6*]

[5*] The structure formula is assigned from morphology and crystallochemical consideration of synthetic phase. There is no chemical analyses for original mineral.
[6*] The structure formula is assigned by A. A. Godovikov from physical properties, initial chemical analysis and crystallochemical consideration.

Ursilite (Calciumursilite)[7*] $[(UO_2)_4(Si_5O_{13})^{\infty2}_2]^{\infty3}Ca_4(OH)_4 \cdot 15H_2O$ [6*]
Calcioursilite $[(UO_2)_4(Si_2O_5)_5]Ca_4(OH)_6 \cdot 15H_2O$

3.1a.3.6b.1.1.2. Neutral
Weeksite family (x = 0.8; y = 2)
  Haiweeite group
  Metahaiweeite $[(UO_2)_2(Si_5O_{13})^{\infty2}]^{\infty3}Ca \cdot nH_2O$ [8*]
  Haiweeite $[(UO_2)_2Si_5O_{12}(OH)_2]Ca \cdot 6H_2O$
  Weeksite $[(UO_2)_2(Si_5O_{13})^{\infty2}]^{\infty3}(K,Na)_2(H_2O)_4$
 *Coutinhoite $[(UO_2)_2(Si_5O_{13})]Th_xBa_{1-2x} \cdot 3H_2O$
*3.1a.3.66.2. Uranylo-disilicates
*3.1a.3.66.2.1. Hydrates
*Carlosbarbosaite $[(UO_2)_2(Nb^{5+}Si)O_6(OH)_2]Ca_{0.5}\square_{0.5} \cdot 2H_2O$

3.1a.3.6b.2. Uranylo-tetrasilicates
3.1a.3.6b.2.1. Hydrates
3.1a.3.6b.2.1.1. Neutral
  Kasolite group (x = 1; y = 1)
  Oursinite $(UO_2)_2(SiO_3OH)_2]^{\infty2}(Co,Mg) \cdot 6H_2O$
  Kasolite $[(UO_2)(SiO_4)]^{\infty2}Pb(H_2O)$
3.1a.3.6b.2.1.2. Acid
  Swamböite group (x = 1; y = 2)
  Swamböite $[(UO_2)_6(SiO_3OH)_6]^{\infty2}_3U^{6+} \cdot 30H_2O$
Sklodowskite family (x = 1; y = 2)
Sklodowskite $[(UO_2)_2(SiO_4H)_2]^{\infty2}Mg(H_2O)_4 \cdot 2H_2O$
Cuprosklodowskite $[(UO_2)_2(SiO_3OH)_2]^{\infty2}Cu(H_2O)_4 \cdot 2H_2O$
Uranophane $[(UO_2)_2(SiO_3OH)_2]^{\infty2}Ca(H_2O)_4 \cdot H_2O$
Beta-uranophane $[(UO_2)_2(SiO_3OH)_2]^{\infty2}Ca(H_2O)_4 \cdot H_2O$
*Uranophane mon. $[(UO_2)_2(SiO_3OH)_2]^{\infty2}Ca(H_2O)_4 \cdot H_2O$
Sodium boltwoodite $[(UO_2)(SiO_3OH)]^{\infty2}(Na,K)(H_2O)$
Boltwoodite $[(UO_2)(SiO_3OH)]^{\infty2}K(H_2O)$

3.1a.3.7a. Uranylo-phosphoric acids (hydrates)
Vanmeersscheite family (x = 1.5; y = 1.5)
Vanmeersscheite $[(UO_2)_3(PO_4)_2]U(OH)_6 \cdot 4H_2O$
Metavanmeersscheite $[(UO_2)_3(PO_4)_2]U(OH)_6 \cdot 2H_2O$

3.1a.3.7b. Uranylo-phosphates
3.1a.3.7b.1. Uranylo-phosphates f-cations
3.1a.3.7b.1.1. Basic
Althupite (x = 1.75; y = 2) $[(UO_2)_7O_2(OH)_5(PO_4)_4]AlTh \cdot 15H_2O$

3.1a.3.7b.1.2. Hydrates
3.1a.3.7b.1.2.1. Basic (x = 1.5; y = 2)
Francoisite-(Nd) $[(UO_2)_3O(OH)(PO_4)_2]^{\infty2}Nd \cdot 6H_2O$
*Francoisite-(Ce) $[(UO_2)_3O(OH)(PO_4)_2]^{\infty2}Ce \cdot 6H_2O$

[7*] The name does not aprooved by the CNMMN IMA and is used after A. A. Chernikov.
[8*] The structure formula is assigned from morphology which is close to morphology of weeksite, initial chemical analysis and crystallochemical consideration.

3.1a.3.7b.2. Uranylo-phosphates s-, d_s- и p_s-cations
3.1a.3.7b.2.1. Actually uranylo-phosphates ($x = UO_2 : PO_4$, $y = UO_2 : Me^{2+}$)
3.1a.3.7b.2.1.1. Hydrates
3.1a.3.7b.2.1.1.1. Basic

Kamitugaite ($x = 2.5$; $y = 2$) $[(UO_2)_5((P,As)O_4)_2(OH)_9]^{\infty 2}PbAl·9,5H_2O$
Renardite ($x = 2$; $y = 2$) $[(UO_2)_2(PO_4)(OH)_2]_2^{\infty 2}Pb·7H_2O$

Mundite family
Mundite ($x = 1.5$; $y = 2$) $[(UO_2)_3O(OH)(PO_4)_2]^{\infty 2}Al·6,5H_2O$
Upalite ($x = 1.5$; $y = 2$) $[(UO_2)_3O(OH)(PO_4)_2]^{\infty 2}Al(H_2O)_5·2H_2O$

Dumontite family
  Dumontite group ($x = 1.5$; $y = 1.5$)
    Phurcalite $[(UO_2)_3O_2(PO_4)_2]^{\infty 2}Ca_2(H_2O)_7$
    Dumontite $[(UO_2)_3O_2(PO_4)_2]^{\infty 2}Pb_2(H_2O)_5$
    Bergenite $[(UO_2)_3O_2(PO_4)_2]^{\infty 2}(Ba_{1.33}Ca_{0.67})_2·16H_2O$ (Ba: Ca=2)
  Dewindtite group ($x = 1.5$; $y = 1$)
    Dewindite $[H(UO_2)_3(PO_4)_2O_2)]^{\infty 2}_2Pb_3(H_2O)_9·3H_2O$

Phosphuranylite family
Phosphuranylite ($x = 1.75$; $y = 2.(3)$) $\{[(UO_2)_3(PO_4)_2O_2]^{\infty 2}_2(UO_2)\}^{\infty 3}(H_3O)_3KCa(H_2O)_8$

Yingjiangite ($x = 1.75$; $y = 3.5$) $[(UO_2)_7(PO_4)_4(OH)_6]K_2Ca·6H_2O$
Threadgoldite ($x=1$; $y = 1.(3)$) $[(UO_2)_2(PO_4)_2]^{\infty 2}Al(OH)(H_2O)_4·4H_2O$
Phuralumite ($x = 1.5$; $y = 1$) $[(UO_2)_3O(OH)(PO_4)_2]^{\infty 2}Al_2(OH)_3·11H_2O$
Triangulite ($x = 1$; $y = 0.(8)$) $[(UO_2)_4(PO_4)_4(OH)_5]^{\infty 2}Al_3·5H_2O$
Vochtenite ($x = 1$; $y = 1.6$) $[(UO_2)_4(PO_4)_4(OH)]^{\infty 2}(Fe,Mg)Fe^{3+}·(12-13)H_2O$
*Lakebogaite ($x = 0.5$; $y = 0.4$) $[(UO_2)_2(PO_4)_4(OH)_2]CaNaFe^{3+}_2H·8H_2O$
Moreauite ($x = 0.(3)$; $y = 0.(2)$) $[(UO_2)(PO_4)_2]^{\infty 2}Al_3(PO_4)(OH)_2·13H_2O$
Ranunculite ($x = 1$; $y = 0.5$) $[(UO_2)(PO_4)]^{\infty 2}HAl(OH)_3·4H_2O$
Furongite ($x = 1$; $y = 0.(3)$) $[(UO_2)_7(PO_4)_{13}]^{\infty 2}Al_{13}(OH)_{14}·58H_2O$

3.1a.3.7b.2.1.1.2. Neutral
Torbernite family ($y = 2$)
  Torbernite group
  α-Torbernite (tetrag.) $[(UO_2)_2(PO_4)_2]^{\infty 2}Cu(H_2O)_4·8H_2O$
  β-Torbernite (tricl.) $[(UO_2)_2(PO_4)_2]^{\infty 2}Cu(H_2O)_4·8H_2O$
  *Metasaleeite $[(UO_2)_2(PO_4)_2]Mg·8H_2O$
  Saleeite $[(UO_2)_2(PO_4)_2]^{\infty 2}Mg(H_2O)_4·6H_2O$
  Autunite $[(UO_2)_2(PO_4)_2]^{\infty 2}Ca\ 11H_2O$
  *Uranospatite $[(UO_2)(PO_4)]_2Al_{1-x}\square_x(H_2O)_{20+3x}F_{1-x}$, $0< x <0,33$
  Metatorbernite group
  Metatorbernite $[(UO_2)_2(PO_4)_2]^{\infty 2}Cu(H_2O)_8$
  Przhevalskite group
  Przhevalskite $[(UO_2)(PO_4)]^{\infty 2}_2Pb·4H_2O$

Sabugalite family ($y = 2$)
Sabugalite $[(UO_2)_4(PO_4)_4]^{\infty 2}HAl(H_2O)_4·8H_2O$
Bassetite $[(UO_2)_2(PO_4)_2]^{\infty 2}Fe^{2+}(H_2O)_4\ (H_2O)_4$
Lehnerite $[(UO_2)_2(PO_4)_2]^{\infty 2}Mn^{2+}(H_2O)_4(H_2O)_4$
Meta-uranocircite I $[(UO_2)_2(PO_4)_2]^{\infty 2}Ba(H_2O)_8$
Meta-uranocircite II $[(UO_2)_2(PO_4)_2]^{\infty 2}Ba(H_2O)_6$

(synthetic phase)

Uranocircite group

Uranocircite $[(UO_2)_2(PO_4)_2]^{\infty 2}Ba \cdot 10H_2O$

Meta-autunite group

Meta-autunite $[(UO_2)_2(PO_4)_2]^{\infty 2}Ca(H_2O)_{6-8}$

*Metanatroautunite = natroautunite $[(UO_2)_2(PO_4)_2]Na_2 \cdot 7H_2O$

Meta-ankoleite $[(UO_2)(PO_4)]^{\infty 2}K(H_2O)_3$

Uramphite $[(UO_2)(PO_4)]^{\infty 2}(NH_4)(H_2O)_3$

*Metauramphite $[(UO_2)_2(PO_4)_2](NH_4)_2 \cdot 6H_2O$

Chernikovite $[(UO_2)(PO_4)]^{\infty 2}(H_3O)(H_2O)_3$

Parsonsite family (y = 0.5)

Ulrichite $[(UO_2)(PO_4)_2]^{\infty 2}CaCu(H_2O)_4$

Parsonsite $[(UO_2)(PO_4)_2]^{\infty 2}Pb_2(H_2O)$

3.1a.3.7b.2.2. Uranylo-phosphato-sulfates

3.1a.3.7b.2.2.1. Hydrates

3.1a.3.7b.2.2.1.1. Basic

Xiangjiangite (y = 2.(6)) $[(UO_2)_4(PO_4)_2(SO_4)_2(OH)]^{\infty 2}(Fe^{3+},Al) \cdot 22H_2O$

Coconinoite (y=0.(3)) $[(UO_2)_2(PO_4)_4(SO_4)](Al,Fe^{3+})_4(OH)_2 \cdot 18\text{-}20H_2O$

*3.1a.3.7c Uranylo-bismuthilo-phosphates

*Phosphowalpurgite $[(UO_2)(BiO)_4(PO_4)_2] \cdot 2H_2O$

*Šreinite $[(UO_2)_4(BiO)_3(PO_4)_2]Pb(OH)_7 \cdot 4H_2O$

3.1a.3.8a. Uranylo-arsenic acids

Trögerite $[(UO_2)(AsO_4)](H_3O) \cdot 3H_2O$

3.1a.3.8b. Uranylo-arsenates

3.1a.3.8b.1. Neutral

*Chistyakovaite (y=1.3) $[(UO_2)_2(AsO_4)_2(F,OH)]Al \cdot 6H_2O$

Hallimondite (y=0.5) $[(UO_2)(AsO_4)_2]^{\infty 2}Pb_2$

3.1a.3.8b.2. Hydrates

3.1a.3.8b.2.1. Basic

Arsenuranylite (y=4) $[(UO_2)_4(AsO_4)_2]Ca(OH) \cdot 6H_2O$

Hüegelite (y=1.5) $[(UO_2)_3O_2(AsO_4)_2]^{\infty 2}Pb_2(H_2O)_5^{9*}$

3.1a.3.8b.2.2. Neutral

*Uramarsite $[(UO_2)(AsO_4)]NH_4 \cdot 3H_2O$

Uranospinite family (y = 2;)

Novacekite group

Novacekite-II $[(UO_2)_2(AsO_4)_2]^{\infty 2}Mg \cdot 10H_2O$

Uranospinite $[(UO_2)_2(AsO_4)_2]^{\infty 2}Ca(H_2O)_4 \cdot 6H_2O$

Heinrichite $[(UO_2)_2(AsO_4)_2]^{\infty 2}Ba(H_2O)_8 \cdot (2\text{-}4)H_2O$

Zeunerite family(y=2)

Zeunerite $[(UO_2)_2(AsO_4)_2]^{\infty 2}Cu(H_2O)_4 \cdot 8H_2O$

Kirchheimerite $[(UO_2)_2(AsO_4)_2]^{\infty 2}Cu(H_2O)_4 \cdot 8H_2O$

Meta-uranospinite $[(UO_2)_2(AsO_4)_2]^{\infty 2}Ca(H_2O)_4 \cdot 4H_2O$

Meta-Na-uranospinite $[(UO_2)_2(AsO_4)_2]^{\infty 2}Na_2 \cdot 5H_2O$

9* The structure formula is assigned on the analogy of dumontite.

Metalodevite $[(UO_2)_2(AsO_4)_2]^{\infty 2}Zn(H_2O)_4 \cdot 6H_2O$
Metaheinrichite $[(UO_2)_2(AsO_4)_2]^{\infty 2}Ba(H_2O)_8$
Metazeunerite family (y=2)
Metazeunerite $[(UO_2)_2(AsO_4)_2]^{\infty 2}Cu(H_2O)_4 \cdot 4H_2O$ or
                                      $[(UO_2)(AsO_4)]^{\infty 2}_2Cu\ (H_2O)_8$
Novacekite-I $[(UO_2)_2(AsO_4)_2]^{\infty 2}Mg(H_2O)_4 \cdot 8H_2O$
Metanováčekite $[(UO_2)_2(AsO_4)_2]^{\infty 2}Mg(H_2O)_4 \cdot 2\text{-}4H_2O$
Kahlerite $[(UO_2)_2(AsO_4)_2]^{\infty 2}Fe^{2+}(H_2O)_4 \cdot 8H_2O$
*Metarauchite $[(UO_2)_2(AsO_4)_2]Ni \cdot 8H_2O$
*Rauchite $[(UO_2)_2(AsO_4)_2]Ni \cdot 10H_2O$
Arsenuranospathite $[(UO_2)_2(AsO_4)_2]AlF \cdot 20H_2O$
Abernathyite family (y=2)
Metakirchheimerite $[(UO_2)_2(AsO_4)_2]^{\infty 2}Co(H_2O)_4 \cdot 4H_2O$
Metakahlerite $[(UO_2)_2(AsO_4)_2]^{\infty 2}Fe^{2+}(H_2O)_4 \cdot 4H_2O$
Sodium uranospinite $[(UO_2)_2(AsO_4)_2]^{\infty 2}(Na_2,Ca)(H_2O)_4 \cdot 6H_2O$
Abernathyite $[(UO_2)(AsO_4)]^{\infty 2}K(H_2O)_3$
*Nielsbohrite $[(UO_2)_3(AsO_4)]K(OH)_4 \cdot H_2O$
H-metauranospinite $[(UO_2)(AsO_4)]^{\infty 2}(H_3O)(H_2O)_3$

*3.1a.3.8в. Uranylo-arsenato-arsenites
*3.1a.3.8в.1. Hydrates
*Seelite $[(UO_2)(AsO_3)_x(AsO_4)_{1\text{-}x}]^{\infty 2}_2Mg \cdot 7H_2O$ (x = ~ 0,7)

*3.1a.3.8г. Uranylo-arsenous acids
*Chadwikite $(UO_2)H(AsO_3)$
*3.1a.3.8г.1. Hydrates
*Mineral D $(UO_2)H(AsO_3) \cdot H_2O$
*Štěpite $(UO_2)H_2(AsO_3)_2 \cdot 4H_2O$

*3.1a.3.8д. Uranylo-arsenites
*3.1a.3.8д.1. Hydrates
*Dymkovite $[(UO_2)_2(As^{3+}O_3)_2]Ni \cdot 7H_2O$

3.1a.3.9a. Uranylo-bismuthilo-arsenic acids (hydrates)
Walpurgite $[(UO_2)(BiO)_4(AsO_4)_2]^{\infty 2} \cdot 2H_2O$

3.1a.3.9b. Uranylo-bismuthilo-arsenates
3.1a.3.9b.1. Hydrates
3.1a.3.9b.1.1. Basic
Asselbornite (y=4) $[(UO_2)_4(BiO)_3(AsO_4)_2]Pb(OH)_7 \cdot 4H_2O$

3.1a.3.10a. Uranylo-carbonic acids
Rutherfordine family
 Rutherfordine $[(UO_2)(CO_3)]^{\infty 2}$
*Blatonite $[(UO_2)(CO_3)]^{\infty 2} \cdot H_2O$
Joliotite $[(UO_2)(CO_3)]^{\infty 2} \cdot 1,5\text{-}2H_2O$
*Oswaldpeetersite $(UO_2)_2(CO_3)(OH)_2 \cdot 4H_2O$

3.1a.3.10 b. Uranylo-carbonates

3.1a.3.10 b.1. Uranylo-carbonates **f**-cations
3.1a.3.10 b.1.1. Hydrates
3.1a.3.10 b.1.1.1. Oxido-uranylo-carbonates
Kamotoite-(Y) (y=1.(3)) $[(UO_2)_4O_4(CO_3)_3]^{\infty 2}(Y,REE)_2 \cdot 14H_2O$
3.1a.3.10 b.1.1.2. Basic
Bijvoetite -(Y) (y=0.(6)) $[(UO_2)_{16}(OH)_8O_8(CO_3)_{16}(H_2O)_{25}(Y,REE)_8](H_2O)_{14}$
Wyartite (y=2.(3)) $[(UO_2)_2O_4(OH)(CO_3)]^{\infty 2}CaU^{5+} \cdot 7H_2O$
Astrocyanite-(Ce) (y=0.2) $[(UO_2)(CO_3)_3]^{\infty 2}Cu_2Ce_2(CO_3)_2(OH)_2 \cdot 1,5H_2O$
Shabaite -(Nd) (y=0.25) $[(UO_2)(CO_3)_3]^{\infty 2}CaNd_2(CO_3)(OH)_2 \cdot 6H_2O$

*3.1a.3.10б.1.1.3. Silicato-uranilo-carbonates
*Lepersonnite-(Gd) $[(UO_2)_{24}(CO_3)_8(Si_4O_{28})]CaGd_2 \cdot 60H_2O$
3.1a.3.10 b.2. Uranylo-carbonates **s**-, d_s- и p_s-cations
3.1a.3.10 b.2.1. Actually uranilo-carbonates
3.1a.3.10 b.2.1.1. Neutral
Widenmannite (y=0.5) $[(UO_2)(OH)_2(CO_3)_3]Pb_2$
*Čejkaite (y = 0,5) $[(UO_2)(CO_3)_3]Na_4$
*Agricolaite $[(UO_2)(CO_3)_3]K_4$
3.1a.3.10 b.2.1.2. Hydrates
3.1a.3.10 b.2.1.2.1. Basic
Sharpite (y=6) $[(UO_2)_6(OH)_4(CO_3)_5]Ca \cdot 6H_2O$
Urancalcarite (y=3) $[(UO_2)_3(OH)_6(CO_3)]^{\infty 3}Ca(H_2O)_3$
Roubaultite (y=1.5) $[(UO_2)_3O_2(OH)_2(CO_3)_2]^{\infty 2}Cu_2(H_2O)_4$
3.1a.3.10 b.2.1.2.2. Neutral
Zellerite family (y = 1)
Fontanite $[(UO_2)_3O_2(CO_3)_2]Ca\,(H_2O)_6$
Zellerite $[(UO_2)(CO_3)_2]^{\infty}Ca(H_2O)_3 \cdot 2H_2O$
Metazellerite $[(UO_2)(CO_3)_2]^{\infty}Ca(H_2O)_3^{\cdot}$
Liebigite family (y = 0.5)
Bayleyite $[(UO_2)(CO_3)_3](Mg(H_2O)_6)_2 \cdot 6H_2O$
Swartzite $[(UO_2)(CO_3)_3]^{[8]}Ca(H_2O)_6 \cdot ^{[6]}Mg(H_2O)_6$
Liebigite $[(UO_2)(CO_3)_3]Ca_2(H_2O)_8 \cdot 3H_2O$
Andersonite $[(UO_2)(CO_3)_3]Na_2Ca(H_2O)_5 \cdot H_2O$
Grimselite $[(UO_2)(CO_3)_3]K_3Na \cdot H_2O$
*Synthetical $[(UO_2)(CO_3)_3]K_2Ca(H_2O)_6$
*Braunerite $[(UO_2)(CO_3)_3]K_2Ca \cdot 6H_2O$
*Linekite $[(UO_2)(CO_3)_3]_2K_2Ca_3 \cdot 7H_2O$
Voglite (y = 0.(3)) $[(UO_2)(CO_3)_4]Ca_2Cu(H_2O)_6$
Rabbittite family
Albrechtschraufite (y = 0.4) $[(UO_2)(CO_3)_3]_2Ca_4MgF_2(H_2O)_{13} \cdot 17H_2O$
Rabbittite (y = 0.(3)) $[(UO_2)_2(CO_3)_6]^{\infty 2}Ca_3Mg_3(OH)_4 \cdot 18H_2O$
Znucalite (y=0.08) $[(UO_2)(CO_3)_3]^{\infty 2}CaZn_{11}(OH)_{20}(H_2O)_4$

*3.1a.3.10б.2.2. Uranylo-carbonato-sulfates
                    *3.1a.3.10б.2.2.1. Hydrates
*Ježekite $[(UO_2)(CO_3)_3](SO_4)_2Na_8(H_2O)_4$

3.1a.3.10 b.2.3. Uranylo-carbonato + sulfato-fluorides
3.1a.3.10 b.2.3.1. Neutral

Schröckingerite (y=0.3) $\{[(UO_2)(CO_3)_3]^{[6]}Na(H_2O)_3^{[8]}Ca_3(SO_4)F(H_2O)_3\}^{\infty2}\cdot4H_2O$

3.1a.3.11. Uranylo-selenites
3.1a.3.11.1. Hydrates
3.1a.3.11.1.1. Basic
Marthozite (y =3) $[(UO_2)_3O_2(Se^{4+}O_3)_2]^{\infty2}Cu^{2+}(H_2O)_8$
Guilleminite (y = 3)* $[(UO_2)_3O_2(SeO_3)_2]^{\infty2}Ba\cdot3H_2O$
*Larisaite $[(UO_2)_3O_2(Se^{4+}O_3)_2]^{\infty2}Na(H_3O)\cdot4H_2O$
3.1a.3.11.1.2. Basic
*Piretite (y = 3) $[(UO_2)_3(Se^{4+}O_3)_2(OH)_4]^{\infty2}Ca\cdot4H_2O$
Demesmaekerite   (y = 0.28) $[(UO_2)_2(SeO_3)_6]^{\infty1}[Cu_5(OH)_6(H_2O)_2Pb_2]^{\infty2}$
Derriksite (y = 0.25) $[(UO_2)(SeO_3)_2]^{\infty1}[Cu_4(OH)_6]^{\infty2}$
Haynesite $[(UO_2)_3(OH)_2(SeO_3)_2]^{\infty2}5H_2O$ [1]*

*3.1a.3.12a. Uranylo-sulfuric acids (hydrates) *3.1a.3.12a.1. Neutral
*Shumwayite $[(UO_2)_2(SO_4)_2]\cdot5H_2O$
                                 *3.1a.3.12a.2. Basic
*Jachymovite $[(UO_2)_8(SO_4)(OH)_{14}]^{\infty2}\cdot13H_2O$

*3.1a.3.126. Uranylo-sulfates
*3.1a.3.126.1. Uranylo-sulfates *s-, d_s-, u p_s*-cations
*3.1a.3.126.1.1. Neutral   *3.1a.3.126.1.1.1. Hydrates
*Beshtauite $[(UO_2)(SO_4)_2](NH_4)_2\cdot2H_2O$
*Geschieberite $[(UO_2)(SO_4)_2]K_2\cdot2H_2O$
*Klaprothite monoclinic $[(UO_2)(SO_4)_4]Na_6\cdot4H_2O$
*Peligotite triclinic $[(UO_2)(SO_4)_4]Na_6\cdot4H_2O$
*Ottohahnite $[(UO_2)_2(SO_4)_5]Na_6(H_2O)_7\cdot1.5H_2O$
*Bobcookite $[(UO_2)_2(SO_4)_4]NaAl\cdot18H_2O$
*Wetherillite $[(UO_2)_2(SO_4)_4]Na_2Mg\cdot18H_2O$
*Svornostite $[(UO_2)(SO_4)_2]_2K_2Mg\cdot8H_2O$
*Oppenheimerite $[(UO_2)(SO_4)_2] Na_2\cdot3H_2O$
*Fermiite $[(UO_2)(SO_4)_3]Na_4\cdot3H_2O$
*3.1a.3.126.1.1.1.1. Basic
*Adolfpateraite (y = 2) $[(UO_2)(SO_4)(OH)]K\cdot H_2O$
*3.1a.3.126.1.1.1.1.1. Hydrates
Metauranopilite
*Plašilite $[(UO_2)_2(SO_4)(OH)]Na\cdot2H_2O$
Rabejacite (y = 2) $[(UO_2)_4O_4(SO_4)_2]^{\infty2}Ca_2\cdot8H_2O$
Meta-uranopilite $[(UO_2)_6(SO_4)(OH)_{10}]^{\infty2}\cdot5H_2O$
  Nickel-zippeite group (y = 3)
   Zinczippeite $[(UO_2)_2O_2(SO_4)]^{\infty2}Zn\cdot3.5H_2O$
   Magnesiozippeite $[(UO_2)_6(SO_4)_3(OH)_{10}]^{\infty2}Mg_2\cdot16H_2O$
   Nickelzippeite $[(UO_2)_6(SO_4)_3(OH)_{10}]^{\infty2}Ni_2\cdot16H_2O$
Deliensite $[(UO_2)_2(SO_4)_2(OH)_2]^{\infty2}Fe\cdot7H_2O$
Johannite $[(UO_2)_2(OH)_2(SO_4)_2]^{\infty2}Cu(H_2O)_4\cdot4H_2O$ (y = 2)

*3.1a.3.126.2. Oxido-uranylo-sulfates *s-, d_s-, u p_s*-cations
*3.1a.3.126.2.1. Hydrates
  Zippeite group

Sodium zippeite (y = 1,5)	$[(UO_2)_8O_5(SO_4)_4]Na_5(OH)_3 \cdot 12H_2O$
Zippeite (y = 1,5)	$[(UO_2)_4O_3(SO_4)_2]K_3(OH) \cdot 3H_2O$
Cobaltzippeite	$[(UO_2)_2O_2(SO_4)]Co \cdot 3.5H_2O$
*Plavnoite	$[(UO_2)_2O_2(SO_4)]K_{0.8}Mn_{0.6} \cdot 3.5H_2O$
*Pseudojohannite (y = 1,2)	$[(UO_2)_4O_4(SO_4)_2(OH)_2]Cu_3 \cdot 12H_2O$
Uranopilite	$[(UO_2)_6(SO_4)O_2(OH)_6(H_2O)_6]^{\infty 2} (H_2O)_8$
*Marécottite	$[(UO_2)_8O_6(SO_4)_4]Mg_3(OH)_2 \cdot 28H_2O$

*3.1a.3.126.3. Uranylo-sulfato-sulfites
*3.1a.3.126.3.1. Basic   *3.1a.3.126.3.1.1. Hydrates

*Meisserite	$[(UO_2)(SO_4)_3(SO_3OH)]Na_5(H_2O)$
*Belakovskiite	$[(UO_2)(SO_4)_4(SO_3OH)]Na_7(H_2O)_3$

*3.1a.3.126.4. Uranylo-sulfato-chlorides *3.1a.3.126.4. 1. Hydrates

*Bluelizardite	$[(UO_2)(SO_4)_4]Na_7Cl(H_2O)_2$

*3.1a.3.126.3. Uranylo-sulfates f-cations
*3.1a.3.12.1.3. Basic   *3.1a.3.12.1.3.1. Hydrates

*Sejkoraite-(Y)	$[(UO_2)_8O_6(SO_4)_4(OH)_2]Y_2 \cdot 26H_2O$

3.1b. QUASISUBTIPE AND HYDROXIDES LITHOPHYLIC CATION WITH MIDDLE FC
3.16.1. **Overclass***: Oxides of Zr
3.1b.1a. *Class*: Simple oxides of Zr
3.1b.1a.1. Neutral
Baddeleyite family

Baddeleyite	ZrO_2

3.1b.1b. *Class:* Complex oxides of Zr^{4+} →titanates of Zr^{4+} → zirconotitanates
3.1b.1b.1. Neutral

*Lakargiite	$Ca(Zr,Ti,Sn)O_3$ или $CaZrO_3$
Srilankite	$ZrTi_2O_6$
*Zirconolite-2M	$CaZrTi_2O_7$
*Zirconolite-3O	$CaZrTi_2O_7$
*Zirconolite-3T	$CaZrTi_2O_7$
Tazheranite	$Ca_2Zr_5Ti_2O_{16}$
Calzirtite	$Ca_2Zr_5Ti_2O_{16} \rightarrow {}^{(8)}Ca^{(8)}(CaZr)_{\Sigma 2}^{(7)}Zr_4^{(6)}Ti_2O_{16}$
*Calzirtite orth.	$Ca_2Zr_5Ti_2O_{16}$
*Hiärneite	$(Ca,Mn^{2+},Na)_2(Zr,Mn^{3+})_5(Sb^{5+},Ti,Fe^{3+})_2O_{16}$
Zirkelite	$(Ti,Ca,Zr)O_{2-x}$
*Laachite	$(Ca,Mn)_2Zr_2Nb_2TiFeO_{14}$
*Elbrusite-(Zr)	$Ca_3(Zr_{1.5}U^{6+}{}_{0.5})Fe_3^{3+}O_{12}$
*Polymignite	$(Ca,Fe,Y,Th)(Nb,Ti,Ta,Zr)O_4$
*Unnamed	$(Gd,Ce,Ca,La,U)_4ZrTi_2O_{12}$
*3.16.16.1.1. Hydrates	
*Menesezite	$Ba_2MgZr_4(BaNb_{12}O_{42}) \cdot 12H_2O$
*Allendeite	$Sc_4Zr_3O_{12}$

3.1b.2. **Overclass***: Oxides of Sn^{4+} and Ti^{4+}
3.1b.2a. *Class:* Simple oxides and hydroxides of Sn^{4+} and Ti^{4+}
3.1b.2a.1. Oxides of Sn^{4+} 3.1b.2a.1.1. Neutral

Cassiterite SnO_2

3.1b.2a.2. Hydroxides of Sn^{4+}
3.1b.2a.2.1. Simple   3.1b.2a.2.1.1. Oxido-hydroxides
Varlamoffite $(Sn,Fe)^{2+}(O,OH)_2$

3.1b.2a.2.2. Complex (hydrostannates)
                                                                 3.1b.2a.2.2.1. Neutral
Schoenfliesite $Mg[Sn^{4+}(OH)_6]$
  Stottite group
  Jeanbandyite $(Fe^{3+},Mn^{2+})[Sn^{4+}(OH,O)_6]$
  Tetrawickmanite $Mn^{2+}[Sn^{4+}(OH)_6]$
  Stottite $Fe^{2+}[Ge^{4+}(OH)_6]$
  Wickmanite group
  Vismirnovite $Zn[Sn^{4+}(OH)_6]$
  Mushistonite $(Cu,Zn,Fe)[Sn^{4+}(OH)_6]$
  Natanite $Fe^{2+}[Sn^{4+}(OH)_6]$
  Wickmanite $(Mn,Ca)[Sn^{4+}(OH)_6]$
  Burtite $Ca[Sn^{4+}(OH)_6]$

3.1b.2a.3. Simple oxides of Ti^{4+}   3.1b.2a.3.1. Neutral
  Rutile family
  Rutile TiO_2
  Brookite TiO_2
  Anatase TiO_2
*Monoclinic TiO_2
*Orthorhombic. TiO_2 with structure α-PbO_2 TiO_2
*Akaogiite TiO_2
  Ilmenorutile series
  Ilmenorutile $Fe_x(Nb,Ta)_{2x} 4Ti_{1-x}O_2$
  Strüverite $(Ti,Ta,Fe^{3+})O_2$
                                                                 *3.1б.2a.3.1.2. Basic $(Ti,Cr,Fe)(O,OH)_2$
*Carmaichaelite $(Ti,Cr,Fe)(O,OH)_2$

3.1b.2b. *Class*: Complex oxides of Ti^{4+} (Sn^{4+}) → titanates (stannates) (only (6)-titanates, (6)- stannates)
3.1b.2b.1. Titanates of *s*-, d_s- and p_s-cations
3.1b.2b.1.1. $M^{3+}(Fe^{3+}, Cr^{3+}, V^{3+}, Al^{3+})$
                                                                 3.1b.2b.1.1.1. Neutral
  Pseudorutile family
  Pseudorutile $Fe^{3+}_2Ti_3O_9$
  Schreyerite $V^{3+}_2Ti_3O_9$
  Kyzylkumite $V^{3+}_2Ti_3O_9$
*Olkhonskite $(Cr,V)_2Ti_3O_9$
*Unnamed $(Cr,V)_2Ti_4O_{11}$
*Unnamed $(Cr,V)_2Ti_2O_7$
  Pseudobrookite family
  Pseudobrookite $Fe^{3+}_2TiO_5$
*Panguite $(Ti,Sc,Al,Mg,Zr,Ca)_{1.8}O_3$ or $(Ti,Sc,Al,Mg,Zr,Ca)_3O_5$
*Kangite $(Sc,Ti,Al,Zr,Mg,Ca,□)_2O_3$

Berdesinskiite	$V^{3+}_2TiO_5$
*Oxyvanite	$V^{3+}_2V^{4+}O_5$

Priderite family (compare with cryptomelane (group))

Priderite	$K(Ti_7{}^{4+}Fe^{3+})O_{16}$
*Batiferrite	$(Ti_2Fe^{3+}{}_8Fe^{2+}{}_2)O_{19}\cdot Ba$
*Haggertyite	$[(Ti^{4+}{}_5Fe^{2+}{}_4Fe^{3+}{}_2Mg)O_{19}]Ba$
*Henrymeyerite	$(Ti_7Fe)O_{16}\cdot Ba$
*Ankangite (discredited)	$(Ti,V^{3+},Cr^{3+})_8O_{16}Ba$

  Mannardite group

Mannardite	$|V^{3+}{}_2Ti_6O_{16}|^{\infty 3}Ba$
Redledgeite	$|Cr^{3+}{}_2Ti_6O_{16}|^{\infty 3}Ba$ or
	$Ba_x[(Cr,Fe,V)^{3+}{}_{2x}Ti_{8-2x}]O_{16}$

3.1b.2b.1.1.2. Basic

Tivanite	$V^{3+}Ti(OH)O_3$

3.1b.2b.1.2. M^{3+} and M^{2+}

3.1b.2b.1.2.1. Neutral

Senaite series (compare with crichtonite (series))

Landauite	$(Na,Pb)(Mn^{2+},Y)(Zn,Fe)_2(Ti,Fe^{3+},Nb)_{18}(O,OH,F)O_{38} \rightarrow$ or
	$^{(12)}Na\{^{(6)}(Ti^{4+}{}_{15}Fe^{3+}{}_3Mn^{2+})_{\Sigma19}(O,OH,F)O_{30}|^{(4)}ZnO_4|_2\}^{\infty3}$
Senaite	$Pb(Mn,Y,U)(Fe,Zn)_2(Ti,Fe,Cr,V)_{18}O_{38} \rightarrow$ or
	$Pb\{(Ti,Fe,Mn)_{19}O_{30}|^{(4)}(Fe,Mn)O_4|_2\}^{\infty3}$
Lindsleyite	$(Ba,Sr)(Zr,Ca)(Fe,Mg)_2(Ti,Cr,Fe)_{18}O_{38}$
Mathiasite	$(K,Ba)(Zr,Ca)(Fe,Mg)_2)(Ti,Cr)_{18}O_{38} \rightarrow$ or
	$(K,Ba)\{^{(6)}(Ti,Cr)_{18}O_{30}|^{(4)}(Fe,Mg)O_4|_2\}^{\infty3}$

Hibonite group (compare with magnetoplumbite (group))

Hibonite	$(Ca,TR)(Al,Mg,Ti)_{12}O_{19}$
*Hibonite-(Fe)	$(Fe,Mg)Al_{12}O_{19}$
Hawthorneite	$Ba(MgCr^{3+}{}_4Fe^{2+}{}_2Fe^{3+}{}_2Ti^{4+}{}_3)_{\Sigma12}O_{19}$
Yimengite	$K(Cr,Ti,Fe,Mg)_{12}O_{19}$
*Shulamitite	$Ca_3TiFe^{3+}AlO_8$
Uhligite	$Ca_3(Ti,Al,Zr)_9O_{20}$ (?)
Jeppeite	$(K,Ba)_2(Ti,Fe^{3+})_6O_{13}$
Kennedyite	$MgFe^{3+}{}_2Ti_3O_{10}$

*3.16.2b.1.2.2. Oxido-hydroxides

*Almeidaite	$PbZn_2(Mn,Y)(Ti,Fe^{3+})_{18}O_{36}(OH,O)_2$

Polysomatic series of **magnesiohögbomite**

*Magnesiohögbomite-2N2S	$Mg_6(Al_{14}Ti_2)O_{30}(OH)_2$
*Magnesiohögbomite-2N3S	$Mg_8(Al_{18}Ti_2)O_{38}(OH)_2$
*Magnesiohögbomite-6N6S	$Mg_{18}(Al_{42}Ti_6)O_{90}(OH)_6$

Högbomit series

Högbomit-10T= *magnesiohögbomite-2N2S

Högbomit-15R = *magnesiohögbomite-6N6S

Högbomit-18R= *magnesiohögbomite-6N6S

* *Högbomit-24R*= *magnesiohögbomite6N6S

*Zincohögbomit-2N2S	$Zn_6(Al_{14}Ti_2)_{\Sigma16}O_{30}(OH)_2$
* Zincohögbomi-2N6S	$Zn_{14}(Al_{30}Ti_2)_{\Sigma32}O_{62}(OH)_2$

* Ferrohögbomit-2N2S $(Fe^{2+}_3ZnMgAl)_{\Sigma 6}(Al_{14}Fe^{3+}Ti^{4+})_{\Sigma 16}O_{30}(OH)_2$

3.1b.2b.1.2.2. Basic

Nigerite family
*Magnesionigerite-2N1S = пенчжичжуанит $Mg_4(Al_{10}Sn_2)_{\Sigma 12}O_{22}(OH)_2$
*Magnesionigerite-6N6S $Mg_{18}(Al_{42}Sn_6)_{\Sigma 48}O_{90}(OH)_6$
Ferronigerite-2N1S $(Fe,Mg)_4(Al_{10}Sn_2)_{\Sigma 12}O_{22}(OH)_2$
*Ferronigerite-6N6S $Fe_8(Al_{42}Sn_6)O_{90}(OH)_6$
*Pengzhizhongite = Magnesionigerite-2N1S $Mg_4(Al_{10}Sn_2)_{\Sigma 12}O_{22}(OH)_2$

3.1b.2b.1.2.3. Hydrates

Cafetite $Ca(Fe^{3+},Al)_2Ti_4O_{12}\cdot 4H_2O$ or
                                           $Ca[Ti_2O_5](H_2O)$

3.1b.2b.1.3. M^{2+} ($2\,M^{2+} \rightarrow M^+M^{3+}$)
3.1b.2b.1.3.1. Neutral
Armalcolite $(Mg,Fe^{2+})Ti^{4+}_2O_5$
   Ilmenite series
   Geikielite $MgTiO_3$
   Ecandrewsite $(Zn,Fe,Mn)TiO_3$
   Ilmenite $Fe^{2+}TiO_3$
   Pyrophanite $MnTiO_3$
Perovskite family (compare with latrappite)
Perovskite $CaTiO_3$
*Barioperovskite $BaTiO_3$
*Megoite $CaSnO_3$
Loparite-(Ce) $(Na,Ce,Sr)(Ce,Th)(Ti,Nb)_2O_6$
Tausonite $SrTiO_3$
*K-Sr-loparite $(Sr,La,K,Ce,Ca,Th,Na)(Ti,Cr,Nb)O_3$
Macedonite $PbTiO_3$

   Ulvöspinel series (compare with oxispinelides (series); sulfospinelides (series); selenospinelides (series)
   Qandilite $MgTi^{4+}[MgO_4] \rightarrow Mg_2TiO_4$
   Ulvöspinel (ulvite) $Fe^{2+}Ti^{4+}[Fe^{2+}O_4] \rightarrow Fe^{2+}_2TiO_4$
3.1b.2b.1.3.2. Basic
Kassite $CaTi_2(OH)_2O_4$

3.1b.2b.1.4. Titanates of M^{3+} and M^+
                                           3.1b.2b.1.4.1. Basic $\rightarrow$ acids
Freudenbergite $Na|Fe^{3+}Ti_3O_8|^{\infty 3} \rightarrow$
                                           $^{(12)}Na_{1-y}|Fe^{3+}_{1-x}Ti_{3-x}Si_xO_8H_{3x+y}|^{\infty 3}$

3.1b.2b.2. Titanates of s-, d_s- and p_s-cations with unknown structure
 and questionable
Kleberite $FeTi_6O_{11}\cdot(OH)_5$
Manganbelyankinite $(Mn,Ca)(Ti,Nb)_5O_{12}\cdot 9H_2O$
Belyankinite $Ca(Ti,Zr,Nb)_6O_{13}\cdot 14H_2O$ (?)

3.1.3.2.2.2. Titanates of f-cationsa

                                           3.1.3.2.2.2.1. Neutral$\rightarrow$ Basic
   Crichtonite series (compare with senaite (series))

*Davidite-(Ce)	$Ce(Y,U)Fe_2(Ti,Fe,Cr,V)_{18}(O,OH,F)_{38}$
Davidite-(La)	$La(Y,U)Fe_2(Ti,Fe,Cr,V)_{18}(O,OH,F)_{38}$
*Unnamed	$(Ca,Ce)Sc(Ti,Fe,Al)_{20}(O,OH)_{38}$
Loveringite	$(Ca,Ce,La)(Zr,Fe)(Mg,Fe)_2(Ti,Fe,Cr,Al)_{18}O_{38}$
*Dessauite-(Y)	$SrYFe^{3+}_2(Ti_{11}Fe_7)_{\Sigma18}O_{38}$
Crichtonite	$(Sr,La,Ce,Y)|(Ti,Fe^{3+},Mn)_{21}O_{38}|^{\infty3}$
*Gramaccioliite – (Y)	$(Pb,Sr)(Y,Mn)(Ti,Fe^{3+})_{18}Fe^{3+}_2O_{38}$
Brannerite family	
Brannerite group	
Lucasite-(Ce)	$(Ce,La)Ti_2(O,OH)_6$
Brannerite	$(U,Ca,Ce)(Ti,Fe^{3+})_2O_6$
Thorutite	$(Th,U,Ca)Ti_2(O,OH)_6$
Yttrocrasite-(Y)	$(Y,Th,Ca,U)(Ti,Fe^{3+})_2(O,OH)_6$

3.1b.3. **Overclass***: Oxides and hydroxides of Nb^{5+} and Ta^{5+}
3.1b.3a. *Class:* Simple oxides and hydroxides of Nb^{5+} and Ta^{5+}
3.1b.3a.1. Neutral

Tantite	$(Ta,Nb)_2O_5$
Ixiolite family	
Ixiolite group	
Ixiolite	$(Ta,Fe,Sn,Nb,Mn)_4O_8$
*Unnamed	$(Sc,Fe^{3+})(Nb,Ta)O_4$
Wodginite group	
*Titanowodginite	$Mn^{2+}TiTa_2O_8$
*Ferrowodginite	$Fe^{2+}SnTa_2O_8$
*Fe^{2+}-Ti-wodginite	$(Fe,Mn)_4(Ti,Sn,Ta)_4(Ta,Nb,W)_8O_{32}$
*Tantalowodginite	$(Mn_2,▨)_4Ta_4Ta_8O_{32}$
*Wodginite	$MnSnTa_2O_8$

3.1b.3a.2.Hydroxides

Kimrobinsonite	$Ta(OH)_3(O,CO_3)$

3.1b.3b. *Class*: Complex oxides of Nb^{5+} and Ta^{5+} ((6)-tantaloniobates→ (4)-tantaloniobates)
3.1b.3b.1. Tantaloniobates of **s**-, **d$_s$**- and **p$_s$**-cations
3.1b.3b.1.1. Tantaloniobates of **s**-, **d$_s$**- and **p$_s$**-cations (without Li and Be)
3.1b.3b.1.1.1. Proper tantaloniobates
3.1b.3b.1.1.1.1. $M^{4+} = Sn^{4+}(Ti^{4+})$
3.1b.3b.1.1.1.1.1. Neutral
3.1b.3b.1.1.1.2. $M^{3+} = Al^{3+}$
3.1b.3b.1.1.1.2.1. Neutral

Alumotantite	$AlTaO_4$
*Heftetjernite	$ScTaO_4$

3.1b.3b.1.1.1.2.2. Basic

Simpsonite	$Al_4(Ta,Nb)_3O_{13}(OH,F)$

3.1b.3b.1.1.1.3. M^{2+} (Sn^{2+}, Mg, Fe^{2+}, Mn^{2+}, Ca^{2+})
3.1b.3b.1.1.1.3.1. Neutral → acids → basic

Thoreaulite series	
Foordite	$Sn|Nb_2O_6|^{\infty2}$
Thoreaulite	$Sn|Ta_2O_6|^{\infty2}$
Tapiolite series	

Tapiolite-Fe $FeTa_2O_6$
Tapiolite-Mn $MnTa_2O_6$
Columbite series
Columbite-(Mg) = Magnocolumbite $(Mg,Fe,Mn)(Nb,Ta)_2O_6$
Columbite-(Fe) = Ferrocolumbite $FeNb_2O_6$
Columbite-(Mn) = Manganocolumbite $(Mn,Fe)(Nb,Ta)_2O_6$
Ferrotantalite = tantalite-(Fe)_ $FeTa_2O_6$
Manganotantalite = tantalite(Mn)- $MnTa_2O_6$
*Magnesiotantalite = tantalite-(Mg) $MgTa_2O_6$
Calciotantite $CaTa_4O_{11}$
Rynersonite family
Rynersonite $Ca(Ta,Nb)_2O_6$
Changbaiite $PbNb_2O_6$

Microlite family (compare with pyrochlore (series))
Microlite series
Bariopyrochlore $Ba_2Nb_2O_7$
*Hydropyrochlore $(H_2O□)_2Nb_2(O,OH)_6·H_2O$
*Hydroxycalciopyrochlore $(Ca,Na,U,□)_2(Nb,Ti)_2O_6(OH)$
*Aspedamite $□_{12}(Fe^{3+},Fe^{2+})_3Nb_4[Th(Nb,Fe^{3+})_{12}O_{42}](H_2O,OH)_{12}$
*Oxystibiomicrolite $(Sb,Ca)_2Ta_2O_6O$
*Fluorcalciomicrolite $(Ca,Na,□)_2Ta_2O_6F$
Bariomicrolite = *hydrokenomicrolite $(□,H_2O)_2Ta_2(O,OH)_6(H_2O)$
Parabariomicrolite $BaTa_4O_{10}(OH)_2·2H_2O$
*Hydrokenomicrolite $(□,H_2O)_2Ta_2(O,OH)_6(H_2O)$
Bismutomicrolite series
*Hydroxykenomicrolite $(□,Na,Sb^{3+})_2|Ta_2O_6(OH)$
*Fluorsodicmicrolite $(Na,Ca)Ta_2O_6F$
Stibiobetafite = oxycalciopyrochlore $Ca_2Nb_2O_6O$
Stannomicrolite = oxystannomicrolite $Sn_2Ta_2O_6O$
Cesplumtantite $(Cs,Na)_2(Pb, Sb^{3+}·Sn^{2+})_3Ta_8O_{24}$

3.1b.3b.1.1.1.3.2. Hydrates
Gerasimovskite $(Mn,Ca)(Nb,Ti)_{5-6}O_{12-16}·8-9H_2O$ (?)
Franconite $(Na,Ca)_2(Nb,Ti)_4O_{11}·9H_2O$
*Hochelagaite $(Ca,Na,Sr)Nb_4O_{11}·8H_2O$
*Ternovite $(Mg,Ca)Nb_4O_{11}·10H_2O$
*Peterandresenite $Mn_4Nb_6O_{19}·14H_2O$

3.1b.3b.1.1.1.4. M^{3+} and M^+

                                        3.1b.3b.1.1.1.4.1. Neutral
Sosedkoite $(K,Na)_5Al_2(Ta,Nb,Sb)_{22}O_{60}$
3.1b.3b.1.1.1.5. M^+

                                        3.1b.3b.1.1.1.5.1. Neutral $\rightarrow$ basic
Rankamaite $(Na,K,Pb)(Ta,Nb,Al)_4(O,OH)_{10}$
Natrotantite $Na_2Ta_4O_{11}$
Irtyshite $Na_2(Ta,Nb)_4O_{11}$
Latrappite $(Ca,Na)(Nb,Ti,Fe)O_3$ (compare perovskite (series);
                                        loparite (group)) macedonite (group));

Lueshite family

Lueshite	$NaNbO_3$
*Isolueshite	$(Na,La,Ca,)(Nb,Ti)O_3$

3.1b.3b.1.1.2. Tantaloniobato-tungstenates of Mg, Fe^{2+}, Mn^{2+}

　　　　　　　　　　3.1b.3b.1.1.2.1. Neutral

Qitianlingite	$(Fe,Mn)^{2+}_2(Nb,Ta)_2W^{6+}O_{10}$
*Koragoite	$Mn_2^{2+}Mn^{3+}Nb_2(Nb,Ta)_3W_2O_{20}$

3.1b.3b.1.2. Tantaloniobates of Li　3.1b.3b.1.2.1. Neutral

Lithiotantite family

Lithiotantite	$Li(Ta,Nb)_3O_8$
Lithiowodginite	$Li(Ta,Nb)_3O_8$

3.1b.3b.2. Tantaloniobates of f-elements
3.1b.3b.2.1. Tantaloniobates of U
3.1b.3b.2.1.1. Neutral

Liandratite	$U^{6+}(Nb,Ta)_2O_8$
Petscheckite	$U^{4+}Fe^{2+}(Nb,Ta)_2O_8$

3.1b.3b.2.2. Tantaloniobates of TR
3.1b.3b.2.2.1. Neutral → basic

Euxenite series

Euxenite-(Y)	$(Y,Ca,Ce,U,Th)|(Nb,Ta,Ti)_2O_6|^{\infty3}$
Vigezzite	$(Ca,Ce)|(Nb,Ta,Ti)_2O_6|^{\infty3}$
*Titanvigezzite	$(Ca,Ce)|(Ti,Nb,Si,Ta)_2O_6|^{\infty3}$
Fersmite	$(Ca,Ce,Na)|(Nb,Ti,Fe,Al)_2(O,OH,F)_6|^{\infty3}$
Tanteuxenite-(Y)	$(Y,Ca,Ce)|(Ta,Nb,Ti)_2(O,OH)_6|^{\infty3}$

Aeschynite series

Niobo-aeschynite-(Ce)	$(Ce,Ca,Th)|(Nb,Ti)_2(O,OH)_6|^{\infty3}$
*Niobo-aeschynite-(Nd)	$(Nd,Ce)|(Nb,Ti)_2(O,OH)_6|^{\infty3}$
*Niobo-aeschynite-(Y)	$(Y,Ca,Ce,Nd,Th)(Nb,Ta,Ti,Fe)_2(O,OH)_6$
*Yttroniobo-aeschynite-(Ce)	$(Ce,Y,Ca,Th)|(Nb,Ti,Ta)_2O_6$
Tantal-aeschynite-(Y)	$(Y,Ce,Ca)|(Ta,Ti,Nb)_2O_6|^{\infty3}$
Aeschynite-(Nd)	$(Nd,Ce,Ca,Th)|(Ti,Nb)_2(O,OH)_6|^{\infty3}$
Aeschynite-(Ce)	$(Ce,Ca,Fe,Th)|(Ti,Nb)_2(O,OH)_6|^{\infty3}$
Aeschynite-(Y)	$(Y,Ca,Fe,Th)|(Ti,Nb)_2(O,OH)_6|^{\infty3}$

Polycrase series

Polycrase-(Y)	$(Y,Ca,Ce,U,Th)|(Ti,Nb,Ta)_2O_6|^{\infty3}$
*Uranopolycrase	$(U,Y)(Ti,Nb)_2O_6$
Kobeite-(Y)	$(Y,U)(Ti,Nb)_2(O,OH)$
Loranskite-(Y)	$(Y,Ce,Ca)ZrTaO_6$

Samarskite series

*Calciosamarskite	$(Ca,Fe,Y)(Nb,Ta,Ti)O_4$
Samarskite-(Y)	$(Y,Ce,U,Fe^{3+})_3(Nb,Ta,Ti)_5O_{16}$
*Samarskite-(Yb)	$(Yb,Y,REE,U,Th,Ca,Fe^{2+})(Nb,Ta,Ti)O_4$
Pyrochlore series	(compare with microlite (series))
Yttropyrochlore-(Y) discredited	
Yttrobetafite-(Y) discredited	
Betafite	$Ca_2(Ti,Nb)O_6(OH)$

Uranpyrochlore discredited
Ceriopyrochlore-(Ce) $Ce_2Nb_2O_6(OH)$
Plumbopyrochlore $Pb_2Nb_2O_7$
Pyrochlore $NaCaNb_2O_6F$
*Fluornatropyrochlore $(Na,Pb,Ca,REE,U)_2Nb_2O_6F$
*Hydroxymanganopyrochlore $(Mn,Th,Na,Ca,REE)_2(Nb,Ti)_2O_6(OH)$
Uranmicrolite $(U_{0.5}Ca_{0.5})Ta_2O_6(OH)$
Calciobetafite $Ca_2(NbTi)O_6(OH)$
*Bismutopyrochlore discredited
Murataite-(Y) $((Y,Na)_6Zn(Zn,Fe^{3+})_4(Ti,Nb,Na)_{12}O_{29}(O,F,OH)_{10}F_4$
Beta-fergusonite series
Beta-fergusonite-(Ce) $(Ce,La,Nd)[NbO_4]$
Beta-fergusonite-(Nd) $(Nd,Ce)[NbO_4]$
Beta-fergusonite-(Y) $Y[NbO_4]$
Ishikawaite $(U,Fe,Y)NbO_4$
 Fergusonite series
 *Iwashiroite-(Y) $YTaO_4$
 *Fergusonite-(Ce) $CeNbO_4 \cdot 0.3H_2O$
 Fergusonite-(Y) $(Y,Er,Ce,Fe)[(Nb,Ta,Ti)O_4]$
 Formanite-(Y) $Y[(Ta,Nb)O_4]$
Yttrotantalite-(Y) $(Y,U,Fe^{2+})[(Ta,Nb)O_4]$

3.1b.3b.2.3. Tantaloniobates of Sb^{3+} and Bi^{3+} 3.1b.3b.2.3.1. Neutral
Stibiotantalite family
 Stibiotantalite group
 Stibiotantalite $Sb|(Ta,Nb)O_4|^{\infty 2}$
 Stibiocolumbite $Sb|NbO_4|^{\infty 2}$
 Bismutocolumbite $Bi|NbO_4|^{\infty 2}$
 *Yttrocolumbite-(Y) $YNbO_4$
 Bismutotantalite $Bi|(Ta,Nb)O_4|^{\infty 2}$
Zimbabweite $^{(8)}Na\{^{(6)}(PbNa_{0.5}K_{0.5})_{\Sigma 2}As^{3+}_4|^{(6)}(Ta_3Nb_{0.5}Ti_{0.5})_{\Sigma 4}O_{18}|^{\infty 2}\}^{\infty 2}$
*Unnamed $Mn^{2+}_3U^{4+}As^{3+}_2Sb^{3+}_2Ta^{5+}_2Ti^{4+}_2O_{20}$

*3.1б.3b.2.3.4. Tantaloniobato-tungstenates Sb^{3+} и Bi^{3+}
*Billwiseite $Sb^{3+}_5(Nb,Ta)_3WO_{18}$

3.1b.4. **Overclass***: Oxides and hydroxides of Mo and W
3.1b.4a. *Class:* Simple oxides and hydroxides of Mo and W
                                3.1b.4a.1.Neutral
Tugarinivite MoO_2
Molybdite MoO_3
*Cupromolybdite $Cu^{2+}_3Mo^{6+}_2O_9$
                                3.1b.4a.2. Oxido-hydroxides
Tungstite $WO_3 \cdot H_2O \rightarrow H_2WO_4 \rightarrow W(OH)_2O_2$
 Alumotungstite series
 Alumotungstite $\square_2W_2O_6(H_2O)$
 Ferritungstite = hydrokenoelsmoreite $\square_2W_2O_6(H_2O)$
                                3.1b.4a.3. Hydrates (molybdenum and
                                            tungsten acids)
Hydrotungstite family

Sidwellite	$MoO_3 \cdot 2H_2O \rightarrow H_2MoO_4 \cdot H_2O \rightarrow Mo(OH)_2O_2 \cdot H_2O$
Hydrotungstite	$WO_3 \cdot 2H_2O \rightarrow H_2WO_4 \cdot H_2O \rightarrow W(OH)_2O_2 \cdot H_2O$
Meymacite	$WO_3 \cdot 2H_2O \rightarrow H_2WO_4 \cdot H_2O \rightarrow W(OH)_2O_2 \cdot H_2O$
*Elsmoreite	$WO_3 \cdot 5H_2O$

3.1b.4b.*Class:* Complex oxides and hydroxides of Mo and W ((6)-molybdates and tungstenates → (4)-molybdates and tungstenates)
3.1b.4b.1. Molybdates and tungstenates of *s*-, *d_s*- and *p_s*-cations
3.1b.4b.1.1. Proper molybdates and tungstenates

3.1b.4b.1.1.1. $M^{3+}(Al^{3+}, Fe^{3+})$	3.1b.4b.1.1.1.1. Oxido-hydroxides →
	→ hydroxides → hydrates

Anthoinite family

Anthoinite	$AlWO_3(OH)_3$
Mpororoite	$AlWO_3(OH)_3 \cdot H_2O$
*Bamfordite	$Fe^{3+}Mo_2(OH)_3O_6 \cdot H_2O$
*Ferrimolybdite	$Fe^{3+}_2[MoO_4]_3 \cdot 8H_2O$ (?)
*Ophirite	$Ca_2Mn_4[Zn_2Mn^{3+}_2(H_2O)_2(Fe^{3+}W_9O_{34})_2] \cdot 46H_2O$

3.1b.4b.1.1.2. M^{3-} and M^{2+}	3.1b.4b.1.1.2.1. Oxido-hydroxides
Jixianite	$Pb(W, Fe^{3+})_2(O, OH)_7$

3.1b.4b.1.1.3. M^{3+}, M^{2+} and M^+	3.1b.4b.1.1.3.1. Hydrates
Phyllotugstite	$HCaFe^{3+}_3[WO_4]_6 \cdot 10H_2O$
*Pittongite	$(Na \cdot H_2O)_{0.7}(W, Fe^{3+})(O, OH)_3$

3.1b.4b.1.1.4. M^{2+}	3.1b.4b.1.1.4.1. Neutral
Wolframite series	
*Huanzalaite	$MgWO_4$
Ferberite	$FeWO_4$
Huebnerite	$MnWO_4$
Sanmartinite	$(Zn, Fe, Ca, Mn)WO_4$
Scheelite series	
Powellite	$Ca[MoO_4]$
Scheelite	$Ca[WO_4]$

* 3.1б.4б.1.1.5 M^{2+} и Mo^{4+}

*Kamiokite	$Fe^{2+}_2Mo^{4+}_3O_8 \rightarrow Fe^{2+}_2Mo^{4+}[MoO_4]_2$

3.1b.4b.1.2. Molybdato (tungstenato)-vanadates
3.1b.4b.1.2.1. Hydrates

Rankachite	$Ca_{0.5}(V^{4+}, V^{5+})(W^{6+}, Fe^{3+})_2O_8(OH) \cdot 2H_2O$

3.1b.4b.1.3. Molybdato (tungstenato)-phosphates
3.1b.4b.1.3.1. Hydrates

Melkovite	$[Ca_2(H_2O)_{15}Ca(H_2O)_6][Mo_8^{6+}P_2Fe_3^{3+}O_{36}(OH)]$

3.1b.4b.1.4. Phospho- molybdato- phosphates
3.1b.4b.1.4.1. Hydrates

Mendozavilite family

Paramendozavilite	$NaAl_4Fe^{3+}_7(OH)_{16}[PMo_{12}O_{40}][PO_4]_5 \cdot 56H_2O$

Mendozavilite-NaFe $[Na_2(H_2O)_{15}Fe^{3+}(H_2O)_6][Mo_8^{6+}P_2Fe_3^{3+}O_{35}(OH)_2]$
*Mendozavilite-NaCu $[Na_2(H_2O)_{15}Cu(H_2O)_6][Mo^{6+}_8P_2Fe^{3+}_3O_{34}(OH)_3]$
*Mendozavilite-KCa $[K_2(H_2O)_{15}Ca(H_2O)_6][Mo^{6+}_8P_2Fe^{3+}_3O_{34}(OH)_3]$

*3.1b.4b.1.5. Molybdato (tungstenato)-arsenates
*3.1b.4b.1.5.1. Hydrates
   ***Betpakdalite** group
*Betpakdalite-CaCa $[Ca_2(H_2O)_{17}Ca(H_2O)_6][Mo^{6+}_8As^{5+}_2Fe^{3+}_3O_{36}(OH)]$
*Betpakdalite-CaMg $[Ca_2(H_2O)_{17}Mg(H_2O)_6][Mo^{6+}_8As^{5+}_2Fe^{3+}_3O_{36}(OH)]$
*Betpakdalite-NaCa $[Na_2(H_2O)_{17}Ca(H_2O)_6][Mo^{6+}_8As^{5+}_2Fe^{3+}_3O_{34}(OH)_3]$
*Betpakdalite-NaNa $[Na_2(H_2O)_{16}Na(H_2O)_6][Mo^{6+}_8As^{5+}_2Fe^{3+}_3O_{33}(OH)_4]$
*Obradovičite-KCu $[K_2(H_2O)_{17}Cu^{2+}(H_2O)_6][Mo_8^{6+}As_2Fe^{3+}_3O_{34}(OH)_3]$
*Obradovičite-NaCu $[Na_2(H_2O)_{17}Cu^{2+}(H_2O)_6][Mo_8^{6+}As_2Fe^{3+}_3O_{34}(OH)_3]$
*Obradovičite-NaNa $[Na_2(H_2O)_{16}Na(H_2O)_6][Mo_8^{6+}As_2Fe_3^{3+}O_{33}(OH)_4]$

3.1b.4b.2. Molybdates and tungstenates of *f*-elements
*3.1b.4b.2.1. Proper molybdates and tungstenates
                                               3.1b.4b.2.1.1. Neutral
Sedovite $U^{4+}[MoO_4]_2$
                                               3.1b.3.4b.2.2. Basic
Mourite $U^{4+}Mo_5(OH)_{10}O_{12}$
   Yttrotungstite -(Y) series
   Cerotungstite-(Ce) $(Ce,Nd)W_2(OH)_3O_6$
   *Yttrotungstite-(Ce) $(Ce,REE)W_2(OH)_3O_6$
   Yttrotungstite-(Y) $(Y,REE)W_2(OH)_3O_6$

*3.1b.4b.2.2. Molybdato- and tungstenato - arsenates
*3.1b.4b.2.2.1. Neutral
*Paraniite-(Y) $Ca_2(Y,REE)(AsO_4)(WO_4)_2$

3.1b.4b.3. Molybdates and tungstenates of *d*-cations
3.1b.4b.3.1. Molybdates of Mo^{4+}
3.1b.4b.3.1.1. Hydrates
Ilsemannite $(Mo^{6+}_2Mo^{4+})O_8 \cdot H_2O \rightarrow MoMo_2O_8 \cdot H_2O \rightarrow$
                                               $Mo^{4+}[MoO_4]_2 \cdot H_2O$ (?)

3.1b.4b.3.2. Molybdates and tungstenates of I*b*-cations - Cu^{2+}
3.1b.4b.3.2.1. Proper molybdates and tungstenates   3.1b.4b.3.2.1.1. Basic
Lindgrenite family
Lindgrenite $Cu^{2+}_3(OH)_2[MoO_4]_2$
*Markascherite $Cu^{2+}_3(OH)_4[MoO_4]$
*Szenicsite $Cu^{2+}_3(OH)_4[MoO_4]$
Cuprotungstite $Cu^{2+}_3(OH)_2[WO_4]_2$

*3.1b.4b.3.2.2. Oxido-molybdato (tungstenato)-sulfates
*Vergasovaite $Cu^{2+}_3O[MoO_4][SO_4]$

3.1b.4b.4. Molybdates and tungstenates of *p*-metals
3.1b.4b.4.1. Molybdates and tungstenates of IV*a*-cations – Pb^{2+}
3.1b.4b.4.1.1. Proper molybdates and tungstenates

3.1b.4b.4.1.1.1. Neutral

Wulfenite family
  Wulfenite group
  Wulfenite $Pb[MoO_4]$
  Stolzite $Pb[WO_4]$
  Raspite $Pb[WO_4]$
 *Raspite, beneficiate Te $Pb[(W_{0.56}Te_{0.44})O_4]$
                     3.1b.4b.4.1.1.2. Oxido-molybdato-halogenides
Pinalite $Pb_3O[WO_4]Cl_2$
*Parkinsonite $Pb_7MoO_9Cl_2$

3.1b.4b.5. Molybdates and tungstenates of semimetals
3.1b.4b.5.1. Molybdates and tungstenates of Bi^{3+}
3.1b.4b.5.1.1. Proper molybdates and tungstenates
                     3.1b.4b.5.1.1.1. Oxido-molybdates (tungstenates)
*Biehlite $(Sb,As)^{3+}_2Mo^{6+}O_6$

  Koechlinite group
  Koechlinite $Bi_2MoO_6 \rightarrow [Bi_2O_2]^{\infty 2}[MoO_4]^{\infty 2}$
  Russellite $Bi_2WO_6 \rightarrow [Bi_2O_2]^{\infty 2}[WO_4]^{\infty 2}$

*3.1б.4б.5.1.1.1.1 Oxido-hydroxido-molybdates (hydrates)
*Gelosaite $BiMo^{6+}_{2-5x}Mo^{5+}_{6x}O_7(OH)\cdot H_2O$ $(0 \leq x \leq 0.4)$ or $Bi^{3+}Mo^{6+}_{2+x}O_7(OH)\cdot H_2O$

*3.1б.4б.5.1.1.2. Oxido –molybdato-arsenates
*Schlegelite $Bi_7O_4[MoO_4]_2[AsO_4]_3$
                     *3.1б.4б.5.1.1.1.1. Hydrates
*Vajdakite $[(Mo^{6+}O_2)_2(H_2O)_2As^{3+}_2O_5]\cdot H_2O$

3.1b.4b.5.1.2. Molybdato-tellurates of Bi^{3+}
                     3.1b.4b.5.1.2.1. Neutral
Chiluite $Bi_6Mo_2Te_2O_{21}$

3.1b.5. Overclass*: Oxides and hydroxides of Mn^{4+}
3.1b.5a. *Class:* Simple oxides and hydroxides of $Mn^{4+} \rightarrow$ complex oxides and hydroxides of Mn^{4+}
                     3.1b.5a.1. Neutral oxido-hydroxides
Pyrolusite family (compare with rutile (family); cassiterite (group))
Pyrolusite β -MnO_2
Ramsdellite γ-MnO_2
 *Akhtenskite ϵ-MnO_2
Nsutite $Mn^{2+}_xMn^{4+}_{1-x}(OH)_{2x}O_{2-2x}$

3.1b.5b. *Class:* Complex oxides and hydroxides of Mn^{4+}
*3.1б.5б.1. Neutral
*Strontiomelane $SrMn^{4+}_6Mn^{3+}_2O_{16}$
*Zenzénite $Pb_3(Fe^{3+},Mn^{3+})_4Mn^{4+}_3O_{15}$
                     3.1b.5b.1. Oxido-hydroxodes $\rightarrow$ hydrates
Cryptomelane family (compare with priderite (family))
  Coronadite group

Manjiroite $Na(Mn_7^{4+}Mn^{3+})O_{16}$
Coronadite $Pb(Mn^{4+}_6Mn^{3+}_2)O_{16}$
Cryptomelane group
Hollandite $Ba(Mn^{4+}_6Mn^{3+}_2)O_{16}$
Cryptomelane $K(Mn^{4+}_7Mn^{3+})_8O_{16}$
Romanèshite $(Ba,\cdot H_2O)_2(Mn^{4+},Mn^{3+})_5O_{10}$
Todorokite family
*Jianshuiite $MgMnO_7\cdot3H_2O$
Todorokite $(Na,Ca,K,Ba,Sr)_{1-x}(Mn,Mg,Al)_6O_{12}\cdot3\text{-}4H_2O$
Woodruffite $Zn_2Mn_5^{4+}O_{12}\cdot4H_2O$
 Rancieite series
Takanelite $(Mn^{2+},Ca)_{2x}Mn^{4+}_{1-x}O_2\cdot0.7H_2O$
Rancieite $(Ca,Mn^{2+})_{0.2}(Mn^{4+},Mn^{3+})O_2\cdot0.6H_2O$
 Chalcophanite series
Chalcophanite $(Zn,Fe^{2+},Mn^{2+})|Mn^{4+}_3O_7|^{\infty2}\cdot3H_2O$
*Ni-chalcophanite $(Ni,Cu,Co^{3+})|Mn^{4+}_3O_7|^{\infty2}\cdot5H_2O$
*Ernienickelite $Ni|Mn^{4+}_3O_7|^{\infty2}\cdot3H_2O$
Aurorite $(Mn,Ag,Ca)|Mn^{4+}_3O_7|^{\infty2}\cdot3H_2O$
*Jianshuiite $(Mg,Mn)^{2+}|Mn^{4+}_3O_7|^{\infty2}\cdot7H_2O$
Cesarolite $Pb(OH)|Mn^{4+}_3O_6(OH)|^{\infty2}$
 Lithiophorite family
 Lithiophorite group
Lithiophorite $|(Al,Li)(OH)_2|^{\infty2}|MnO_2|^{\infty2}$
Mn^{2+}-lithiophorite $|(Al,Mn^{2+},Li)(OH)_2|^{\infty2}|MnO_2|^{\infty2}$
Janggunite $|(Mn^{2+},Fe^{3+})_{1+x}(OH)_4|^{\infty2}|Mn^{4+}_{5-x}(OH)_2O_8|^{\infty2}$
 Birnessite series $R_{2x}(OH,\cdot H_2O)_{6x}|(Mn^{4+},Mn^{3+},Mg,Ca)_{1-x}(OH)_2O|^{\infty2}$;
                                  $R = Na, K; 1/2Ca, 1/2Mg; 1/3Mn^{3+}$

Mn-Birnessite
Mg-Birnessite
Ca-Birnessite (or simple birnessite)
Na-Birnessite
*Clinobirnessite
Asbolane series $M_{1-y}(OH)_{2-2y+x}Mn^{4+}_2(OH)_{2x}O_{4-2x}$;
                                  $M = Al, Fe^{3+}; Ni^{2+}, Co^{2+}, Fe^{2+}, Ca$

Al-Asbolane
Ni-Asbolane
Co-Asbolane
Vernadite $MnO_2\cdot nH_2O$

3.1b.6. **Overclass***: Oxides and hydroxides of V^{4+}
3.1b.6a. *Class:* Simple oxides and hydroxides of V^{4+}
                                  3.1b.6a.1. Neutral
Paramontroseite VO_2

                                  3.1b.6a.2. Oxido-hydroxides (hydrates ?)
Doloresite $6VO_2\cdot4H_2O \rightarrow V^{4+}_6(OH)_4O_4$
 Duttonite family
Duttonite $VO_2\cdot H_2O \rightarrow V^{4+}(OH)_2O$
Lenoblite $VO_2\cdot H_2O \rightarrow V^{4+}(OH)_2O$

3.1b.6b.**Class:** Complex oxides and hydroxides of $V^{4+} \rightarrow$ vanadites
3.1b.6b.1. Oxides of s-cations and V^{4+}
*3.1b.6b.1.1. Neutral
*Cavoite $CaV^{4+}_3O_7$
                                           3.1b.6b.1.2. Hydrates
*Bassoite $SrV^{4+}_3O_7 \cdot 4H_2O$
Simplotite $CaV^{4+}_4O_9 \cdot 5H_2O$

3.1b.6b.2. Oxides of d-cations and V^{4+}
                                           3.1b.6b.2.1. Neutral
Nolanite $(^{(6;4)}Fe^{2+}_6V^{3+}_4)_{\Sigma 10}V^{4+}_{13}O_{38}$
                                           3.1b.6b.2.2. Oxido-hydroxides
Häggite $V_2(OH)_3O_2 \rightarrow V^{3+}V^{4+}(OH)_3O_2$

3.1b.6b.3. Oxides of p-cations and V^{4+}
                                           3.1b.6b.3.1. Neutral
Stibivanite $Sb^{3+}_2VO_5$

3.1b.7. **Overclass***: Oxides and hydroxides of V^{5+}
3.1b.7a. *Class:* Simple oxides and hydroxides of V^{5+}
                                           3.1b.7a.1. Neutral
Shcherbinaite family
Shcherbinaite $^{(5)}V_2O_5$
Bannermanite $(Na,K)_xV^{4+}_xV^{5+}_{6-x}O_{15}$ (при $x = 0 \rightarrow 3V_2O_5$)
                                           3.1b.7a.2. Hydrates
Homologous series of **navajoite** -
Navajoite $(V^{5+},Fe^{3+})_{10}O_{24} \cdot 12H_2O$
Bariandite $Al_{0.6}(V^{4+}V^{5+})_8O_{20} \cdot 9H_2O$
Vanoxite $V^{4+}_4V^{5+}_2O_{13} \cdot 8H_2O$ (?)

3.1b.7b. *Class:* Complex oxides and hydroxides of $V^{5+} \rightarrow$ (6)-vanadates $\rightarrow$ (5)-vanadates $\rightarrow$ (4)-vanadates
3.1b.7b.1. Quasiclass: (6)-Vanadates
3.1b.7b.1.1. (6)- Vanadates of s-, d_s- and p_s-cations
                                           3.1b.7b.1.1.1. Hydrates
Pascoite family
Pascoite $Ca_3(H_2O)_{17}[V_{10}O_{28}]^{\infty}$
*Magnesiopascoite $Ca_2Mg[V_{10}O_{28}] \cdot 16H_2O$
*Rakovanite $Na_3(H_2O)_{15}H_3[V_{10}O_{28}]$
*Kokinosite $Na_2Ca_2(H_2O)_{24}[V_{10}O_{28}]$
*Gunterite $(Na_{4-x}Ca_x)_{\Sigma 4}(H_2O)_{16}(H_{2-x}V_{10}O_{28}] \cdot 6H_2O$
Hummerite $K_2Mg_2(H_2O)_{16}[V_{10}O_{28}]^{\infty}$
Huemulite $Na_4Mg(H_2O)_{24}[V_{10}O_{28}]^{\infty}$
*Lasalite $Na_2Mg_2(H_2O)_{20}[V_{10}O_{28}]^{\infty}$
*Hughesite $Na_3Al(H_2O)_{22}[V_{10}O_{28}]^{\infty}$
*Postite $Mg(H_2O)_6Al_2(OH)_2(H_2O)_8[V_{10}O_{20}] \cdot 13H_2O$
*Wernerbaurite $\{(NH_4)_2[Ca_2(H_2O)_{14}](H_2O)_2\}[V_{10}O_{28}]\}$
*Schindlerite $(NH_4)_4Na_2(H_2O)_{10}[V_{10}O_{28}]$

3.1b.7b.1.2. (6)-Vanadates of d-cations
3.1b.7b.1.2.1. (6)-Vanadates of V^{4+}
                                3.1b.7b.2.1.1. Hydrates

Corvusite series
Corvusite $(Na,Ca,K)_x(V^{5+},V^{4+},Fe^{2+})_8O_{20}\cdot 4H_2O$ x = 0.8-1.2
Grantsite $(Na,Ca)_x(V^{5+},V^{4+})_6O_{16}\cdot 4H_2O$ x = □ 2,5
Straczekite $(Ca,K,Ba)(V^{5+}V^{4+})_8O_{20}\cdot 3H_2O$
*Nashite $Na_3Ca_2[(V^{5+}{}_9V^{4+})O_{28}]\cdot 24H_2O$

3.1b.7b.1.2.2. (6)-Vanadates of p-cations
                                3.1b.7b.1.2.2.1. Neutral
Dreyerite family (compare with pucherite)
Dreyerite $Bi[^{(6)}VO_4]^{\infty}$
Clinobisvanite $Bi[^{(6)}VO_4]^{\infty}$

3.1b.7b.2. Quasiclass: (5)-Vanadates
3.1b.7b.2.1. (5)-Vanadates of s-, d_s- and p_s-cations
3.1b.7b.2.1.1. Proper (5)-vanadates
3.1b.7b.2.1.1.1. (5)-Vanadates with $[^{(5)}V_6O_{16}]^{\infty}$- radicals
                                3.1b.7b.2.1.1.1.1. Hydrates (neutral)
Hewettite family
  Hewettite group
  Hewettite $Ca(H_2O)_9[V_6O_{16}]^{\infty}$
  Barnesite $Na_2(H_2O)_3[V_6O_{16}]^{\infty}$
  Metahewettite $Ca(H_2O)_3[V_6O_{16}]^{\infty}$
*Phosphovanadylite-Ba $Ba[V^{4+}{}_4P_2O_{12}(OH)_4]\cdot 12H_2O$
*Phosphovanadylite-Ca $Ca[V^{4+}{}_4P_2O_{12}(OH)_4]\cdot 12H_2O$

3.1b.7b.2.1.1.2. (5)-Vanadates with $[^{(5)}V_2O_6]^{\infty}$- radicals
                                3.1b.7b.2.1.1.2.1. Basic → hydrates (basic)
Delrioite family
Metadelrioite $SrCa(OH)_2[V_2O_6]^{\infty}$
Delrioite $Sr[V^{5+}{}_2O_6]\cdot 4H_2O$
*Calciodelrioite $Ca[V_2O_6]^{\infty}\cdot 4H_2O$

                                3.1b.7b.2.1.1.2.2. Hydrates (neutral)
Rossite family
Metarossite $Ca(H_2O)_2[V_2O_6]^{\infty}$
Rossite $Ca(H_2O)_4[V_2O_6]^{\infty}$
*Ansermetite $Mn[V_2O_6]\cdot 4H_2O$

3.1b.7b.2.1.2. (5)-Vanadates-(4)-vanadates
                                3.1b.7b.2.1.2.1. Hydrates (basic)
Vanalite $NaAl_8V_{10}O_{38}\cdot 30H_2O \to$
                                $NaAl_8(H_2O)_{25}(OH)_{11}[V^{5+}{}_6O_{16}]^{\infty}[VO_4]_4$

3.1b.7b.2.2. (5)-Vanadates of d-cations
3.1b.7b.2.2.1. (5)-Vanadates of V^{4+}

3.1b.7b.2.2.1.1. Proper (5)-vanadates

$\qquad$ 3.1b.7b.2.2.1.1.1. Hydrates

Sherwoodite $\qquad$ $Ca_9Al_2V^{4+}_4V^{5+}_{24}O_{80}\cdot56H_2O$

Satpaevite $\qquad$ $Al_{12}V^{4+}_2V^{5+}_6O_{37}\cdot30H_2O \rightarrow$

$\qquad$ $Al_{12}V^{4+}_2(OH)_{42}[V^{5+}_6O_{16}]^\infty\cdot9H_2O$

Melanovanadite $\qquad$ $Ca(V^{4+}_2V^{5+}_2)O_{10}\cdot5H_2O$

3.1b.7b.2.2.1.2. (5)-Vanadates-(4)-vanadates

$\qquad$ 3.1b.7b.2.2.1.2.1. Hydrates

Hendersonite $\qquad$ $Ca_3(V^{4+},V^{5+})_{12}O_{32}\cdot12H_2O$

Bokite $\qquad$ $KAl_3Fe^{3+}_6V^{4+}_6V^{5+}_{20}O_{76}\cdot15H_2O$

*3.16.76.2.2.1.3. (5)-Vanadato-(4)-vanadao-arsenito-arsenstes

*3.16.76.2.2.1.3.1. Hydrates

*Vanarsite $\qquad$ $NaCa_{12}(As^{3+}V^{5+}_{8.5}V^{4+}_{3.5}As^{5+}_6O_{51})_2\cdot78H_2O$

*Morrisonite $\qquad$ $Ca_{11}(As^{3+}V^{5+}_{10}V^{4+}_2As^{5+}_6O_{51})_2\cdot78H_2O$

*Gatewayite $\qquad$ $Ca_6(As^{3+}V^{5+}_9V^{4+}_3As^{5+}_6O_{51})\cdot31H_2O$

3.1b.7b.3. Quasiclass: (4)-Vanadates

3.1b.7b.3.1. (4)-Vanadates of s-, d_s- and p_s- cations

3.1b.7b.3.1.1. Divanadates

$\qquad$ *3.1b.7b.3.1.1.1. Neutral

*Metamunirite $\qquad$ β-$NaV^{5+}O_3$

*Ronneburgite $\qquad$ $K_2Mn^{2+}V^{5+}_4O_{12}$

$\qquad$ 3.1b.7b.3.1.1.2. Hydrates

Munirite $\qquad$ $Na_2[V_2O_6]^\infty\cdot4H_2O$

*Dickthomssenite $\qquad$ $Mg[V_2O_6[\cdot7H_2O$

Alvanite $\qquad$ $(Zn,Ni)Al_4[V_2O_6](OH)_{12}\cdot2H_2O$

*Ankinovichite $\qquad$ $(Ni,Zn)Al_4[V_2O_6](OH)_{12}\cdot(H_2O)_2$

3.1b.7b.3.1.2. Trivanadates

$\qquad$ 3.1b.7b.3.1.2.1. Hydrates

Pintadoite $\qquad$ $Ca_2[V_2O_7]\cdot9H_2O$

3.1b.7b.3.1.3. Tetravanadates

3.1b.7b.3.1.3.1. $M^{3+} = Al^{3+}$, Fe^{3+} ($2M^{3+}\rightarrow M^{2+}M^{4+}$)

3.1b.7b.3.1.3.1.1. Proper tetravanadates

 *3.1b.7b.3.1.3.1.1.1. Neutral

*Ziminaite $\qquad$ $(Fe^{3+},Al)_6[V^{5+}O_4]_6$

*Koksharovite $\qquad$ $CaMg_2Fe_4^{3+}[V^{5+}O_4]_6$

*Reppiaite $\qquad$ $Mn^{2+}_5(OH)_4[V^{5+}O_4]_2$

*Argandite $\qquad$ $Mn^{2+}_7(OH)_8[VO_4]_2$

3.1b.7b.3.1.3.1.1.3. Hydrates

$\qquad$ 3.1b.7b.3.1.3.1.1.3.1. Basic

Santafeite $\qquad$ $(Ca,Sr,Na)_3(Mn^{2+},Fe^{3+})_2Mn_2^{4+}(VO_4)_4(OH,O)_5\cdot2H_2O$

$\qquad$ 3.1b.7b.3.1.3.1.1.3.2. Neutral

Schubnelite family

Steigerite $\qquad$ $Al[VO_4]\cdot3H_2O$

Fervanite $\qquad$ $Fe_4[VO_4]_4\cdot5H_2O$

Schubnelite $\qquad$ $Fe[VO_4]\cdot H_2O$

Rusakovite $(Fe^{3+},Al)_5(OH)_9[(VO_4),(PO_4)]_2 \cdot 3H_2O$
*Kolovratite $(Ni,Zn)_xVO_4 \cdot nH_2O$

3.1b.7b.3.1.3.1.2. Vanadato-phosphates
                                                3.1b.7b.3.1.3.1.2.1. Hydrates
   Schoderite family
   Schoderite $Al_2[VO_4][PO_4] \cdot 8H_2O$
   Metaschoderite $Al_2[VO_4][PO_4] \cdot 6H_2O$

*3.16.76.3.1.3.1.3. Vanadato-arsenates
*Gottlobite $CaMg(OH)[(VO_4),(AsO_4)]$
*Nabiasite $BaMn_9(OH)_2[(V,As)O_4]_6$
*Fianelite $Mn^{2+}_2V^{5+}(V^{5+},As^{5+})O_7 \cdot 2H_2O$

3.1b.7b.3.1.3.2. M^{3+} and M^{2+} 3.1b.7b.3.1.3.2.1. Neutral
Lyonsite $Cu^{2+}_3Fe^{3+}_4[VO_4]_6$
*Grigorievite $Cu^{2+}_3Fe^{3+}_2Al_2[VO_4]_6$
                                                3.1b.7b.3.1.3.2.2. Basic
Mounanaite $PbFe^{3+}_2(OH)_2[VO_4]_2$
*Krettnichite $PbMn^{3+}_2(OH)_2[VO_4]_2$
*Tokyoite $Ba_2Mn^{3+}(OH)[VO_4]_2$

3.1b.7b.3.1.3.3. M^{3+}, M^{2+} and M^+
3.1b.7b.3.1.3.3.1. Neutral
Howardevansite $Na_2Cu^{2+}_2Fe^{3+}_4[VO_4]_6$

3.1b.7b.3.1.3.4. M^{2+} 3.1b.7b.3.1.3.4.1. Oxido-tetravanadates
                                                and tetravanadato-chlorides
Heyite $Pb_5Fe^{2+}_2O_4[VO_4]_2$
                                                3.1b.7b.3.1.3.4.2. Basic
   Descloizite family
      Calciovolborthite group
      Calciovolborthite $CaCu(OH)[VO_4]$
      Calcium mottramite $Pb_2CaCu_3(OH)_3[VO_4]_3$
      Descloizite group
      Descloizite $Pb(Zn,Cu)(OH)[VO_4]$
      Mottramite $PbCu(OH)[VO_4]$
      Čechite $Pb(Fe^{2+},Mn)(OH)[VO_4]$
      Pyrobelonite $PbMn(OH)[VO_4]$
   Vesignieite $BaCu_3(OH)_2[VO_4]_2$
   Leningradite $PbCu_3Cl_2[VO_4]_2$
      Brackebuschite series
      Gamagarite $Ba_2(Fe,Mn)^{3+}(OH)[VO_4]_2$
      Brackebuschite $Pb_2(Mn,Fe)^{3+}(OH)[VO_4]_2$
      *Calderonite $Pb_2Fe^{3+}(OH)[VO_4]_2$
      *Fe-brackebuschite $(Pb_{1,8}Zn_{0,2})(Fe^{3+}_{0,75}Mn^{3+}_{0,15}Al_{0,1})(OH)[VO_4]_2$

3.1b.7b.3.1.3.5. Vanadates of M^{2+} and M^+
3.1b.7b.3.1.3.5.1. Polyvanadates
                                                3.1b.7b.3.1.3.5.1.1. Hydrates

Huemulite $Na_4Mg(H_2O)_{24}[V_{10}O_{28}]^{\infty}$
3.1b.7b.3.1.3.5.2. Tetravanadates

                       3.1b.7b.3.1.3.5.2.1. Neutral
Palenzonaite $NaCa_2Mn^{2+}{}_2[VO_4]_3$
*Schäferite $NaCa_2Mg^{2+}{}_2[VO_4]_3$

3.1b.7b.3.1.3.5.3. Vanadates with unknown structure
                       3.1b.7b.3.1.3.5.3.1. Hydrates
Kazakhstanite $Fe^{3+}{}_5(OH)_9V^{4+}{}_3V^{5+}{}_{12}O_{39}\cdot 9H_2O$

3.1b.7b.3.2. (4)-Vanadates of f-cations
3.1b.7b.3.2.1. Tetravanadates   3.1b.7b.3.2.1.1. Neutral
  Wakefieldite series
  Wakefieldite-(Ce) $Ce[VO_4]$
  Wakefieldite-(Y) $Y[VO_4]$
  *Wakefieldite-(La) $La[VO_4]$
  *Wakefieldite-(Nd) $Nd[VO_4]$

3.1b.7b.3.3. (4)-Vanadates of d-cations
3.1b.7b.3.3.1. Vanadates of **I**b-cations
3.1b.7b.3.3.1.1. Cu^{2+}
3.1b.7b.3.3.1.1.1. Trivanadates   3.1b.7b.3.3.1.1.1.1. Neutral
  Ziesite family
  Ziesite $\beta\text{-}Cu_2[V_2O_7]$
  Blossite $\alpha\text{-}Cu_2[V_2O_7]$
                     3.1b.7b.3.3.1.1.1.2. Hydrates (basic)
Volborthite $Cu_3(OH)_2(H_2O)_2[V_2O_7]$
*Karpenkoite $(Co,Zn)_3(OH)_2(H_2O)_2[V_2O_7]$

3.1б.7б.3.3.1.1.1.3. Оксидо-триванадато-хлориды
3.1б.7б.3.3.1.1.1.3.1. Кристаллогидраты (основные)
*Engelhauptite $KCu_3(OH)_2[V_2O_7]Cl$

*3.1б.7б.3.3.1.1.2. Триванадато-тетраванадаты
*3.1б.7б.3.3.1.1.2.1. Оксидо-триванадато-тетраванадаты
*Kainotropite $Cu_4Fe^{3+}O_2[V_2O_7][VO_4]$

3.1b.7b.3.3.1.1.3. Tetravanadates   3.1b.7b.3.3.1.1.3.1. Neutral
Mcbirneyite $Cu_3[VO_4]_2$
*Pseudolyonsite $Cu_3[VO_4]_2$
*Borisenkoite $Cu_3[(V,As)O_4]_2$

                     3.1b.7b.3.3.1.1.3.2. Oxido-tetravanadates
Stoiberite family
  Stoiberite $Cu_5O_2[VO_4]_2$
  *Starovaite $KCu_5O[VO_4]_3$
  Fingerite $Cu_{11}O_2[VO_4]_6$

*3.1б.7б.3.3.1.1.3.3. Oxido-tetravanadato-chlorides
*Averievite $Cu_5O_2[VO_4]_2\cdot CuCl_2\cdot MCl$, M=Cs,K,Rb

*Yaroshevskite $Cu_9O_2[VO_4]_4Cl_2$

3.1b.7b.3.3.1.1.3.4. Hydroxides
Turanite $Cu_5(OH)_4[VO_4]_2$

*3.1б.7б.3.3.1.2. M^+ и M^{2+}
*3.1б.7б.3.3.1.2.1. Ag^+, Hg^{2+}
*Tillmannsite $(Ag_3Hg)(V,As)O_4$

3.1b.7b.3.3.1.3. Cu^{2+}, Pb^{2+} and Bi^{3+}
3.1b.7b.3.3.1.3.1. Tetravanadates
                           3.1b.7b.3.3.1.3.1.1. Hydrates (basic)
Duhamelite $Pb_2Cu_4Bi(OH)_3[VO_4]_4·8H_2O$

*3.1б.7б.3.3.2. Vanadates of II*b*-cations
*3.1б.7б.3.3.2.1. Zn^{2+}
*3.1б.7б.3.3.2.1.1. Trivanadates
*3.1б.7б.3.3.2.1.1.1. Hydrates
*Martyite $Zn_3(OH)_2[V_2O_7]·2H_2O$

3.1b.7b.3.4. (4)-Vanadates of *p*-cations
3.1b.7b.3.4.1. Vanadates of **IV***a*-cations
3.1b.7b.3.4.1.1. Pb^{2+}
3.1b.7b.3.4.1.1.1. Trivanadates   3.1b.7b.3.4.1.1.1.1. Neutral
Chervetite $Pb_2[V_2O_7]$
3.1b.7b.3.4.1.1.2. Tetravanadates
                           3.1b.7b.3.4.1.1.2.1. Tetravanadato-chlorides
                           3.1b.7b.3.4.1.1.2.1.1. Oxido-chlorido-vanadates
Kombatite $Pb_{14}O_9[VO_4]_2Cl_4$
                           3.1b.7b.3.4.1.1.2.1.2. Neutral
Vanadinite $Pb_5[VO_4]_3Cl$ (compare with apatite (family);
                           pyromorphite (group))
3.1b.7b.3.4.1.1.2.2. Tetravanadato-chromates
                           3.1b.7b.3.4.1.1.2.2.1. Hydrates
Cassedanneite $Pb_5(VO_4)_2(CrO_4)_2 · H_2O$

3.1b.7b.3.4.1.2. Pb^{2+} and Bi^{3+}
3.1b.7b.3.4.1.2.1.Hydrovanadato-tetravanadates
                           3.1b.7b.3.4.1.2.1.1. Hydrates
Pottsite $PbBi[VO_3OH][VO_4]·2H_2O$

3.1b.7b.3.4.2. Vanadates of **V***a*-cations
3.1b.7b.3.4.2.1. Tetravanadates (orthovanadates)
                           3.1b.7b.3.4.2.1.1. Neutral
Pucherite $Bi[^{(4)}VO_4]^∞$ (compare with dreyerite (family))
                           3.1b.7b.3.4.2.1.2. Oxido-hydroxides
Schumacherite $Bi^{3+}_3(OH)O[VO_4]_2$
*Hechtsbergite $Bi_2O(OH)[VO_4]$

3.1c. **Quasisubtype***: Oxides and hydroxides of chalcophylic cations (without **Va**- and **VIa**- cations)

3.1c.1. **Overclass***: oxides and hydroxides of **Ib**-cations

3.1c.1.1. Cu^+

3.1c.1.1.1. Simple

Cuprite Cu_2O

3.1c.1.2. Cu^+ and Fe^{3+} (Cu^{2+})

3.1c.1.2.1. Complex

Delafossite family

  Delafossite group

   Delafossite $Cu^+Fe^{3+}O_2$

   Mcconnellite $Cu^+Cr^{3+}O_2$

  Crednerite $CuMnO_2$

  Paramelaconite $Cu^+_2Cu^{2+}_2O_3$

3.1c.1.3. Cu^{2+}

3.1c.1.3.1. Proper oxides

3.1c.1.3.1.1. Simple

Tenorite CuO

3.1c.1.3.2. Oxido-halogenides   3.1c.1.3.2.1. Simple

Melanothallite Cu_2OCl_2

3.1c.1.3.2.2.Complex

Murdochite $Pb^{4+}_2Cu^{2+}_{12}O_{15}(Cl,Br)_2$

3.1c.1.3.3. Hydroxides and hydroxido-halogenides

3.1c.1.3.3.1. Simple

Spertiniite $Cu(OH)_2$

                                       3.1c.1.3.3.2. Hydrates

Calumetite family

Calumetite $Cu(OH,Cl)_2 \cdot 2H_2O$

Anthonyite $Cu(OH,Cl)_2 \cdot 3H_2O$

3.1c.2. **Overclass***: Oxides and hydroxides of **IIb**-cations

3.1c.2.1. Oxides and hydroxides of Hg

3.1c.2.1.1. Hg^+

3.1c.2.1.1.1. Oxido-halogenides

3.1c.2.1.1.1.1. Simple

Poyarkovite family

Poyarkovite Hg_3OCl

Kadyrelite $Hg_6^{1+}Br_3O_{1.5}$

3.1c.2.1.1.2. Hydroxido-oxido-halogenides

3.1c.2.1.1.2.1. Simple

Eglestonite $Hg^+_6(OH)OCl_3$

3.1c.2.1.2. Hg^+ and Hg^{2+}

3.1c.2.1.2.1. Oxido-halogenides

3.1c.2.1.2.1.1. Complex

Terlinguaite $Hg_2OCl \rightarrow Hg^+Hg^{2+}OCl$
*Aurivilliusite $Hg^+Hg^{2+}OI$
*Tedhadleyite $Hg^+_{10}Hg^{2+}O_4I_2(Cl,Br)_2$

*3.1c.2.1.2.2. Oxido-sulfides
*Deanesmithite $Hg^+_2Hg^{2+}_3Cr^{6+}O_5S_2$

3.1c.2.1.3. Hg^{2+}
3.1c.2.1.3.1. Proper oxides
3.1c.2.1.3.1.1. Simple
Montroydite   HgO

3.1c.2.1.3.2. Oxido-halogenides
3.1c.2.1.3.2.1. Simple
Comancheite family
Pinchite $Hg_5O_4Cl_2$
*Terkinguacreekite $Hg^{2+}_3O_2Cl_2$
Comancheite $Hg_{13}O_9(Cl,Br)_8$

*3.1c.2.1.3.3. Oxido-halogenido-carbonates
*Vasilyevite $Hg^{2+}_{10}O_6I_3(Br,Cl)_3(CO_3)$

3.1c.2.2. Oxides and hydroxides of Zn^{2+} and Cd^{2+}
3.1c.2.2.1. Simple oxides and hydroxides
3.1c.2.2.1.1. Proper oxides
3.1c.2.2.1.1.1. Simple
Zincite   ZnO (compare with bromellite (group))
Monteponite   CdO (compare with periclase (series)

3.1c.2.2.1.2. Hydroxides
3.1c.2.2.1.2.1. Simple
Sweetite family
Sweetite $Zn(OH)_2$
Ashoverite $Zn(OH)_2$
Wülfingite $Zn(OH)_2$

3.1c.3.**Overclass***: Oxides and hydroxides of **IIIa**-cations
3.1c.3.1. Proper oxides
3.1c.3.1.1. Simple
Avicennite Tl_2O_3

3.1c.3.2. Hydroxides
3.1c.3.2.1. Simple
Dzhalindite $In(OH)_3$

3.1c.4. **Overclass***: Oxides and hydroxides of **IVa**-cations
3.1c.4.1. Oxides and hydroxides of Pb
3.1c.4.1.1. Oxides and hydroxides of Pb^{2+}
3.1c.4.1.1.1. Proper oxides
3.1c.4.1.1.1.1. Simple

Litharge family
Litharge PbO
Massicot $\beta\text{-}PbO$
*Unnamed $\alpha\text{-}PbO$

3.1c.4.1.1.2. Oxido-halogenides
3.1c.4.1.1.2.1. Simple
Mendipite $Pb_3O_2Cl_2$
*Damaraite $Pb_3O_2(OH)Cl$

3.1c.4.1.1.3. Oxido-silicato-chlorides
3.1c.4.1.1.3.1. Simple
Asisite $Pb_7SiO_8Cl_2 \rightarrow Pb_7O_4[SiO_4]Cl_2$

3.1в.4.1.1.4. Oxido-borato-carbonato- chlorides
3.1в.4.1.1.4.1. Simple
*Mereheadite $Pb_{47}O_{24}(OH)_{13}Cl_{25}(BO_3)_2(CO_3)$

3.1c.4.1.2. Oxides and hydroxides of Pb^{2+} and Pb^{4+}
3.1c.4.1.2.1. Proper oxides
3.1c.4.1.2.1.1. Complex
Minium $Pb_3O_4 \rightarrow Pb^{2+}_2Pb^{4+}O_4$

3.1c.4.1.3. Oxides and hydroxides of Pb^{4+}
3.1c.4.1.3.1. Proper oxides
3.1c.4.1.3.1.1. Simple
Plattnerite family
 Plattnerite PbO_2
 Scrutinyite $\alpha\text{-}PbO_2$

*3.1в.4.1.3.1.1. Complex
*Lindqvistite $Pb^{2+}_2Mn^{2+}Fe^{3+}_{16}O_{27}$

3.1c.4.2 Oxides and hydroxides of Ge и Sn
3.1c.4.2.1. Oxides and hydroxides of Sn^{2+}
3.1c.4.2.1.1. Proper oxides
3.1c.4.2.1.1.1. Simple
Romarchite SnO

3.1c.4.2.1.2. Oxido-hydroxides and oxido-hydroxido-halogenides
3.1c.4.2.1.2.1. Simple
Hydroromarchite family
 Hydroromarchite $Sn_3(OH)_2O_2$
 Abhurite $Sn_3(OH)_2OCl_2$
 *Unnamed $Sn_4O(OH,F)_6$

*3.1c.4.2.1.2.2. Complex
*Eyselite $Fe^{3+}Ge^{4+}_3O_7(OH)$

*3.1в.4.2.1.2.2. Oxides and hydroxides of Ge^{4+}

*Krieselite $(Al,Ga)_2(Ge,C)O_4(OH)_2$

3.1d. Quasisubtype*: Oxides and hydroxides of Va-катионов
3.1d.1. Overclass*: Oxides and hydroxides of As^{3+},Sb^{3+},Bi^{3+}
3.1d.1a. Class: Simple oxides and hydroxides of $As^{3+}, Sb^{3+}, Bi^{3+}$
3.1d.1a.1. Proper oxides
 Arsenolite family
  Arsenolite group
   Arsenolite $2As_2O_3 \rightarrow As_4O_6$
   Senarmontite $2Sb_2O_3 \rightarrow Sb_4O_6$
  Valentinite $|Sb_2O_3|^{\infty}$
  Claudetite $|As_2O_3|^{\infty 2}$
 *Stibioclaudetite $AsSbO_3$
  Bismite $\alpha\text{-}|Bi_2O_3|^{\infty 3}$
 *Sphaerobismoite $Bi_2O_3|$

3.1d.1a.2. Oxido (hydroxido)-halogenides
  Bismoclite group
   Zavaritskite $BiOF$
   Daubreeite $BiO(OH,Cl)$
   Bismoclite $BiOCl$
  Onoratoite $Sb_8O_{11}Cl_2$
 *Torrecillasite $Na(As,Sb)^{3+}_4O_6Cl$

3.1d.1a.3. Oxido-silicates
Sillenite $\gamma\text{-}Bi_{12}SiO_{20} \rightarrow \gamma\text{-}Bi_{12}O_{16}[SiO_4]$

3.1d.1b. *Class*: Complex oxides and hydroxides of $As^{3+}, Sb^{3+}, Bi^{3+} \rightarrow$ (6)-arsenites,
antiminites, bismuthites
3.1d.1b.1. Arsenites of *s*-, d_s- and p_s- cations
3.1d.1b.1.1. Arsenites of M^{3+}

                                                 3.1d.1b.1.1.1. Oxides and hydroxides
Karibibite $Fe^{3+}_2As_4(O,OH)_9$

3.1d.1b.1.2. M^{3+} and M^{2+}
3.1d.1b.1.2.1. Proper arsenites

                                                 3.1d.1b.1.2.1.1. Neutral
Schneiderhöhnite $Fe^{2+}Fe^{3+}_3As_5O_{13}$
Stenhuggarite $CaFe^{3+}SbAs_2O_7$
                                                 *3.1d.1b.1.2.1.2. Basic
*Graeserite $Fe^{3+}_4Ti_3AsO_{13}(OH)$
*Fetiasite $Fe^{2+}Fe^{3+}_2[As^{3+}_2O_5]O_2$

                                                 3.1d.1b.1.2.1.3. Hydrates
Lazarenkoite $(Ca,Fe^{2+})Fe^{3+}As_3O_7 \cdot 3H_2O$
Cafarsite $Ca_8(Ti,Fe^{2+},Fe^{3+},Mn)_{6-7}As_{12}O_{36} \cdot 4H_2O$

3.1d.1b.1.2.2. Arsenito-sulfides
3.1d.1b.1.2.2.1. Neutral
Versiliaite $(Fe^{2+},Fe^{3+},Zn)_8(Sb^{3+},Fe^{3+},As)_{16}O_{32}S_{1.3}$

3.1d.1b.1.3. Arsenites of M^{2+}
3.1d.1b.1.3.1. Proper arsenites

	3.1d.1b.1.3.1.1. Neutral
Schafarzikite	$FeSb_2O_4$
	3.1d.1b.1.3.1.2. Basic

Manganarsite family
Manganarsite $Mn_3(OH)_4As_2O_4$
Magnussonite $Mn_5(OH,Cl)As_3O_9$
                                              3.1d.1b.1.3.1.3. Acid
Trigonite $Pb_3Mn(AsO_3)_2(AsO_2OH)$
                                              3.1d.1b.1.3.1.4. Hydrates
Rouseite $Pb_2Mn(AsO_3)_2 \cdot 2H_2O$

3.1d.1b.1.3.2. Arsenito-halogenides   3.1d.1b.1.3.2.1. Neutral
Nanlingite $NaCa_5LiMg_{12}(AsO_3)_2[Fe^{2+}(AsO_3)_6]F_{14}$
*Lucabindiite $(K,NH_4)(As_4O_6)(Cl,Br)$

3.1d.1b.2. Arsenites of **d-** cations
3.1d.1b.2.1. Arsenites of Ti^{4+}, V^{4+} 3.1d.1b.2.1.1. Basic
 Tomichite family
  Tomichite group
  Tomichite $(V,Fe)^{3+}_4Ti_3(OH)AsO_{13}$
  Derbylite $Fe^{3+}_4Ti_3(OH)SbO_{13}$
  Hemloite $(Ti,V^{3+},Fe^{2+},Fe^{3+})_{12}(OH)(As,Sb)^{3+}_2O_{23}$
                                              *3.1г.1б.2.1.2. Hydrates
*Dukeite $Bi^{3+}_{24}Cr^{6+}_8O_{57}(OH)_6(H_2O)_3$

3.1d.1b.2.2. Arsenites of **I**b-cations
3.1d.1b.2.2.1. Arsenites of Au (only Au^+)
3.1d.1b.2.2.1.1. Antimonito- antimonates
                                              3.1d.1b.2.2.1.1.1. Neutral
*3.1d.1б.2.2.1.2. Antimonites
*Unnamed $Au_2Sb^{3+}O_2(OH)$

3.1d.1b.2.2.2. Arsenites of Cu (only Cu^{2+})
                                              3.1d.1b.2.2.2.1. Neutral
Trippkeite $CuAs_2O_4$

*3.1d.1б.2.2.3. Bismuthites Cu (only Cu^{2+}) и M^{3+}
*Kusachiite $CuBi_2O_4$

*3.1г.1б.2.2.3.1. Bismuthito-oxido-sulfato-halogenides
*Atlasovite $Cu^{2+}_6Fe^{3+}Bi^{3+}O_4[SO_4]_5 \cdot KCl$

3.1d.1b.2.3. Arsenites of **II**b-cations
3.1d.1b.2.3.1. Arsenites of Hg
3.1d.1b.2.3.1.1. Oxido-halogenides
Kelyanite $Hg_{12}SbO_6BrCl_2$

3.1d.1b.2.3.2. Arsenites of Zn
3.1d.1b.2.3.2.1. Neutral
Leiteite family
Leiteite $ZnAs_2O_4$
Reinerite $Zn_3As_2O_6$

3.1d.1b.3. Arsenites of *p*-cations
3.1d.1b.3.1. Arsenites of **IV*a***-cations (Pb^{2+})
3.1d.1b.3.1.1. Proper arsenites
3.1d.1b.3.1.1.1. Neutral
Paulmooreite $Pb_2As_2O_5$

3.1d.1b.3.1.2. Arsenito-halogenites   3.1d.1b.3.1.2.1. Neutral →basic
    Nadorite group
    Nadorite $PbSbO_2Cl$
    Perite $PbBiO_2Cl$
Finnemanite $Pb_5As_3O_9Cl$
Gebhardite $Pb_8As_4O_{11}Cl_6$
Freedite $Pb_8Cu^+(As^{3+}O_3)_2O_3Cl_5$
  Ecdemite family
    Ecdemite group
    Ecdemite $Pb_6As_2O_7Cl_4$
    Thorikosite $Pb_3(OH)(Sb_{0.6}As_{0.4})O_3Cl_2$
    Heliophyllite $Pb_6As_2O_7Cl_4$

3.1d.2. **Overclass:** Oxides and hydroxides of As^{5+}, Sb^{5+}, Bi^{5+} (all complex) → arsenates, antimonates and bismuthates (only (6)-arsenates, (6)-antimonates and (6)-bismuthates)
3.1d.2.1. Quasiclass: (6)-arsenates, (6)-antimonates and (6)-bismuthates
3.1d.2.1.1. (6)-Arsenates, (6)-antimonates and (6)-bismuthates of - *s*-, *d$_s$*- and *p$_s$*- cations
3.1d.2.1.1.1. ((6) - Arsenates, (6) - antimonates and (6) - bismuthates of *s*-, *d$_s$*- and *p$_s$*- cations (without Li and Be)
3.1d.2.1.1.1.1. (6)-Arsenates, (6)-antimonates and (6)-bismuthates of M^{3+}
3.1d.2.1.1.1.1.1. Proper (6)-Arsenates, (6)-antimonates and (6)-bismuthates
                           3.1d.2.1.1.1.1.1.1. Basic
Bahianite $Al_5(OH)_2Sb_3O_{14}$ → $\{Al_5(OH)_2|Sb_3O_{14}|^\infty\}^{\infty 3}$
3.1d.2.1.1.1.2. (6)-Antimonates of M^{3+} and M^{2+}
3.1d.2.1.1.1.2.1. Neutral
Melanostibite $Mn^{2+}_2Fe^{3+}SbO_6$
*Paganoite $NiBi^{3+}As^{5+}O_5$
Cualstibite $Cu_2AlSb^{5+}(OH)_{12}$
*Zincalstibite $(Zn,Cu)_2Al(OH)_6[Sb(OH)_6]$ или $Zn_2AlSb(OH)_{12}$
*Omsite $(Ni,Cu)_2Fe^{3+}(OH)_6[Sb(OH)_6]$

3.1d.2.1.1.1.2.2. (6)-Antimonates-sulfato-hlorides
                           3.1d.2.1.1.1.2.3.1. Basic
Mammothite $Pb_6Cu_4Al(OH)_{16}SbO_2[SO_4]_2Cl_4$

3.1d.2.1.1.1.3. (6)- antimonates M^{2+}
3.1d.2.1.1.1.3.1. Neutral
    Byströmite group (compare with tapiolite (group)) $MgSb_2^{5+}O_6$

Byströmite	$MgSb^{5+}_2O_6$
Tripuhyite	$Fe^{2+}Sb^{5+}_2O_6$ or $FeSbO$

*3.1г.2.1.1.1.3.2. Basic
*Bottinoite	$Ni[Sb^{5+}(OH)_6]_2 \cdot 6H_2O$

3.1d.2.1.1.1.3.3. Oxido-antimonates
Ingersonite	$Ca_3MnSb^{5+}_4O_{14}$

3.1d.2.1.1.1.3.3.1. Hydrates
Brandholzite	$Mg(H_2O)_6[Sb(OH)_6]_2$

3.1d.2.1.1.1.4. (6)-Antimonates M^{2+} and M^+
3.1d.2.1.1.1.4.1. Basic
Roméite series (compare with pyrochlore (series); partzite (series))
*Fluorcalcioroméite	$[Sb^{5+}_2O_6]^{\infty 3}(Ca,Na)_2F$
*Oxycalcioroméite	$Ca_2Sb_2O_7$
*Oxyplumboroméite	$Pb_2Sb_2O_7$
Na-romeite	$[Sb_2O_6]^{\infty 3}(Na,Ca,Mn)_2(OH,F)$
Lewisite	$[(Sb,Ti)_2(O,OH)_6|^{\infty 3}(Ca,Fe,Na)_2(O,OH)$

3.1d.2.1.1.1.5. (6)-Antiminates of M^+ 3.1d.2.1.1.1.5.1. Acid
Mopungite	$Na[Sb(OH)_6]$

*3.1d.2.1.1.1.5.2. Neutral
*Brizziite	$NaSb^{5+}O_3$

3.1d.2.1.1.2. (6) - Antimonates of Be→ beryllo-antimonates
3.1e.2.1.1.2.1. Neutral
Swedenborgite	$NaBe_4SbO_7 \rightarrow$
	$^{(12)}Na[^{(4)}Be_4O(^{(6)}SbO_6)]^{\infty 3}$

3.1d.2.1.2. (6)-Arsenates, (6)-antimonates of d-cations
3.1d.2.1.2.1. (6)-Arsenates, (6)-antimonates of Ib-elements
3.1d.2.1.2.1.1. (6)-Arsenates, (6)-antimonates of Cu^{2+}
3.1e.2.1.2.1.1.1. Basic
Namibite	$Cu(BiO)_2[VO_4](OH)$

3.1d.2.1.2.1.2. (6) Antimonates of Cu^+
3 .1e.2.1.2.1.2.1. Acid-basic
Partzite group (compare with pyrochlore (series); romeite (series)
Partzite	$Cu_2Sb_2O_5(OH)_2$ (?)$\rightarrow$
	$|Sb_2O_5|^{\infty 3} \cdot Cu_2(OH)_2$ (?)
Stetefeldtite	$Ag_2Sb_2(O,OH)_7 \rightarrow$
	$|Sb_2(O,OH)_6|^{\infty 3} \cdot Ag_2(O,OH)$
*Auriacusite	$Fe^{3+}Cu^{2+}[(As,Sb)O_4]O$

3.1d.2.1.2.2. (6)-Arsenates and (6)-antimonates of IIb-elements
3.1d.2.1.2.2.1. (6)-Arsenates and (6)-antimonates of Hg
3.1d.2.1.2.2.1.1. Basic (acid ?)
*Shanovite	$Hg^{2+}_8Sb^{5+}_2O_{13}$
Shakhovite	$Hg^+_4(OH)_3Sb^{5+}O_3$

3.1d.2.1.2.2.2. (6)-Arsenates and (6)-Antimonates of Zn
3.1d.2.1.2.2.2.1. Neutral

Ordoñezite $ZnSb_2O_6$ (compare with tapiolite (group); byströmimite (group)
                                       *3.1d.2.1.2.2.2.2. Basic
*Sabelliite $(Cu,Zn)_2Zn[(As,Sb)O_4](OH)_3$

3.1d.2.1.3. (6) - Arsenates and and (6) - Antimonates of *p*-metals
3.1d.2.1.3.1. (6)- Arsenates and (6) - antimonates of **IVa**-elements (Pb^{2+})
*Rosiaite $PbSb_2O_6$
                                       3.1d.2.1.3.1.1. Neutral
Monimolite $(Pb,Ca)_3Sb_2O_8$
                                       3.1d.2.1.3.1.2. Basic (acid)
Bindheimite $Pb_2Sb_2O_7$

3.1d.2.1.3.2. (6)-Arsenates, (6)-antimonates of **Va**-elements
3.1d.2.1.3.2.1. (6)-Arsenates, (6)-antimonates of Sb^{3+} and Bi^{3+}
                                       3.1d.2.1.3.2.1.1. Neutral $\rightarrow$ basic (acid)
Cervantite (compare with stibiotantalite (group)) $Sb_2O_4 \rightarrow {}^{(4)}Sb^{3+}|^{(6)}Sb^{5+}O_4|^{\infty 2}$
*Clinocervantite β-Sb_2O_4
Stibiconite group (compare with romeite (series); partzite (group); bindheimite)
  Stibiconite   questionable $Sb_3O_6(OH) \rightarrow |Sb^{5+}_2O_6|^{\infty 3} \cdot Sb^{3+}(OH)$
  Bismutostibiconite questionable $Bi(Sb,Fe)_2O_7 \rightarrow |(Sb,Fe)_2O_6|^{\infty 3} \cdot BiO$

3.1e. Quasisubtype*: Oxides and hydroxides of VIa-cations (Te)
3.1e.1. Overclass*: Oxides and hydroxides of Te^{4+}
3.1e.1a. Class: Simple oxides and hydroxides of Te^{4+}
  Tellurite family
Tellurite TeO_2
Paratellurite TeO_2

3.1e.1b. ***Class:*** Complex oxides and hydroxides $Te^{4+} \rightarrow$ Tellurites
3.1e.1b.1. Tellurites of **s**- and d_s-ations
3.1e.1b.1.1. Tellurites M^{3+}
3.1e.1b.1.1.1. Tellurites Fe^{3+}
3.1e.1b.1.1.1.1. Proper tellurites
                                       3.1e.1b.1.1.1.1.1. Neutral
Blakeite   questionable $Fe^{3+}_2Te_3O_9$
                                       3.1e.1b.1.1.1.1.2. Basic and hydroxido-chlorides
Mackayite $Fe^{3+}Te_2(OH)O_5$
Rodalquilarite $H_3Fe^{3+}_2Te_4O_{12}Cl \rightarrow Fe^{3+}_2Te_4(OH)_3O_9Cl$
                                       3.1e.1b.1.1.1.1.3. Hydrates
Emmonsite $Fe^{3+}_2Te_3O_9 \cdot 2H_2O$
Sonoraite $Fe^{3+}Te(OH)O_3 \cdot H_2O$

3.1e.1b.1.1.1.2. Tellurito-sulfates   3.1e.1b.1.1.1.2.1. Hydrates
Poughite $Fe^{3+}_2Te_2O_6[SO_4] \cdot 3H_2O$

3.1e.1b.1.1.2. Tellurites of Fe^{3+} and Bi^{3+}
3.1e.1b.1.1.2.1. Tellurito-tellurates   3.1e.1b.1.1.2.1.1. Hydrates
Yecoraite $Fe_3^{3+}Bi_5O_9(Te^{4+}O_3)(Te^{6+}O_4)_2 \cdot 9H_2O$
*Pingguite $Bi_6Te^{4+}_2O_{13}$

3.1e.1b.1.1.3. Tellurites of M^{3+} and M^{2+}
3.1e.1b.1.1.3.1. Proper tellurites
3.1e.1b.1.1.3.1.1. Hydrates (basic)
Eztlite $Pb_3Fe^{3+}_6Te^{4+}_3Te^{6+}(OH)_{10}O_{15} \cdot 8H_2O$

3.1e.1b.1.1.3.2. Tellurito-monoalu-monosilicates
                            3.1e.1b.1.1.3.2.1. Hydrates (basic)
Burckhardtite $Pb_2(Fe^{3+}Te^{6+})[AlSi_3O_8]O_6$

3.1e.1b.1.1.4. Tellurites of M^{2+}
3.1e.1b.1.1.4.1. Proper tellurites
                            3.1e.1b.1.1.4.1.1. Neutral
Spiroffite $(Mn,Zn)_2Te_3O_8$
*Zincospiroffite $Zn_2Te_3O_8$
Denningite $(Mn,Ca)Te^{4+}_4O_{10}$
               3.1e.1b.1.1.4.1.2. Neutral$\rightarrow$ Hydrates (compounds inclusions)
Zemannite family
Kinichilite $Mg_{0.5}Mn^{2+}Fe^{3+}(Te^{4+}O_3)_3\ 4.5H_2O$
Zemannite $Mg_{0.5}ZnFe^{3+}(Te^{4+}O_3)_3\ 4.5H_2O$
*Ilirneyite $Mg_{0.5}\{ZnMn^{3+}(TeO_3)_3\}^{\infty 3} \cdot 4.5H_2O$

3.1e.1b.1.1.4.2. Tellurito-tellurates
                            3.1e.1b.1.1.4.2.1. Neutral
Carlfriesite $CaTe^{4+}_2Te^{6+}O_8$
*Walfordite $(Fe^{3+}Te^{6+})Te^{4+}_3O_8$

3.1e.1b.1.1.4.3. Tellurito-carbonates
                            3.1e.1b.1.1.4.3.1. Neutral
Mroseite $CaTeO_2[CO_3]$

3.1e.1b.1.1.5. Tellurites of M^{2+} and M^+
3.1e.1b.1.1.5.1. Proper tellurites
                            3.1e.1b.1.1.5.1.1. Hydrates
Keystoneite $Mg_{0,5}(Ni^{2+}Fe^{3+})_{\Sigma 2}(TeO_3)_3 \cdot 4.5H_2O$

3.1e.1b.1.1.5.2. Tellurito-sulfato-halogenides
3.1e.1b.1.1.5.2.1. Oxido-tellurito-sulfato-halogenides
Nabokoite $KCu^{2+}_7Te^{4+}O_4[SO_4]_5Cl$

3.1e.1b.2. Tellurites of d-cations
3.1e.1b.2.1. Tellurites of **IV**b-cations (Ti^{4+})
                            3.1e.1b.2.1.1. Neutral
Winstanleyite $TiTe_3O_8$

3.1e.1b.2.2. Tellurites of **I**b-cations·
3.1e.1b.2.2.1. Tellurites of Cu^{2+}
*3.1e.1b.2.2.1. Proper tellurites
                            3.1e.1b.2.2.1.1.1. Neutral
Rajite $CuTe_2O_5$
Balyakinite $CuTeO_3$

3.1e.1b.2.2.1.1.2. Hydrates
3.1e.1b.2.2.1.2.1.1. Basic
Cesbronite $Cu_5Te_2O_6(OH)_6 \cdot 2H_2O$
*Brumadoite $Cu_3(Te^{6+}O_4)(OH)_4 \cdot 5H_2O$
3.1e.1b.2.2.1.1.2.2. Neutral

Graemite family
Graemite $CuTeO_3 \cdot H_2O$
Teineite $Cu(Te,S)O_3 \cdot 2H_2O$

*3.1e.1b.2.2.1.2. Tellurito-arsenates
*Juabite $CaCu^{2+}_{10}(TeO_3)_4[AsO_4]_4(OH)_2(H_2O)_4$

*3.1e.1b.2.2.1.3. Tellurito-halogenides
*3.1e.1b.2.2.1.3.1. Basic
*Mojaveite $Cu^{2+}_6[T^{6+}O_4(OH)_2](OH)_7Cl$

3.1e.1b.2.2.2. Tellurites of Cu^{2+} and Zn
3.1e.1b.2.2.2.1. Proper tellurites
3.1e.1b.2.2.2.1.1. Basic
Quetzalcoatlite $Cu^{2+}_3Zn_6Te^{6+}_2O_{12}(OH)_6 \cdot Ag_xPb_yCl_{x+2y}$

*3.1e.16.2.2.3. Tellurito-arsenato-halogenides
*3.1e.1b.2.2.3.1.Hydrates
*Eurekadumpite $(Cu,Zn)_{16}(TeO_3)_2(AsO_4)_3Cl(OH)_{18} \cdot 7H_2O$

3.1e.1b.2.2.2.2. Tellurito-tellurato-halogenides
3.1e.1b.2.2.2.2.1. Hydrates
3.1e.1b.2.2.2.2.1.1. Basic
Tlalocite $Cu_{10}Zn_6Te^{4+}Te^{6+}_2(OH)_{25}O_{11}Cl \cdot 27H_2O$

3.1e.1b.2.2.2.3. Tellurito-tellurato-sulfates
3.1e.1b.2.2.2.3.1. Neutral
Tlapallite $H_6Ca_2Cu_3Te^{4+}_4Te^{6+}O_{18}[SO_4]$

3.1e.1b.2.2.3. Tellurites of Cu^{2+} and Pb
Choloalite $(Pb,Ca)_3(Cu,Sb)_3Te_6O_{18}Cl$

3.1e.1b.2.3. Tellurites of **II**b-elements
3.1e.1b.2.3.1. Tellurites of Hg^+
3.1e.1b.2.3.1.1. Neutral
Magnolite $Hg^+_2TeO_3$

3.1e.1b.3. Tellurites of p-cations (Pb^{2+})
3.1e.1b.3.1. Proper tellurites
3.1e.1b.3.1.1. Neutral

Plumbotellurite family
Plumbotellurite α-$PbTeO_3$
Fairbankite $PbTeO_3$

*3.1e.16.3.2. Tellurato-halogenides

*Telluroperite $Pb_3Te^{4+}O_4Cl_2$

3.1e.1b.3.2. Tellurito-tellurates   3.1e.1b.3.2.1. Acid
                                      3.1e.1b.3.2.2. Hydrates
Oboyerite $H_6Pb_6Te^{4+}_3Te^{6+}_2O_{21}\cdot2H_2O$

3.1e.1b.4. Tellurites of semimetals
3.1e.1b.4.1. Tellurites of Bi^{3+}
                                      3.1e.1b.4.1.1. Neutral
Smirnite Bi_2TeO_5
Chekhovichite $Bi_2Te_4O_{11}$

3.1e.2. **Overclass:** Oxides and hydroxides of Te^{6+} (all complex) → tellurates (all (6)-tellurates)

3.1e.2.1. Tellurites of *s*- and *d_s*- cations (M^{2+})
                                      3.1e.2.1.1. Neutral
Yafsoanite $Ca_3Zn_3[TeO_6]_2 \rightarrow$
                                      $Ca_3Te_2|ZnO_4|_3$
                                      3.1e.2.1.2. Hydrates
Cuzticite $Fe_2TeO_6\cdot3H_2O$

3.1e.2.2. Tellurates of *d*-cations
3.1e.2.2.1. Tellurates of Mn^{4+}
3.1e.2.2.1.1. Tellurates of Mn^{4+} and Pb
                                      3.1e.2.2.1.1.1. Neutral
Kuranakhite $PbMn^{4+}TeO_6$
                                      *3.1д.2.2.1.1.2. Hydrates
*Xocolatlite $Ca_2Mn^{4+}_2Te^{6+}_2O_{12}\cdot H_2O$

3.1e.2.2.2. Tellurates of I*b*-cations
3.1e.2.2.2.1. Tellurates of Cu^{2+} 3.1e.2.2.2.1.1. Basic
Xocomecatlite $Cu_3Te^{6+}O_4(OH)_4$
*Frankhawthorneite $Cu_2Te^{6+}O_4(OH)_2$
*Mcalpineite $Cu^{2+}_3Te^{6+}O_6$
                                      *3.1д.2.2.1.1.1. Hydrates
*Raisaite $Cu^{2+}Mg[Te^{6+}O_4(OH)_2]\cdot6H_2O$
*Utahite $Cu^{2+}_5Zn_3(Te^{6+}O_4)_4(OH)_8\cdot7H_2O$
*Leisingite $Cu^{2+}(Mg,Cu,FeZn)_2Te^{6+}O_6\cdot6H_2O$

3.1e.2.2.2.2. Tellurates of Cu^{2+} and Pb
3.1e.2.2.2.2.1. Basic
Khinite family
Khinite $PbCu_3Te^{6+}(OH)_2O_6$
Parakhinite = khinit-3T $PbCu_3Te^{6+}(OH)_2O_6$
*Housleyite $Pb_6Cu^{2+}Te^{6+}_4(OH)_2O_{18}$
*Timroseite $Pb_2Cu^{2+}_5(Te^{6+}O_6)_2(OH)_2$
*Paratimroseite $Pb_2Cu^{2+}_4(Te^{6+}O_6)_2(H_2O)_2$
*Andycristiite $PbCu^{2+}(Te^{6+}O_5)(H_2O)$
*Eckhardite $(Ca,Pb)Cu^{2+}Te^{6+}O_5(H_2O)_2$

*3.1e.2.2.2.2.3. Tellurato-carbonates
*3.1e.2.2.2.2.3.1. Basic
*Agaite $Pb_3Cu^{2+}Te^{6+}O_5(OH)_2[CO_3]$

*3.1e.2.2.2.2.4. Tellurato-sulfates
*3.1e.2.2.2.2.4.1. Basic
*3.1e.2.2.2.2.4.1.1.Hydrates
*Bairdite $Pb_2Cu^{2+}_4(Te^{6+}O_5)_2(OH)_2[SO_4](H_2O)$

3.1e.2.2.2.3. Tellurates of **II**b-cations
3.1e.2.2.3.1. Tellurates of Zn
3.1e.2.2.3.1.1. Tellurates of Zn and Pb
3.1e.2.2.3.1.1.1. Tellurato-arsenates   3.1e.2.2.3.1.1.1.1. Basic
Dugganite $Pb_3Zn_3Te^{6+}O_6[AsO_4]_2$
*Joëlbruggerite $Pb_3Zn_3(Sb^{5+},Te^{6+})O_5[AsO_4]_2(OH,O)$

*3.1e.2.2.3.1.1.2. Tellurato-phosphates
*3.1e.2.2.3.1.1.2.1.Basic
*Kuksite $Pb_3Zn_3Te^{6+}O_6[PO_4]_2$

*3.1e.2.2.3.1.1.2. Tellurato -vanadates
*3.1e.2.2.3.1.1.2.1.Basic
*Cheremnykhite $Pb_3Zn_3Te^{6+}O_6[VO_4]_2$

3.1e.2.3. Tellurates of p-metals
3.1e.2.3.1. Tellurates of **IV**a-metals   3.1e.2.3.1.1. Hydrates
*Ottoite $Pb_2Te^{6+}O_5$
Schieffelinite $Pb_{10}Te_6O_{20}(OH)_{14}[SO_4]\cdot5H_2O$
*Chromschieffelinite $Pb_{10}Te_6O_{20}(OH)_{14}[CrO_4](H_2O)_5$
*Thorneite $Pb_6[Te^{6+}_2O_{10}][CO_3]Cl_2(H_2O)$

3.1e.2.4. Tellurates of semimetals (Bi^{3+})   3.1e.2.4.1. Hydrates
Montanite $Bi_2TeO_6\cdot2H_2O$

3.1f. Quasisubtype*: Oxides and hydroxides of nonmetals (lithophylic) elements
3.1f.1. **Class:** Oxides and hydroxides of Si and Ge (silicic and germanium anhydrides,
silicic and germanium asids)
3.1f.1.1. Ionic-covalent crystals   3.1f.1.1.1. Neutral
  Stishovite group (compare with rutile, cassiterite)
  Stishovite $^{(6)}SiO_2$
  *Mon. (mineral high pressure) SiO_2
  Argutite $^{(6)}GeO_2$
Silice family
Coesite $\{SiO_2\}^{\infty3}$
β-Tridymite $\{SiO_2\}^{\infty3}$
α-Tridymite $\{SiO_2\}^{\infty3}$
*Orthorhombic SiO_2
*Seifertite orth. SiO_2
Cristobalite $\{SiO_2\}^{\infty3}$
β-Quartz $\{SiO_2\}^{\infty3}$

α-Quartz	$\{SiO_2\}^{\infty 3}$
Lutecine	$\{SiO_2\}^{\infty 3}$

3.1f.1.2. Globular crystals	3.1f.1.2.1. Hydrates
Opal family	
Opal	$SiO_2 \cdot nH_2O$
Silhydrite	$3SiO_2 \cdot H_2O$

3.1f.1.3. Silico-organic substances

Melanophlogite	$Si_{46}O_{92} \, C_2H_{17}O_5$
*Chibaite	$SiO_2 \cdot n(CH_4, C_2H_6, C_3H_8, C_4H_{10})$; ($n_{max} = 3/17$)

3.1f.2. **Class:** Oxides and hedroxides of B (boric anhydride and boric asids)
Sassolite family

Metaborite	HBO_2
*Clinometaborite	$\beta\text{-}HBO_2$
Sassolite	$B(OH)_3 \rightarrow H_3BO_3$

3.1f.3 **Class:** Oxides and hedroxides of Se (selenium anhydrite)

Downeyite	SeO_2

3.2. Subtype: Oxosalts (anisodesmical)
3.2.1. **Class:** Silicates
Subclass: Silicates of cations with low FC
Silicates and alumosilicates of **s-**, **d_s-** and **p_s-**cations
Silicates and alumosilicates of **s-**, **d_s-** and **p_s-**cations (without Li^+ and Be^{2+})
Proper silicates and alumosilicates
Zero-alumosilicates (K = 0 Neutral, with compounds inclusions which contain H_2O
  Cordierite family
   Cordierite series

Cordierite	$Mg_2\{Al_3[(Si_5Al)_{\Sigma 6}O_{18}]\}^{\infty 3}$
Sekaninaite	$Fe^{2+}_2\{Al_3[(Si_5Al)_{\Sigma 6}O_{18}]\}^{\infty 3}$
Indialite	$(Mg,Fe)_2\{Al_3[(Si_5Al)_{\Sigma 6}O_{18}]\}^{\infty 3}$
*Ferroindialite	$(Fe,Mg)_2\{Al_3[(Si_5Al)_{\Sigma 6}O_{18}]\}^{\infty 3}$

Feldspar family
Plagioclase subfamily (**Ca-Na-feldspar**)

Anorthite	$Ca[Al_2Si_2O_8]^{\infty 3}$ (An); $An_{90\text{-}100}$
*Dmisteinbergite	$Ca[Al_2Si_2O_8]$
*Svyatoslavite	$Ca[Al_2Si_2O_8]$
*Lisetite	$Na_2Ca[Al_2Si_2O_8]_2$
Bytownite	$An_{70\text{-}90}$
*Maskelynite – it is a glass whis bytownite composition	
Labradorite	$An_{50\text{-}70}$
Andesine	$An_{30\text{-}50}$
Oligoclase	$An_{10\text{-}30}$
*Lingunite	$Na[AlSi_3O_8]$

Na-K-feldspar subfamily

Albite	$Na[AlSi_3O_8]$ (Ab); $An_{0\text{-}10}$
*Albite tetragonal	$Na[AlSi_3O_8]$
*Kumdykolite	$Na[AlSi_3O_8]$

Anorthoclase $\qquad$ $(Na,K)[AlSi_3O_8]^{\infty3}$

Sanidine $\qquad$ $(K,Na)[AlSi_3O_8]^{\infty3}$

Microcline $\qquad$ $K[AlSi_3O_8]^{\infty3}$

*Rubicline $\qquad$ $Rb[AlSi_3O_8]^{\infty3}$

Orthoclase $\qquad$ $K[AlSi_3O_8]^{\infty3}$

*Mineral mon., pseudotet. whis hollandite structure. $K[AlSi_3O_8]$

*Kokchetavite $\qquad$ $K[AlSi_3O_8]$

K-Ba-feldspar subfamily

Hyalophane $\qquad$ $(K,Ba)[(Si,Al)Si_3O_8]^{\infty3}$

Celsian $\qquad$ $Ba[Al_2Si_2O_8]^{\infty3}$

Paracelsian $\qquad$ $Ba[Al_2Si_2O_8]^{\infty3}$

*Filatovite $\qquad$ $K[(Al,Zn)_2(As,Si)_2O_8]^{\infty3}$

Slawsonite $\qquad$ $(Sr,Ca)[Al_2Si_2O_8]^{\infty3}$

Banalsite $\qquad$ $BaNa_2[Al_2Si_2O_8]^{\infty3}{}_2$

Buddingtonite $\qquad$ $NH_4[AlSi_3O_8]$

Feldspathoid family

Nepheline subfamily

Stronalsite $\qquad$ $Na_2Sr[AlSiO_4]^{\infty3}{}_4$

Nepheline $\qquad$ $4(Na,K)[AlSiO_4]^{\infty3} \to KNa_3[AlSiO_4]^{\infty3}{}_4$

Trikalsilite $\qquad$ $(K_{0.67}Na_{0.33})[AlSiO_4]^{\infty3}$

Panunzite $\qquad$ $(K_{0.7}Na_{0.3})[AlSiO_4]^{\infty3}$

Kalsilite $\qquad$ $K[AlSiO_4]^{\infty3}$

*Megakalsilite $\qquad$ $K[AlSiO_4]$

Kaliophilite $\qquad$ $K[AlSiO_4]^{\infty3}$

Leucite subfamily

Leucite $\qquad$ $K[AlSi_2O_6]^{\infty3}$

Ammonioleucite $\qquad$ $NH_4[AlSi_2O_6]^{\infty3}$

Pollucite $\qquad$ $(Cs,Na)[AlSi_2O_6]^{\infty3} \cdot nH_2O$

Scapolite series $(Ca,Na)_4(Si,Al)_{12}O_{24}(CO_3,SO_4,Cl)$

*Silvialite $\qquad$ $[Al_6Si_6O_{24}]\,Ca_4(SO_4)$ or

$\qquad$ $[Al_2Si_2O_8]^{\infty3}{}_3\,Ca_4(SO_4)$

Meionite $\qquad$ $[(Si,Al)_4O_8]_3(Ca,Na)_4(CO_3,SO_4,Cl),\quad Me_{75\text{-}100}$

Mizzonite $\qquad$ $Me_{50\text{-}75}$

Dipyre $\qquad$ $Me_{25\text{-}50}$

Marialite $\qquad$ $[AlSi_3O_8]^{\infty3}{}_3Na_4Cl; Me_{0\text{-}25}$

Cancrinite series

*Balliranoite $\qquad$ $[Al_6Si_6O_{24}][(Na,K)_6(CO_3)_2][Ca_2Cl_2]$

Cancrinite $\qquad$ $[Al_6Si_6O_{24}][(Na,K)_6(CO_3)_2][Ca_2(H_2O)_2]$

Cancrisilite $\qquad$ $[Al_5Si_7O_{24}][Na_5CO_3(H_2O)][Na_2(H_2O)_2]$

Davyne $\qquad$ $[Al_6Si_6O_{24}][(Na,K)_6(SO_4)_{0.5}Cl][Ca_2Cl_2]$

*Depmeierite $\qquad$ $[Al_6Si_6O_{24}][Na_6(PO_4),(CO_3))_{0.5\text{-}1}(H_2O)][Na_2(H_2O)_2]$

*Hydroxicancrinite $\qquad$ $[Al_6Si_6O_{24}][Na_6(OH)(CO_3)][Na_2(H_2O)_2]$

*Kyanoxalite $\qquad$ $[Al_{5\text{-}6}Si_{6\text{-}7}O_{24}][Na_6(C_2O_4)(H_2O)_3)(OH)][Na_2(H_2O)_2]$

Microsommite $\qquad$ $[Al_6Si_6O_{24}][(Na,K)_6(SO_4)][Ca_2Cl_2]$

*Pitiglianoite $\qquad$ $[Al_6Si_6O_{24}][Na_4K_2(SO_4)][Na_2(H_2O)_2]$

*Qudridavyne $\qquad$ $[Al_6Si_6O_{24}][(Na,K)_6Cl_2)][Ca_2Cl_2]$

*Carbobystrite $\qquad$ $[Al_6Si_6O_{24}]Na_8(CO_3)\cdot4H_2O$

*Farneseite $\qquad$ $[Al_6Si_6O_{24}]_7Na_{46}Ca_{10}(SO_4)_{12}\cdot6H_2O$

*Alloriite $\qquad$ $[Al_6Si_6O_{24}][Na_5K_{1.5}Ca](SO_4)(OH)_{0.5}\cdot H_2O$

*Biachellaite $[Al_6Si_6O_{24}](Na,Ca,K)_8(OH)_{0.5}(SO_4)_2 \cdot H_2O$

*Kircherite $[Al_{12}Si_{12}O_{48}](Na_{10}Ca_4K_2)(SO_4)_4 \cdot 2/3H_2O$

*Tounkite $[Al_6Si_6O_{24}](Na,Ca,K)_8(SO_4)_2Cl \cdot 0.5H_2O$

Afghanite $[Al_{24}Si_{24}O_{96}][(Na,K)_{22}Ca_{10}](SO_4)_6Cl_6$

Giuseppettite $[Si_{48}Al_{48}O_{192}]Na_{42}K_{16}Ca_6](SO_4)_{10}Cl_2 \cdot 5H_2O$

*Marinellite $[Al_{36}Si_{36}O_{144}]Na_{42}Ca_6(SO_4)_8Cl_2 \cdot 6H_2O$

Franzinite $[Si_{30}Al_{30}O_{120}](Na,K)_{30}Ca_{10}(SO_4)_{10} \cdot 2H_2O$ or
                              $[SiAlO_4]_{30}(Na,K)_{30}Ca_{10}(SO_4)_{10} \cdot 2H_2O$

*Fantappieite $[Al_{99}Si_{99}O_{396}]Na_{82.5}Ca_{33}K_{16.5}(SO_4)_{33} \cdot 6H_2O$

*Bystrite $[Al_6Si_6O_{24}]Ca(Na,K)_7(S_3)_{1.5} \cdot H_2O$

Vishnevite series

Wenkite $[(Si,Al)_{20}O_{41}]Ba_4Ca_6(OH)_2(SO_4)_3 \cdot H_2O$

Liottite $[Si_{18}Al_{18}O_{72}]Na_{16}Ca_8(SO_4)_5Cl_4$

Sacrofanite $[(Si,Al)_{12}O_{24}](Na,Ca)_9(OH,SO_4,CO_3,Cl)_4 \cdot nH_2O$

Vishnevite $[(Al_6Si_6)O_{24}]Na_8(SO_4) \cdot 2H_2O$

Sodalite series

Sodalite $[AlSiO_4]^{\infty3}_6 Na_8Cl_2$

Haüyne $[(Si_3Al_3)O_{12}]Na_3Ca(SO_4)$

*Vladimirivanovite $[Al_6Si_6O_{24}]Na_6Ca_2[SO_4,S_3,S_2,Cl]_2 \cdot H_2O$

Lazurite series

Lazurite $[(Al_6Si_6O_{24})]Na_6Ca_2(SO_4,S,S_2,S_3,Cl,OH)_2$

Nosean $[(Si_6Al_6)O_{24}]Na_8(SO_4) \cdot H_2O$

Zeolites family Low-silicic zeolites (Si : Al up 1 to 1,(6))

Parthéite $[(Si_4Al_4O_{15}]Ca_2(OH)_2 \cdot 4H_2O$

Thomsonite subfamily(Si : Al = 1)

Gismondine $[AlSiO_4]^{\infty3}_2Ca(H_2O)_4$

Thomsonite-Ca $[AlSiO_4]^{\infty3}_5NaCa_2(H_2O)_6$

*Thomsonite-Sr $[AlSiO_4]^{\infty3}_5(Sr,Ca)_2Na \cdot 6\text{-}7H_2O$

Willhendersonite $[AlSiO_4]^{\infty3}_3KCa(H_2O)_5$

*Willhendersonite-Ca $[AlSiO_4]^{\infty3}_3CaK(H_2O)_5$

Amicite $[AlSiO_4]^{\infty3}_4 K_2Na_2(H_2O)_5$

*Flörkeite $[AlSiO_4]_8 K_3Ca_2Na \cdot 12H_2O$

*Bellbergite $[AlSiO_4]^{\infty3}_{18} (K,Ba,Sr)_2Sr_2Ca_2(Ca,Na)_4 \cdot 30(H_2O)$

*Tschörtnerite $[AlSiO_4]^{\infty3}_{12} Ca_4(K,Ca,Sr,Ba)_3Cu_3(OH)_8 \cdot nH_2O$ n ≈ 20

Scolecite-natrolite subfamily (Si : Al = 1,5)

Scolecite $[Al_2Si_3O_{10}]^{\infty3} Ca(H_2O)_3$

Cowlesite $[Al_2Si_3O_{10}]^{\infty3} Ca(H_2O)_{5\text{-}6}$

Edingtonite $[Al_2Si_3O_{10}]^{\infty3} Ba(H_2O)_4$

Mesolite $[Al_2Si_3O_{10}]^{\infty3}_3 Na_2Ca_2(H_2O)_8$

Gonnardite $[(Si,Al)_5O_{10}](Na,Ca)_2 \cdot 3H_2O$

Natrolite $[Al_2Si_3O_{10}]^{\infty3} Na_2(H_2O)_2$

Paranatrolite $[Al_2Si_3O_{10}]^{\infty3} Na_2(H_2O)_3$

Garronite subfamily (Si:Al = 1,(6))

Garronite-Ca $[Al_3Si_5O_{16}]^{\infty3}_2 NaCa_{2.5}(H_2O)_{14}$

Garronite-Na $[Al_3Si_3O_{16}]^{\infty3}_2 Na_6 \cdot 8.5H_2O$

Phillipsite-Ca $[(Si_{10}Al_6)O_{32}]Ca_3 \cdot 12H_2O$

*Phillipsite-K $[(Si_{10}Al_6)O_{32}]K_6 \cdot 12H_2O$

*Phillipsite-Na $[(Si_{10}Al_6)O_{32}]Na_6 \cdot 12H_2O$

*Unnamed (Si:Al = 1,9) $[Al_{11}Si_{21}O_{64}]Ca_5K_2 \cdot 18.4H_2O$

Middle-silicic zeolites (Si : Al up 2 to 2,2)
  Gobbinsite   (Si:Al =2.2) $[(Si_{11}Al_5)O_{32}]Na_5 \cdot 11H_2O$
  Wairakite subfamily (Si : Al = 2)
  Wairakite $[AlSi_2O_6]^{\infty 3}_2 Ca(H_2O)_2$
  Laumontite $[AlSi_2O_6]^{\infty 3}_2 Ca(H_2O)_4$
   Chabazite series
   *Chabazite-Ca $[AlSi_2O_6]^{\infty 3}_4 Ca_2 \cdot 13H_2O$
   *Chabazite-K $[AlSi_2O_6]^{\infty 3}_4 K_2NaCa_{0.5} \cdot 11H_2O$
   *Chabazite-Na $[Al_4Si_8O_{24}](Na_3K) \cdot 11H_2O$
   *Chabazite-Mg $[Al_3Si_9O_{24}]^{\infty 3}Mg_{0.7}K_{0.5}Ca_{0.5}Na_{0.1} \cdot 10H_2O$
   *Chabazite-Sr $[Al_4Si_8O_{24}](Sr,Ca)_2 \cdot 11H_2O$
   Lévyne-Ca $[AlSi_2O_6]^{\infty 3}_6 Ca_3 \cdot 18H_2O$
   *Lévyne-Na $[AlSi_2O_6]^{\infty 3}_6 Na_6 \cdot 18H_2O$
   Herschelite $[Al_2Si_4O_{12}]_2(Na_2,K_2,Ca,Sr,Mg)_2 \cdot 12H_2O$
  Gmelinite subfamily (Al : Si = 2)
  Gmelinite-Ca $[AlSi_2O_6]^{\infty 3}_4 Ca_2(H_2O)_{11}$
  *Gmelinite-Na $[AlSi_2O_6]^{\infty 3}_4 Na_4(H_2O)_{11}$
  *Gmelinite-K $[AlSi_2O_6]^{\infty 3}_4 K_4 \cdot 22H_2O$
  Faujasite $[AlSi_2O_6]^{\infty 3}_4 Na_2Ca(H_2O)_{16}$
  α-Leongardite $[AlSi_2O_6]^{\infty 3}_4(K,Na)_2Ca(H_2O)_7$
  Analcime $[AlSi_2O_6]^{\infty 3}Na(H_2O)$
  Perlialite $[AlSi_2O_6]^{\infty 3}_{12} K_9Na(Ca,Sr)(H_2O)_{15}$
  Harmotome $[Si_{12}Al_4O_{32}]Ba_2(H_2O)_{12}$
  *Meierite $Ba_{44}Si_{66}Al_{30}O_{192}Cl_{25}(OH)_{33}$ or $[AlSi_{2.2}O_{6.4}] Ba_{1.5}Cl_{0.8}(OH)_{1.1}$

High-silicic zeolites (Si : Al up 2,5 to 3,5)
Merlinoite (Si : Al = 2,(5)) $[Al_9Si_{23}O_{64}]^{\infty 3}K_5Ca_2(H_2O)_{24}$
*Montesommaite $[Al_9Si_{23}O_{64}]^{\infty 3}K_9 \cdot 10H_2O$
Offretite (Si : Al = 2,6) $[Al_5Si_{13}O_{36}]^{\infty 3} KCaMg(H_2O)_{15}$
Erionite-Ca $[Al_{10}Si_{26}O_{72}]Ca_5 \cdot 30H_2O$
*Erionite-K $[(Al_{10}Si_{26}O_{72})]K_{10} \cdot 30H_2O$
*Erionite-Na $[(Al_{10}Si_{26}O_{72})]Na_{10} \cdot 30H_2O$
  Stilbite subfamily (Si : Al = 3)
   Yugawaralite $[AlSi_3O_8]^{\infty 3}_2 Ca(H_2O)_4$
  Epistilbite $[AlSi_3O_8]^{\infty 3}_6 Ca_3(H_2O)_{16}$
  *Tschernichite $[AlSi_3O_8]^{\infty 3}_2 Ca \cdot 8H_2O$
  *Brewsterite-Ba $[AlSi_3O_8]^{\infty 3}_4 Ba_2 \cdot 10H_2O$
  Brewsterite-Sr $[AlSi_3O_8]^{\infty 3}_4 Sr_2 \cdot 10H_2O$
  Wellsite $[AlSi_3O_8]^{\infty 3}_2 (Ba,Ca,K_2)(H_2O)_6$
  *Heulandite-Ba $[Al_9Si_{27}O_{72}](Ba,Ca,Sr,K,Na)_5 \cdot 22H_2O$
  *Heulandite-Ca $[Al_9Si_{27}O_{72}](Ca,Na,K)_5 \cdot 26H_2O$
  *Heulandite-K $[Al_9Si_{27}O_{72}](K,Ca,Na,)_5 \cdot 26H_2O$
  *Heulandite-Na $[Al_9Si_{27}O_{72}](Na,Ca,K)_6 \cdot 22H_2O$
  *Heulandite-Sr $[Al_9Si_{27}O_{72}](Sr,Ca,Na)_5 \cdot 24H_2O$
  Mazzite-Mg $[Al_{10}Si_{26}O_{72}]^{\infty 3} Mg_5 \cdot 30H_2O$
  *Mazzite-Na $[Al_8Si_{28}O_{72}]^{\infty 3} Na_8 \cdot 30H_2O$
  *Stilbite-Ca $[(Al_9Si_{27})O_{72}]NaCa_4 \cdot 28H_2O$

*Stilbite-Na $[Al_9Si_{27}O_{72}]Na_9 \cdot 28H_2O$

Goosecreekite $[AlSi_3O_8]^{\infty 3}{}_2Ca(H_2O)_5$

*Maricopaite $[Al_{12}Si_{36}O_{99}]Ca_2Pb_7\ n(H_2O,OH)$

Paulingite-Ca (Si : Al = 3,2) $[Al_5Si_{16}O_{42}]^{\infty 3}{}_2(Ca,K,Na,Ba,\square)_{10} \cdot 34H_2O$

*Paulingite-Na $[Al_{10}Si_{35}O_{90}](Na_2,K_2,Ca,Ba)_5 \cdot 45H_2O$

*Paulingite-K $[(Si,Al)_{42}O_{84}](K,Ca,Na,Ba,\square)_{10} \cdot 34H_2O$

Stellerite subfamily (Si : Al = 3,5)

Stellerite $[Al_2Si_7O_{18}]^{\infty 3}{}_4\ Ca_4 \cdot 28H_2O$

Barrerite $[Al_2Si_7O_{18}]^{\infty 3}Na_2 \cdot 6H_2O$

*Direnzoite (Si : Al = 3,6) $[Al_{13}Si_{47}O_{120}]NaK_6MgCa_2 \cdot 36H_2O$

Ultra-high-silicic zeolites (Si : Al up 3,8 to 6 and above)

Svetlozarite (Si : Al = 3.8) $[Al_{10}Si_{38}O_{96}](Na_2,Ca,K_2)_5 \cdot 25H_2O$

*Boggsite (Si : Al = 4,2) $[(Al_{18}Si_{77})_{\Sigma 96}O_{192}]Na_3Ca_8 \cdot 70H_2O$

  Dachiardite series (Si : Al = 5)

Sodium dachiardite $[Al_8Si_{40}O_{96}](Na_2,Ca,K_2)_{4\text{-}5} \cdot 26H_2O$

Dachiardite-Ca $[Al_4Si_{20})O_{48}]Ca_2 \cdot 13H_2O$

Dachiardite-K $[Al_4Si_{20}O_{48}]K_4 \cdot 13H_2O$

***Clinoptilolite** subfamily (Si : Al = 5)

*Clinoptilolite-Ca $[Al_6Si_{30}O_{72}]Ca_3 \cdot 20H_2O$

*Clinoptilolite-K $[Al_6Si_{30}O_{72}]K_6 \cdot 20H_2O$

*Clinoptilolite-Na $[Al_6Si_{30}O_{72}]Na_6 \cdot 20H_2O$

Tsaregorodtsevite (Si : Al = 5) $[Si_4(SiAl)O_{12}]N(CH_3)_4$

  Mordenite subfamily (Si : Al = 5)

Mordenite $[AlSi_5O_{12}]^{\infty 3}{}_8(\ Na_2,Ca,K_2)_4 \cdot 28H_2O$

*Terranovaite $[Si_{68}Al_{12}O_{160}](Na,Ca)_8 \cdot 29H_2O$

*Gottardite (Si : Al = 6,2) $[Si_{117}Al_{19}O_{272}]Na_3Mg_3Ca_5 \cdot 93H_2O$

 ***Ferrierite** subfamily (Si : Al = 6,2)

*Ferrierite-K $[Al_5Si_{31}O_{72}]K_2NaMg \cdot 18H_2O$

*Ferrierite-Na $[Al_5Si_{31}O_{72}]Na_3KMg_{0,5} \cdot 18H_2O$

*Ferrierite-Mg $[Al_7Si_{29}O_{72}]Mg_{2,5}K_{0,5}Na_{0,5}Ca_{0,5} \cdot 18nH_2O$

*Mutinaite (Si : Al = 7,7) $[Al_{11}Si_{85}O_{192}]Na_3Ca_4 \cdot 60H_2O$

Basic zero-alumosilicates

Bicchulite family

Bicchulite $Ca_2(OH)_2[Al_2SiO_6]^{\infty 3}$

Kamaishilite $Ca_2(OH)_2[Al_2SiO_6]^{\infty 3}$

Hydrates; neutral

Cymrite $Ba[AlSiO_4]^{\infty 2}{}_2 \cdot nH_2O\ \ (n = 0.5\text{-}1)$

Zeroalumosilicato-carbonato-chlorides   *Hydrates

*Kampfite $Ba_{12}[(Si_{11}Al_5)O_{31}](CO_3)_8Cl_5$

Zerosilicates   *Hydrates

*Afwillite $Ca_3[SiO_4][SiO_2(OH)_2] \cdot 2H_2O$

*Hydrates basic

*Makatite $Na_2[Si_4O_8](OH)_2 \cdot 4H_2O$

*Yegorovite $Na_4[Si_4O_8](OH)_4 \cdot 7H_2O$

*Megaciclite $Na_8K[Si_9O_{18}](OH)_9 \cdot 19H_2O$

Zero-monoalumo- and zero-monosilicates $(0 < K < 1)$
*Ellingsenite $Na_5Ca_6[Si_{18}O_{38}](OH)_{13} \cdot 6H_2O$

Zero-monoalumosilicates c $K_\Sigma = 0,3$ and $0,(3)$
                                     Neutral
Naujakasite $Na_6\{Fe^{2+}[Al_4Si_8O_{26}]^{\infty 2}$
*Manganonaujakasite $Na_6(Mn,Fe)^{2+}[Al_4Si_8O_{26}]^{\infty 2}$
                                     Acid
Lithosite $K_3[HAl_2Si_4O_{13}]^{\infty 3}$

*Zero-monosilicates with $K = 0,3$ *Basic $\rightarrow$ Hydrates
*Esquireite $Ba[Si_6O_{13}] \cdot 7H_2O$

Zero-monoalumosilicates with $K_\Sigma = 0,4$ Neutral $\rightarrow$ Hydrates
 Latiumite family
Tuscanite $[(Si,Al)_5O_{11}]^{\infty 2}{}_2KCa_6(SO_4,CO_3)_2(OH) \cdot H_2O$
Latiumite $[(Si,Al)_5O_{11}]^{\infty 2 \cdot} (Ca,K)_4(SO_4,CO_3)$

Zero-monoalumosilicates with $K_\Sigma = 0,45$
*Kenyaite $Na_2[Si_{22}O_{41}(OH)_8] \cdot 6H_2O$
*Günterblassite $(K,Ca,Ba,Na,\square)_3Fe[(Si,Al)_{13}O_{25}(OH,O)_4] \cdot 7H_2O$
*Hillesheimite $(K,Ca,Ba,\square)_2(Mg,Fe,Ca,\square)_2[(Si,Al)_{13}O_{23}(OH)_6](OH) \cdot 8H_2O$

*Zero-monoalumosilicato-halogenides
*Umbrianite $K_7Na_2Ca_2[Al_3Si_{10}O_{29}]F_2Cl_2$

*Zero-monoalumosilicates and zero-monosilicates c $K_\Sigma = 0,5$
*Ertixiite $Na_2[Si_4O_9]$
*Tosudite $Na(Al_4Mg_2)[AlSi_7O_{18}](OH)_{12} \cdot 5H_2O$

Zero-monoalumosilicates with $K_\Sigma = 0,5$
                                     Acid zero-monoalumosilicates
Ussingite $[AlSi_3O_8(OH)]^{\infty 3} Na_2$
                                     *Hydrates
*Franklinphillite $K_4Mn^{2+}{}_{48}[Al_9Si_{63}O_{163}](OH)_{53} \cdot 6H_2O$

*Zero-monoalumosilicates with $K_\Sigma = 0,57$ *Neutral
*Wadalite $Ca_6(Al,Si,Mg,Fe)_7O_{16}Cl_3$ (?)
                                     *Hydrates
*Magadiite $[Si_7O_{13}(OH)_3]Na \cdot 4H_2O$

Zero-monoalumosilicates and zero-monosilicates with $K_\Sigma = 0,6$
                                     Neutral $\rightarrow$ basic$\rightarrow$hydrates
*Parsettensite $(K,Na,Ca)_{7.5}(Mn,Mg)_{49}Si_{72}O_{168}(OH)_{50} \cdot nH_2O$
Zussmanite $[AlSi_{17}O_{42}]^{\infty 2}(K,Na)(Fe,Mg,Mn,Al)_{13}(OH)_{14}$
*Coombsite $[Al_{1,5}Si_{16,5}O_{42}]^{\infty 2}(K,Na)(Mn,Fe,Mg)_{13}(OH)_{14}$
 Stilpnomelane series
 Calciumferristilpnomelane $[Al_2Si_{16}O_{42}]^{\infty 2} Ca_{0,5}(Fe^{3+},Fe^{2+})_{12}(O,OH)_{12}(H_2O)_{6-8}$

Kaliferristilpnomelane

$$[Al_2Si_{16}O_{42}]^{\infty 2}K_2(Fe^{3+},Fe^{2+})_{12}(O,OH)_{12}(H_2O)_{6-8}$$

Calciumferrostilpnomelane $\quad$ $[Al_2Si_{16}O_{42}]^{\infty 2}Ca_{0,5}(Fe^{2+},Fe^{3+})_{12}(O,OH)_{12}(H_2O)_{6-8}$

Stilpnomelane (kaliferrostilpnomelane)

$$[Al_2Si_{16}O_{42}]^{\infty 2}K(Fe^{2+},Fe^{3+},Mg)_{12}(O,OH)_{12}(H_2O)_{6-8}$$

*Lennilenapeite $\quad$ $[(Si,Al)_{18}(O,OH)_{42}]\,K_{1,5-1,75}(Mg,Mn^{2+},Fe^{2+},Zn)_{12}(O,OH)_{12}\cdot4H_2O$

Ganophyllite (manganstilpnomelane))

$$[Al_2Si_{16}O_{42}]^{\infty 2}K(Mn^{2+},Fe^{2+},Fe^{3+})_{12}(O,OH)_{12}(H_2O)_{6-8}$$

*Tamaite $\quad$ $(Ca,K,Ba,Na)_{3-4}Mn_{24}(Si,Al)_{40}(O,OH)_{112}\cdot21H_2O$

*Carlosturanite $\quad$ $[(Si,Al)_{12}O_{28}](Mg,Fe,Ti)_{21}(OH)_{34}\cdot H_2O$

*Kvanfieldite $\quad$ $[Si_6O_{14}]Na_4(Ca,Mn)(OH)_2$

Zero-monoalumosilicates and zero-monosilicates with $K_\Sigma = 0,75$

Neutral → basic→hydrates

Delhayelite family

Fedorite $\quad$ $[Si_8O_{19}]^{\infty 2}(K,Na)_{2-3}(Ca,Na)_7(OH,F)_2\cdot nH_2O)$

Delhayelite $\quad$ $[AlSi_7O_{19}]^{\infty 2}K_7Ca_5Na_3F_4Cl_2$

*Fivegite $\quad$ $[AlSi_7O_{17}(O_{2-x}OH_x)]K_4Ca_2[(H_2O)_{2-x}OH_x]Cl$

*Hydrodelhayelite $\quad$ $[AlSi_7O_{17}(OH)_2]KCa_2\cdot6H_2O$

Macdonaldite $\quad$ $[Si_8O_{18}(OH)]^{\infty 2}BaCa_4\cdot10H_2O$

Mountainite $\quad$ $[Si_8O_{19}(OH)]^{\infty 2}KNa_2Ca_2\cdot6H_2O$

Rhodesite $\quad$ $[Si_8O_{19}]^{\infty 2}KHCa_2\cdot5H_2O$

*Lalondeite $\quad$ $[Si_8O_{19}]_2(Na,Ca)_6(Ca,Na)_3(F,OH)_2\cdot3H_2O$

*Zero-monoalumosilicates and zero-monosilicates with $K_\Sigma = 0,8$ O Hydrates

*Sarcolite $\quad$ $Na_4Ca_{12}[Al_8Si_{12}O_{46}][SiO_4,PO_4](OH,H_2O)_4(CO_3,Cl)$

Zeophyllite $\quad$ $Ca_{13}(OH)_2F_8[Si_{10}O_{28}]\cdot6H_2O$

*Zakharovite $\quad$ $Na_4Mn^{2+}{}_5[Si_{10}O_{24}](OH)_6\cdot6H_2O$

*Akatoreite $\quad$ $Mn^{2+}{}_9[Al_2Si_8O_{24}](OH)_8$

Zero-monoalumosilicates and zero-monosilicates with $K_\Sigma = 0,83$

Hydrates (basic and neutral)

Truscottite family

Truscottite $\quad$ $(Ca,Mn)_{14}(OH)_8[Si_{24}O_{58}]\,2H_2O$

Reyerite $\quad$ $(Na,K)_2Ca_{14}[Al_2Si_{22}O_{58}](OH)_8\,6H_2O$

*Eggletonite $\quad$ $(Na,K,Ca)_2(Mn,Fe)_8(OH)_7[(Si,Al)_{12}O_{29}]\cdot11H_2O$

*Zero-monoalumosilicates and zero-monosilicates with $K_\Sigma = 0,88$

*Armbrusterite $\quad$ $K_5Na_6Mn^{3+}Mn^{2+}{}_{14}[Si_9O_{22}]_4(OH)_{10}\cdot4H_2O$

Monoalumosilicates and monosilicates $(K = 1)$

Monoalumosilicates $\quad$ Neutral

Sillimanite family

Sillimanite $\quad$ $Al[AlSiO_5]^\infty$

Mullite $\quad$ $Al_2[(Al_{2+2x}Si_{2-2x})O_{10-x}]^\infty$

$\quad$ Basic alumosilicates hydrates

*Allofan $\quad$ $Al_2SiO_5\cdot H_2O$

Prehnite $\quad$ $Ca_2Al(OH)_2[AlSi_3O_{10}]^{\infty 2}$

Mica family

Fragile mica subfamily – $^{(4)}$(Al,Fe^{3+}) : Si up 3 : 1 to 1 : 1

Clintonite	Ca$\{$(Mg$_2$Al)(OH)$_2$[Al$_3$SiO$_{10}$]$^{\infty 2}\}^{\infty 2}$
Margarite	Ca$\{$Al$_2\square$(OH)$_2$[Al$_2$Si$_2$O$_{10}$]$^{\infty 2}\}^{\infty 2}$
Chernykhite	BaV$^{3+}_2\square$(OH)$_2$[Al$_2$Si$_2$O$_{10}$]$^{\infty 2}\}^{\infty 2}$

Siderophyllite subfamily

Preiswerkite	Na$\{$Mg$_2$Al(OH)$_2$[Al$_2$Si$_2$O$_{10}$]$^{\infty 2}\}^{\infty 2}$
Siderophyllite	K$\{$Fe$^{2+}_2$Al(OH)$_2$[Al$_2$Si$_2$O$_{10}$]$^{\infty 2}\}^{\infty 2}$

Anandite subfamily

Kinoshitalite	(Ba,K)$\{$(Mg,Mn)$_3$(OH)$_2$[Al$_2$Si$_2$O$_{10}$]$^{\infty 2}\}^{\infty 2}$
*Ferrokinoshitalite	BaFe$^{2+}_3$(OH)$_2$[Al$_2$Si$_2$O$_{10}$]$^{\infty 2}\}^{\infty 2}$
Anandite	BaFe$^{2+}_3$(OH)[(Si$_3$Fe^{3+})O$_{10}$]S
*Oxykinoshitalite	BaMg$_2$Ti^{4+}O$_2$[(Si$_2$Al$_2$)O$_{10}$]

Usual mica subfamily – $^{(4)}$(Al,Fe)$^{3+}$: Si ~ 1 : 3

Paragonite	Na$\{$Al$_2\square$(OH)$_2$[AlSi$_3$O$_{10}$]$^{\infty 2}\}^{\infty 2}$
*Na-Sr mica	Na$_{0,50}$Sr$_{0,25}$Al$_2$(Na$_{0,25}\square_{0,75}$)(OH)$_2$[Al$_{1,25}$Si$_{2,75}$O$_{10}$]

Muscovite series

Muscovite	K$\{$Al$_2\square$(OH)$_2$[AlSi$_3$O$_{10}$]$^{\infty 2}\}^{\infty 2}$
*Ganterite	Ba$_{0,5}$(Na,K)$_{0,5}\{$Al$_2$(OH)$_2$[Al$_{1,5}$Si$_{2,5}$O$_{10}$]$^{\infty 2}\}^{\infty 2}$
*Chromphyllite	K$\{$Cr$_2\square$(OH,F)$_2$[AlSi$_3$O$_{10}$]$^{\infty 2}\}^{\infty 2}$
Roscoelite	K$\{$V$_2$(OH)$_2$[AlSi$_3$O$_{10}$]$^{\infty 2}\}^{\infty 2}$
*Nanpingite	Cs$\{$Al$_2$(OH,F)$_2$[AlSi$_3$O$_{10}$]$^{\infty 2}\}^{\infty 2}$
Tobelite	NH$_4\{$Al$_2$(OH)$_2$[AlSi$_3$O$_{10}$]$^{\infty 2}\}^{\infty 2}$

Phlogopite subfamily

Wonesite	(Na,K,$\square$)(Mg,Fe,Al)$_6$(OH,F)$_4$[(Si,Al)$_8$O$_{20}$]$^{\infty 2}\}^{\infty 2}$
Sodium phlogopite	Na$\{$Mg$_3$(OH)$_2$[AlSi$_3$O$_{10}$]$^{\infty 2}\}^{\infty 2}$
Hydroxyl-phlogopite	K$\{$Mg$_3$(OH)$_2$[AlSi$_3$O$_{10}$]$^{\infty 2}\}^{\infty 2}$
Tetraferriphlogopite	K$\{$Mg$_3$(OH)$_2$[(Fe^{3+}Si$_3$O$_{10}$]$^{\infty 2}\}^{\infty 2}$
Fluorophlogopite	K$\{$Mg$_3$(F,OH)$_2$[AlSi$_3$O$_{10}$]$^{\infty 2}\}^{\infty 2}$
*Oxyphlogopite	K(Mg,Ti,Fe)$_3$(O,F)$_2$[(Si,Al)$_4$O$_{10}$]
*Aspidolite	NaMg$_3$(OH)$_2$[AlSi$_3$O$_{10}$]
Manganophyllite	K(Mn,Mg,Al)$_{2-3}$(OH)$_2$[(Al,Si)$_4$O$_{10}$]
*Schirozulite	KMn$^{2+}_3$(OH)$_2$[AlSi$_3$O$_{10}$]
Annite	KFe$^{2+}_3$(OH)$_2$[(Fe^{3+}Si$_3$O$_{10}$]
Tetraferriannite	K$\{$(Fe^{2+},Mg)$_3$(OH)$_2$[(Fe^{3+},Al)Si$_3$O$_{10}$]$^{\infty 2}\}^{\infty 2}$
Fluorannite	KFe$^{2+}_3$F$_2$[AlSi$_3$O$_{10}$]
*Montdorite	KFe$^{2+}_{1,5}$Mn$^{2+}_{0,5}$Mg$_{0,5}\square_{0,5}$F$_2$[Si$_4$O$_{10}$]
*Yangzhumingite	KMg$_{2,5}$F$_2$[Si$_4$O$_{10}$]$^\infty$

Biotite – micas between, or close to, the annite-phlogopite and siderophellite - eastonite joins; dark micas without lithium

Chlorite family

Diseptochlorite subfamily

Sudoite	Mg$_2$Al$_3$(OH)$_8$[AlSi$_3$O$_{10}$]
*Glagolevite	NaMg$_6$(OH,O)$_8$[AlSi$_3$O$_{10}$]·H$_2$O

Clinochlore series

Corundophilite = Fe-clinochlor	(Mg,Fe)$_3\{$(Mg,Al)$_3$(OH)$_8$[AlSi$_3$O$_{10}$]
Leuchtenbergite = clinochlor	(Mg,Al)$_6$(OH)$_8$[AlSi$_3$O$_{10}$]
Clinochlore	(Mg$_5$Al(OH)$_8$[AlSi$_3$O$_{10}$]
Ripidolite	(Mg,Fe,Al)$_6$(OH)$_8$[AlSi$_3$O$_{10}$]
Prochlorite	Mg$_5$Al(OH)$_8$[AlSi$_3$O$_{10}$]

Chamosite	$(Fe,Al,Mg)_6(OH)_8[AlSi_3O_{10}]$
Orthochamosite	$(Fe^{2+},Mg,Fe^{3+})_5Al(OH,O)_8[AlSi_3O_{10}]$
Gonyerite	$Mn^{2+}_5Fe^{3+}(OH)_8[Fe^{3+}Si_3O_{10}]$
Pennantite	$Mn^{2+}_5Al(OH)_8[AlSi_3O_{10}]$
Nimite	$(Ni,Mg,Al)_6(OH)_8[AlSi_3O_{10}]$

Septechlorites subfamily

Odinite (mon.)	$(Fe^{3+},Mg,Al,Fe^{2+})_{2.5}(OH)_4[(Si,Al)_2O_5]$

 Amesite series

Amesite	$Mg_2Al(OH)_4[AlSiO_5]$
Brindleyite (nimesite)	$(Ni,Al)_3(OH)_4[(Si,Al)_2O_5]$
Fraipontite	$(Zn,Al)_3(OH)_4[(Si,Al)_2O_5]$

 Cronstedtite series

Berthierine	$(Fe^{2+},Fe^{3+},Al)_3(OH)_4[(Si,Al)_2O_5]$
Cronstedtite	$(Fe^{2+}Fe^{3+})_3(OH)_4[(Si,Fe^{3+})_2O_5]$
*Guidottiite	$(Mn_2Fe^{3+})(OH)_4[(Si,Fe^{3+})O_5]$

Hydromica family

Vermiculite subfamily

Vermiculite	$Mg_{0.7}(Mg,Fe^{3+},Al)_6(OH)_4[(Si,Al)_8O_{20}]\cdot8H_2O$
Brammallite	$(Na,H_3O)(Al,Mg,Fe)_2(OH)_2[Si,Al]_4O_{10}]$
Illite	$K_{0.65}Al_{2.0}(OH)_2[Al_{0.65}Si_{3.35}O_{10}]$
Hydrobiotite	$K(Mg,Fe^{2+})_6(OH)_4[(Si,Al)_8O_{20}]\cdot nH_2O$
*Rudenkoite	$Sr_3(OH,O)_8Cl_2[(Si_{3.5}Al_{3.5})O_{10}]\cdot H_2O$

 Celadonite series

Celadonite	$KFe^{3+}(Mg,Fe^{2+})[\square Si_4O_{10}](OH)_2$
*Alumoceladonite	$KAl(Mg,Fe^{2+})[\square Si_4O_{10}](OH)_2$
*Ferroalumoceladonite	$KFe^{2+}Al[\square Si_4O_{10}](OH)_2$
*Ferroceladonite	$KFe^{3+}(Fe^{2+},Mg)[\square Si_4O_{10}](OH)_2$
*Chromceladonite	$KCr^{3+}(Mg,Fe^{2+})[\square Si_4O_{10}](OH)_2$
*Manganiceladonite	$KMgMn^{3+}[Si_4O_{10}](OH)_2$
Glauconite	$(K,Na)(Mg,Fe^{2+},Fe^{3+})(Fe^{3+},Al)[Si,Al]_4O_{10}](OH)_2$

*Corrensite orth., regular interstratification of trioctahedral **chlorite** with either trioctahedral **vermiculite** or trioctahedral **smectite**.

Smectite family

Montmorillonite subfamily

Swinefordite	$Ca_{0.2}(Li,Al,Mg,Fe)_3(OH,F)_2[(Si,Al)_2O_5]_2\cdot nH_2O_2$
Montmorillonite	$(Na,Ca)_{0.3}(Al,Mg)_2(OH)_2[Si_2O_5]_2\cdot nH_2O$
*Montmorillonite-Fe	$(Na,Ca)_{0.3}(Fe,Mg)_2(OH)_2[Si_2O_5]_2\cdot nH_2O$
*Brinrobertsite	$(Na,K,Ca)_{0.3}(Al,Fe,Mg)_4(OH)_4[(Si,Al)_8O_{20}]\cdot3,5H_2O$
Volkonskoite	$(Na,K,Ca)_{0.3}(Al,Mg,Fe)_4(OH)_4[(Si,Al)_8O_{20}]\cdot3,5H_2O$
Beidellite	$(Na,Ca)_{0.3}Al_2(OH)_2[(Si,Al)_4O_{10}]\cdot nH_2O$
Nontronite	$Na_{0.3}Fe_2^{3+}(OH)_2[(Si,Al)_4O_{10}]\cdot nH_2O$
*Rectorite	$(Na,Ca)Al_4(OH)_4[(Si,Al)_8O_{20}]\cdot2H_2O$
*Yakhontovite	$CaCu^{2+}_2(OH)_2[Si_4O_{10}]\cdot3H_2O$

Saponite subfamily

Sobotkite	$(K,Ca_{0.5})_{0.33}(Mg_{0.66}Al_{0.33})_3(OH)_2[(Si_3Al)O_{10}]\cdot1\text{-}5H_2O$
Saponite	$(Ca,Na)_{0.3}(Mg,Fe^{2+})_3(OH)_2[(Si,Al)_4O_{10}]\cdot4H_2O$
Ferrisaponite	$Ca_{0.3}(Fe^{3+}Mg,Fe^{2+})_3(OH)_2[(Si,Al)_4O_{10}]\cdot4H_2O$
*Ferrosaponite	$Ca_{0.3}(Fe^{2+},Mg,Fe^{3+})_3(OH)_2[(Si,Al)_4O_{10}]\cdot4H_2O$

Palygorskite-sepiolite family (alumosilicates → silicates)

 Palygorskite subfamily

Palygorskite $(Mg,Al)_2(OH)[Si_4O_{10}]\cdot4H_2O$
Yofortierite $(Mn^{2+},Mg,Fe^{3+})_5(OH,H_2O)_2[Si_8O_{20}]\cdot7H_2O$
Tuperssuatsiaite $Na_2(Fe^{3+},Mn^{2+})_3(OH)_2[Si_8O_{20}]\cdot4H_2O$
*Raite $Na_3Mn^{2+}_3Ti^{4+}_{0.25}(OH)_2[Si_8O_{20}]\cdot10H_2O$
*Windhoekite $Ca_2Fe^{3+}_{2.67}(OH)_4[(Si,Al)_8O_{20}]\cdot10H_2O$
Sepiolite subfamily
Sepiolite $Mg_4(OH)_2[Si_6O_{15}]\cdot6H_2O$
*Ferrisepiolite $(Fe^{3+},Fe^{2+},Mg)_4(O,OH)_2[(Si,Fe^{3+})_6O_{15}]\cdot6H_2O$
Falcondoite $Ni_4(OH)_2[Si_6O_{15}]\cdot6H_2O$
Loughlinite $Na_2Mg_3[Si_6O_{16}]\cdot8H_2O$
Osumilite family
Armenite $BaCa_2Al_3[Al_3Si_9O_{30}]\cdot2H_2O$
 Osumilite series
 Yagiite $(Na,K)_3Mg_4Al_6[(Si,Al)_{12}O_{30}]_2$
 Osumilite-(Mg) $(K,Na)(Mg,Fe^{2+})_2(Al,Fe^{3+})_3[(Si,Al)_{12}O_{30}]$
 Osumilite $(K,Na)(Fe,Mg,Mn)^{2+}_2(Al,Fe^{3+})_3[(Si,Al)_{12}O_{30}]$
 *Trattnerite $(Mg,Fe^{2+})_3Fe^{3+}_2[Si_{12}O_{30}]$
 Chayesite $K(Mg,Fe^{2+})_2(Mg,Fe^{2+})_2Fe^{3+}[Si_{12}O_{30}]$
 *Unnamed $Fe^{2+}_5Mg^{2+}_5[(Al,Si)_{12.5}O_{30}]_2$

Monosilicates
Proper monosilicates   Neutral
Gillespite family
Gillespite $BaFe^{2+}[Si_4O_{10}]^{\infty2}$
Sanbornite $Ba[Si_2O_5]^{\infty2}$
*Bigcreekite $Ba[Si_2O_5]\cdot4H_2O$
Natrosilite $Na_2[Si_2O_5]^{\infty2}$
Fenaksite $K_2Na_2Fe^{2+}_2[Si_8O_{20}]^{\infty}$
*Manaksite $NaKMn[Si_4O_{10}]$
*Tuhualite $(Na,K)Fe^{2+}Fe^{3+}[Si_6O_{15}]$
*Kalifersite $(K,Na)_5Fe^{3+}_7[Si_{20}O_{50}](OH)_6\cdot12H_2O$
 Roedderite series
 Roedderite $KNa(Mg,Fe)_5[Si_{12}O_{30}]^{\infty}$
 Merrihueite $KNa(Fe,Mg,)_5[Si_{12}O_{30}]^{\infty}$
 Eifelite $KNa_3Mg_4[Si_{12}O_{30}]^{\infty}$
 *Shibkovite $K(Ca,Mn,Na)_2(K_{2-x}\square_x)Zn_3[Si_{12}O_{30}]$
*Shirokshinite $KNaMg_2[Si_4O_{10}]F_2$
Agrellite $NaCa_2F[Si_4O_{10}]^{\infty}$
*Friedrichbeckeite $K(\square_{0.5}Na_{0.5})_2Mg_2(MgBe_2)[Si_{12}O_{30}]\square$
                          Hydrates (basic)
*Cairncrossite (M^{2+} : OH = 4.5) $Sr_2Ca_{7-x}Na_{2x}(OH)_2[Si_4O_{10}]_4(H_2O)_{15-x}$ ($0 \leq x \leq 1$)
*Calcinaksite $KNaCa[Si_4O_{10}]\cdot H_2O$
Canasite   (M^{2+} : OH = 2) $K_3Na_3Ca_5(O,OH,F)_4[Si_{12}O_{30}]^{\infty}$
*Fluorcanasite $K_3Na_3Ca_5(F,OH)_4[Si_{12}O_{30}]\cdot H_2O$
*Frankamenite $K_3Na_3Ca_5(OH)F_3[Si_{12}O_{30}]\cdot H_2O$
Talc-pyrophyllite family (M^{2+} : OH = 1,5)
*Erlianite $Fe^{2+}_4Fe^{3+}_2(OH)_8[Si_2O_5]_3$
 Pyrophyllite series
 Pyrophyllite $\{Al_2(OH)_2[Si_2O_5]^{\infty2}_2\}^{\infty2}$

Ferripyrophyllite $\{Fe^{3+}_2(OH)_2[Si_2O_5]^{\infty2}_2\}^{\infty2}$
Talc series
Talc $\{Mg_3(OH)_2[Si_2O_5]^{\infty2}_2\}^{\infty2}$
*Stevensite $Mg_3(OH)_2[Si_4O_{10}]$
Minnesotaite $\{(Fe,Mg)_3(OH)_2[Si_2O_5]^{\infty2}_2\}^{\infty2}$
Willemseite = nickel-kerolite $\{(Ni,Mg)_3(OH)_2[Si_2O_5]^{\infty2}_2\}^{\infty2}$
Kerolite series
Kerolite $\{Mg_3(OH)_2[Si_2O_5]^{\infty2}_2\}^{\infty2}H_2O$
*__Tungusite__ family (M^{2+} : OH = 1)
*Tungusite $Ca_{14}Fe^{2+}_9(OH)_{22}[Si_6O_{15}]_4$
Pyrosmalite family (M^{2+} : OH = 0,8)
  Pyrosmalite series
  Pyrosmalite-(Fe) $(Fe^{2+},Mn)_8(OH,Cl)_{10}[Si_6O_{15}]^{\infty2}$
  Pyrosmalite-(Mn) $(Mn^{2+},Fe)_8(OH,Cl)_{10}[Si_6O_{15}]^{\infty2}$
Brokenhillite $(Mn,Fe)_8(OH,Cl)_{10}[Si_6O_{15}]^{\infty2}$
Mcgillite $(Mn,Fe)_8(OH)_8Cl_2[Si_6O_{15}]^{\infty2}$
Friedelite $Mn_8(OH,Cl)_{10}[Si_6O_{15}]^{\infty2}$
Bementite $Mn_7(OH)_8[Si_6O_{15}]^{\infty2}$
*Innsbruckite $Mn_{33}(OH)_{38}[Si_2O_5]_{14}$
Serpentine family (M^{2+} : OH = 0,75)
Antigorite $\{Mg_6(OH)_8[Si_4O_{10}]^{\infty2}\}^{\infty2}$
Caryopilite $\{Mn_6(OH)_8[Si_4O_{10}]^{\infty2}\}^{\infty2}$
Clinochrysotile $\{Mg_3(OH)_4[Si_2O_5]^{\infty2}\}^{\infty2}$
Lizardite $\{Mg_3(OH)_4[Si_2O_5]^{\infty2}\}^{\infty2}$
Orthochrysotile $\{Mg_3(OH)_4[Si_2O_5]^{\infty2}\}^{\infty2}$
Greenalite $\{(Fe^{2+},Fe^{3+})_{2-3}(OH)_4[Si_2O_5]^{\infty2}\}^{\infty2}$
Karpinskite $\{(Mg,Ni)_2(OH)_2[Si_2O_5]^{\infty2}\}^{\infty2}$
*Willemseite $\{(Ni,Mg)_3(OH)_2[Si_2O_5]_2\}$
Nepouite $\{Ni_3(OH)_4[Si_2O_5]^{\infty2}\}^{\infty2}$
Pecoraite $\{Ni_3(OH)_4[Si_2O_5]^{\infty2}\}^{\infty2}$
Kaolinite-halloysite family (M^{2+} : OH = 0,75)
  Kaolinite subfamily
  Kaolinite $\{Al_2(OH)_4[Si_2O_5]^{\infty2}\}^{\infty2}$
  Dickite $\{Al_2(OH)_4[Si_2O_5]^{\infty2}\}^{\infty2}$
  Nacrite $\{Al_2(OH)_4[Si_2O_5]^{\infty2}\}^{\infty2}$
  *Kellyite $(Mn^{2+},Mg,Al)_3(OH)_4[(Si,Al)_2O_5]$
  Halloysite subfamily
  Halloysite-10Å. $\{Al_2(OH)_4[Si_2O_5]^{\infty2}\}^{\infty2}(H_2O)_2$
  *Halloysite-7Å $\{Al_2(OH)_4[Si_2O_5]^{\infty2}\}^{\infty2}$
  Endellite = halloysite-10Å $\{Al_2(OH)_4[Si_2O_5]^{\infty2}\}^{\infty2}(H_2O)_2$
  Hisingerite $\{Fe^{3+}_2(OH)_4[Si_2O_5]^{\infty2}\}^{\infty2}(H_2O)_2$
Grumantite $Na[Si_2O_4(OH)]^{\infty2}H_2O$
Kanemite $Na[Si_2O_4(OH)]\,3H_2O$

                        Basic → silicato-fluorides → Hydrates

Apophyllite family
  Apophyllite series
  Natrofluorapophyllite = apophyllite-(NaF) $NaCa_4F[Si_4O_{10}]^{\infty2}_2\cdot8H_2O$
  Hydroxyapophyllite = apophyllite-(KOH) $KCa_4F[Si_4O_{10}]^{\infty2}_2\cdot8H_2O$

Fluorapophyllite = apophyllite-(KF) $KCa_4(F,OH)[Si_4O_{10}]^{\infty 2}_2 \cdot 8H_2O$
Bannisterite $(Ca,K,Na,)(Mn^{2+},Fe^{2+})_{10}(OH)_8[(Si,Al)_{16}O_{38}]^{\infty 2} \cdot nH_2O$
*Gyrolite $(NaCa_2)Ca_{14}(OH)_8[(Si_{23}Al)O_{60}] \cdot (14+x)H_2O$
*Orlymanite $Ca_4Mn^{2+}_3(OH)_6[Si_8O_{20}] \cdot 2H_2O$
*Cryptophyllite $K_2Ca[Si_4O_{10}] \cdot 5H_2O$
*Shlykovite $KCa[Si_4O_9(OH)] \cdot 3H_2O$
*Aklimaite $Ca_4[Si_2O_5(OH)_2](OH)_4 \cdot 5H_2O$
Suolunite $Ca_2[Si_2O_5(OH)_2] H_2O$
Revdite $Na_{16}[Si_{16}O_{27}(OH)_{26}] \cdot 28H_2O$

Neutral

Nekoite $Ca_3[Si_6O_{15}]^{\infty 2} 7H_2O$

Mono-mono-disilicates with mixed silicooxygens radical
Basic
Charoite $(K,Na)_5(Ca,Ba,Sr)_8(OH,F)[Si_6O_{15}]^{\infty 2}_2 [Si_6O_{16}] (H_2O)_n$
Okenite $\{Ca_8(H_2O)_6[Si_6O_{15}]^{\infty}_2[Si_6O_{16}]^{\infty}\}^{\infty 2} Ca_2(H_2O)_{12}$

Mono-disilicates (including isomorphic alumosilicates)
Mono-disilicates with K = 1,1(6)
Hydrates (basic)
Riversideite family
Tacharanite $Ca_8Al_{1,33}(H_2O)_9(OH)_6[Si_{12}O_{31}]^{\infty 2}$ (?)
Riversideite $Ca_{10}(H_2O)_3(OH)_6[Si_{12}O_{31}]^{\infty 2}$
Plombierite $Ca_{10}(H_2O)_{18}(OH)_6[Si_{12}O_{31}]^{\infty 2}$

Mono-disilicates with K = 1,(3)
Basic
*Denisovite $(K,Na)Ca_2(F,OH)[Si_3O_8]$
*Marshallsussmanite $NaCaMn(OH)[Si_3O_8]$
Jimthompsonite family
Jimthompsonite $(Mg,Fe^{2+})_5(OH)_2[Si_6O_{16}]^{\infty}$
Clinojimthompsonite $(Mg,Fe^{2+})_5(OH)_2[Si_6O_{16}]^{\infty}$
 *Ca-jimthompsonite $Ca_2(Mg,Fe)_8(OH)_4[Si_6O_{16}]_2$
*Hydrates (neutral)
*Shafranovskite $(Na,K)_6(Mn,Fe)^{2+}_3[Si_9O_{24}] \cdot 6H_2O$

Mono-disilicates with K = 1,(3) + 1,5
Basic
Chesterite $(Fe,Mg)_{17}(OH)_6[Si_6O_{16}]^{\infty 2}_2[Si_4O_{11}]^{\infty 2}_2$

Mono-disilicates with K=1,4
Acid
Tokkoite $K_2|Ca_4(F,OH)[Si_7O_{18}(OH)]^{\infty}|^{\infty 2}$
(it as isostructural with tinaksite during isomorphism: $\square 2Ca^{2+}(F,OH)^- \rightarrow Na^+Ti^{4+}O^{2-}$)

Mono-disilicates with K = 1,5
Basic
Amphiboles family
 Mg,Fe^{2+}- amphiboles subfamily

Gedrite series
Magnesiogedrite = gedrite
Gedrite $\square Mg_2(Mg_3Al_2)(OH)_2[(Al_2Si_6)O_{22}]^{\infty}$
Ferrogedrite $(Fe^{2+},Mg)_5Al_2(OH)_2[AlSi_3O_{11}]_2$
Sodicgedrite $Na(Mg,Fe^{2+})_6Al(OH)_2[AlSi_3O_{11}]^{\infty}_2$
*Sodic-ferrogedrite $NaFe_2(Fe_4Al)(OH)_2[AlSi_3O_{11}]^{\infty}_2$
Anthophyllite series
Magnesio-anthophyllite = anthophyllite
Anthophyllite $\square Mg_2Mg_5(OH)_2[Si_4O_{11}]^{\infty}_2$
*Proto-anthophyllite $(Mg,Fe^{2+})_7(OH)_2[Si_4O_{11}]^{\infty}_2$
Ferro-anthophyllite $Fe^{2+}_2Fe^{2+}_5(OH)_2[Si_4O_{11}]^{\infty}_2$
Proto-ferro-anthophyllite $\square\{Fe_2^{2+}\}\{Fe_5^{2+}\}(OH)_2[Si_4O_{11}]_2$
*Protomangano-ferro-anthophyllite $\square Mn^{2+}_2Fe^{2+}_5(OH)_2[Si_4O_{11}]^{\infty}_2$
*Sodicanthophyllite $NaMg_7(OH)_2[Al_{0,5}Si_{3,5}O_{11}]^{\infty}_2$
Protomangan-anthophyllite $(Fe,Mn)_7(OH)_2[Si_4O_{11}]^{\infty}_2$
Cummingtonite series
Magnesio-cummingtonite = cummingtonite
Cummingtonite $\square Mg_2Mg_5(OH)_2[Si_4O_{11}]^{\infty}_2$
Grunerite $\square Fe^{2+}_2Fe^{2+}_7(OH)_2[Si_4O_{11}]^{\infty}_2$
*Manganogrunerite $(Mn,Fe^{2+})_7(OH)_2[Si_4O_{11}]^{\infty}_2$
*Permanganogrunerite $Mn^{2+}_4Fe^{2+}_3(OH)_2[Si_4O_{11}]^{\infty}_2$
*Manganocummingtonite = tirodite $\square Mn_2Mg_5(OH)_2[Si_4O_{11}]^{\infty}_2$

Mn-amphiboles subfamily
                                    Basic
Dannemorite = manganogrunerite

***Mg,Fe,Ca-amphiboles** subfamily
Ca-amphiboles subfamily
 Hornblende series
 Tschermakite subseries
 Tschermakite $Ca_2Mg_3Al_2(OH)_2[AlSi_3O_{11}]^{\infty}_2$
 Aluminotschermakite $Ca_2Mg_3Al_2(OH)_2[AlSi_3O_{11}]^{\infty}_2$
 Ferro-tschermakite $\square Ca_2(Fe_3^{2+}Al_2\}(OH)_2[AlSi_3O_{11}]^{\infty}_2$
 *Alumino-ferrotschermakite $\square Ca_2(Fe^{2+}_3Al_2)(OH)_2[AlSi_3O_{11}]^{\infty}_2$
 Ferro-ferri-tschermakite $\square Ca_2(Fe^{2+}_3Fe^{3+}_2)(OH)_2[AlSi_3O_{11}]^{\infty}_2$
 Ferritschermakite $\square Ca_2Mg_3Fe^{3+}_2(OH)_2[AlSi_3O_{11}]^{\infty}_2$
 *Ferri-ferro-tschermakite $\square Ca_2(Fe_3^{2+}Fe_2^{3+})(OH)_2[(AlSi_3O_{11}]^{\infty}_2$
 Hornblende subseries
 Magnesio-hornblende $\square Ca_2Mg_4Al(OH)_2[Al_{0,5}Si_{3,5}O_{11}]^{\infty}_2$
 *Magnesio-ferri-fluoro-hornblende $\square Ca_2(Mg_4Fe^{3+})F_2[Al_{0,5}Si_{3,5}O_{11}]_2$
 *Ferrimagnesiohornblendite $Ca_2[Mg_4(Fe,Al)](OH)_2[Al_{0,5}Si_{3,5}O_{11}]^{\infty}_2$
 Ferro-hornblende $\square Ca_2Fe^{2+}_4Al(OH,F)_2[Al_{0,5}Si_{3,5}O_{11}]^{\infty}_2$
 *Ferro-ferri-hornblende $\square Ca_2(Fe^{2+}_4Fe^{3+})(OH)_2[Al_{0,5}Si_{3,5}O_{11}]^{\infty}_2$
 Tremolite subseries (Na+K) в A < 0,5; Ti < 0,5
 Tremolite $\square Ca_2Mg_5(OH)_2[Si_4O_{11}]^{\infty}_2$
 Actinolite $\square Ca_2Mg_{<4,5}Fe^{2+}_{>0,5}(OH)_2[Si_4O_{11}]^{\infty}_2$
 Ferro-actinolite $\square Ca_2Mg_{<2,5}Fe^{2+}_{>2,5}(OH)_2[Si_4O_{11}]^{\infty}_2$

*Parvo-manganotremolite $\square CaMnMg_5(OH)_2[Si_4O_{11}]^\infty{}_2$

Ti-amphiboles subfamily (Na : Ca = 1 : 2)
*Obertiite $NaNa_2(Mg_3Fe^{3+}Ti)O_2[Si_4O_{11}]$
*Dellaventuraite $NaNa_2(MgMn^{3+}{}_2Ti^{4+}Li)O_2[Si_4O_{11}]_2$
*Ferroobertiite $NaNa_2(Fe^{2+}{}_3Fe^{3+}Ti^{4+})O_2[Si_4O_{11}]_2$
Kaersutite series
Kaersutite $NaCa_2(Mg_3Ti^{4+}Al)O_2[AlSi_3O_{11}]^\infty{}_2$
Ferro-kaersutite $NaCa_2(Fe^{2+}{}_3Ti^{4+}Al)O_2[AlSi_3O_{11}]^\infty{}_2$
Magnesiosadanagaite $NaCa_2(Mg_3Al_2)(Al_3Si_5O_{22})(OH)_2[(Al_{1.5}Si_{2.5}O_{11}]^\infty{}_2$
Sadanagaite $NaCa_2(Mg_3Al_2)(OH)_2[(Al_{1.5}Si_{2.5}O_{11}]^\infty{}_2$
*Potassic-aluminosadanagaite $KCa_2[Fe^{2+}{}_3(Al,Fe^{3+})_2](OH)_2[Al_3Si_5O_{22}]$
*Potassic-sadanagaite $KCa_2(Mg_3Al_2)(OH)_2[Al_3Si_5O_{22}]$
*Potassic-magnesio-sadanagaite $KCa_2(Mg_3Fe^{3+}{}_2)(OH)_2[Al_3Si_5O_{22}]$
*Potassic-ferri-sadanagaite $KCa_2(Mg^{2+}{}_3Fe^{3+}{}_2)(OH)_2[Al_3Si_5O_{22}]$
*Cannilloite $CaCa_2(Mg_4Al)(OH,F)_2[Al_3Si_5O_{22}]$
*Fluorocannilloite $CaCa_2(Mg_4Al)F_2[Al_3Si_5O_{22}]$

Pargasite subfamily (Na : Ca = 1 : 2)
*Fluoro-potassic-pargasite $KCa_2(Mg_4Al)Al_2F_2[AlSi_3O_{11}]_2$
Pargasite series
Pargasite $NaCa_2Mg_4Al(OH)_2[AlSi_3O_{11}]^\infty{}_2$
Potassic-pargasite $KCa_2Mg_5(OH,F)_2[(Si,Al)_4O_{11}]^\infty{}_2$
*Potassic-fluoro -pargasite $KCa_2(Mg_4Al)Al_2F_2[AlSi_3O_{11}]_2$
*Chloro-potassic-pargasite $(K,Na)Ca_2(Fe^{2+},Mg)_4Fe^{3+}Cl_2[AlSi_3O_{11}]_2$
Ferrohydroxylpargasite $NaCa_2(Fe^{2+},Mg)_4Al(OH,F)_2[AlSi_3O_{11}]^\infty{}_2$
*Ferro-pargasite $NaCa_2(Fe^{2+},Mg)_4Al(OH)_2[Si_6Al_2O_{22}]$
*Chromio-pargasite $NaCa_2Mg_4Cr^{3+}(OH)_2[Si_6Al_2O_{22}]$
Fluorpargasite $NaCa_2(Mg,Fe^{2+})_4AlF_2[AlSi_3O_{11}]^\infty{}_2$
Hastingsite series
Magnesio-hastingsite $NaCa_2(Mg,Fe^{2+})_4Fe^{3+}(OH)_2[AlSi_3O_{11}]^\infty{}_2$
*Fluoro-magnesiohastingsite $NaCa_2(Mg,Fe^{2+})_4Fe^{3+}F_2[AlSi_3O_{11}]^\infty{}_2$
Hastingsite $NaCa_2(Fe^{2+},Mg)_4Fe^{3+}(OH)_2[AlSi_3O_{11}]^\infty{}_2$
*Fluoro-potassichastingsite $(K,Na)Ca_2(Fe^{2+}{}_4 Fe^{3+})F_2[Si_6Al_2O_{22}]$
*Chloro-potassic-hastingsite $(K,Na)Ca_2(Fe^{2+}{}_4 Fe^{3+}) Cl_2[Si_6Al_2O_{22}]$
*Potassic-magnesio-hastingsite $KCa_2(Mg_4Fe^{3+})(OH)_2[Si_6Al_2O_{22}]$
*Potassic-hastingsite $KCa_2(Fe^{2+}{}_4Fe^{3+})(OH)_2[Si_6Al_2O_{22}]$
*Oxo-magnesio-hastingsite $NaCa_2(Mg_2Fe^{3+}{}_3)O_2[Si_6Al_2O_{22}]$
Edenite series
Edenite $NaCa_2Mg_5(OH)_2[Al_{0.5}Si_{3.5}O_{11}]^\infty{}_2$
*Fluoro-edenite $NaCa_2Mg_5F_2[Al_{0.5}Si_{3.5}O_{11}]^\infty{}_2$
Ferro-edenite $NaCa_2Fe^2{}_5(OH)_2[Al_{0.5}Si_{3.5}O_{11}]^\infty{}_2$
*Parvo-manganoedenite $NaCaMnMg_5(OH)_2[Al_{0.5}Si_{3.5}O_{11}]_2$
Winchite subfamily (Na : Ca = 1 : 1)
Winchite series
Winchite $\square NaCa(Mg_4Al)(OH)_2[Si_4O_{11}]^\infty{}_2$
*Parvowinchite $(NaMn^{2+})(Mg_4Al)(OH)_2[Si_4O_{11}]_2$
Ferro-winchite $\square NaCaFe^{2+}{}_4Al(OH)_2[Si_4O_{11}]^\infty{}_2$
*Ferri-winchite $\square (NaCa)Mg_4Fe^{3+}(OH)_2[Si_4O_{11}]^\infty{}_2$

Barroisite $\square NaCa(Mg,Fe^{2+})_3Al_2(OH)_2[Al_{0,5}Si_{3,5}O_{11}]^{\infty}_2$

Ferro-barroisite $\square(NaCa)Fe^{2+}_3Al_2(OH)_2[Al_{0,5}Si_{3,5}O_{11}]^{\infty}_2$

Ferri-barroisite $\square(NaCa)(Mg_3Fe^{3+}_2)(OH)_2[Al_{0,5}Si_{3,5}O_{11}]^{\infty}_2$

Ferro-ferri-barroisite $\square(NaCa)(Fe^{2+}_3Fe^{3+}_2)(OH)_2[Al_{0,5}Si_{3,5}O_{11}]^{\infty}_2$

Glaucophane subfamily (Na_2)

Glaucophane series

Glaucophane $\square Na_2(Mg,Fe^{2+})_3Al_2(OH)_2[Si_4O_{11}]^{\infty}_2$

Ferro-glaucophane $\square Na_2(Fe^{2+},Mg)_3Al_2(OH)_2[Si_4O_{11}]^{\infty}_2$

Crossite series

Crossite $Na_2(Mg,Fe^{2+})_3(Al,Fe^{3+})_2(OH)_2[Si_4O_{11}]^{\infty}_2$

Ribeckite series

Hydroxylmagnesioribeckite=magnesioriebeckite

$Na_2(Mg,Fe^{2+})_3Fe^{3+}_2(OH)_2[Si_4O_{11}]^{\infty}_2$

Hydroxylriebeckite = riebeckite $Na_2(Fe^{2+}_3Fe^{3+}_2)(OH,F)_2[Si_4O_{11}]^{\infty}_2$

Fluororibeckite $\square Na_2(Fe^{2+}_3Fe^{3+}_2)F_2[Si_4O_{11}]^{\infty}_2$

*Magnesioribeckite $\square Na_2(Mg,Fe^{2+})_3Fe^{3+}_2(OH)_2[Si_4O_{11}]^{\infty}_2$

Taramite series

*Aluminomagnesiotaramite $Na(Ca,Na)(Mg_3Al_2)(OH)_2[AlSi_3O_{11}]^{\infty}_2$

*Fluoro-alumino-magnesiotaramite $Na(Ca,Na)(Mg_3Al_2)F_2[AlSi_3O_{11}]^{\infty}_2$

Magnesiotaramite $Na_2Ca(Mg_3\ AlFe^{3+}(OH)_2[AlSi_3O_{11}]^{\infty}_2$

Taramite $Na(NaCa)\ (Mg_3Al_2)(OH)_2[AlSi_3O_{11}]^{\infty}_2$

*Aluminotaramite $Na(Ca,Na)(Fe^{2+}_3Al_2)(OH)_2[AlSi_3O_{11}]_2$

*Ferrimagnesiotaramite $Na(Ca,Na)(Mg_3Fe^{3+}_2)(OH)_2[AlSi_3O_{11}]^{\infty}_2$

Ferri-taramite $Na(CaNa)(Mg_3Fe^{3+}_2)(OH)_2[AlSi_3O_{11}]_2$

*Ferro-ferri-taramite $Na(CaNa)(Fe_3^{2+}Fe_2^{3+})(OH)_2[AlSi_3O_{11}]_2$

*Chloro-potassic-ferri-magnesiotaramite $K(Ca,Na)(Mg_3Fe^{3+}_2)Cl_2[AlSi_3O_{11}]^{\infty}_2$

Richterite series

Richterite $Na_2Ca(Mg,Fe^{2+})_5(OH)_2[Si_4O_{11}]^{\infty}_2$

*Potassic-richterite $K(Na,Ca)Mg_5(OH)_2[Si_4O_{11}]^{\infty}_2$

Fluoro-richterite $Na(NaCa)Mg_5F_2[Si_4O_{11}]^{\infty}_2$

Fluoro-potassicrichterite $KNaCaMg_5F_2[Si_4O_{11}]^{\infty}_2$

Ferror-richterite $Na(NaCa)Fe^{2+}_5(OH)_2[Si_4O_{11}]^{\infty}_2$

*Richterite-MgSrK $(K,Na)(Ca,Sr,Mg,Na)Na(Mg,Na)_5(OH)_2[Si_4O_{11}]_2$

Katophorite series

*Magnesiokatophorite $Na(CaNa)(Mg_4Al)(OH)_2[Al_{0,5}Si_{3,5}O_{11}]^{\infty}_2$

*Ferrikatophorite $Na(NaCa)(Fe^{2+}_4Fe^{3+})(OH)_2[Al_{0,5}Si_{3,5}O_{11}]^{\infty}_2$

Aluminokatophorite $*Na_2Ca(Fe^{2+}_4Al)(OH)_2[Al_{0,5}Si_{3,5}O_{11}]^{\infty}_2$

Katophorite $*Na(NaCa)(Mg^{2+}_4Al)(OH)_2[Al_{0,5}Si_{3,5}O_{11}]^{\infty}_2$

Arfvedsonite subfamily

Nybøite $NaNa_2Mg_3Al_2(OH)_2[Al_{0,5}Si_{3,5}O_{11}]^{\infty}_2$

*Ferri-nybøite $NaNa_2(Mg_3Fe^{3+}_2)(OH)_2[Al_{0,5}Si_{3,5}O_{11}]_2$

*Ferro-ferri-nybøite $NaNa_2(Fe^{2+}_3Fe^{3+}_2)(OH)_2[Al_{0,5}Si_{3,5}O_{11}]_2$

*Ferro-nybøite $NaNa_2(Fe^{2+}_3Al_2)(OH)_2[Al_{0,5}Si_{3,5}O_{11}]_2$

*Fluoronybøite $NaNa_2(Mg_3Al_2)F_2[Al_{0,5}Si_{3,5}O_{11}]_2$

Eckermannite series

Eckermannite $*NaNa_2(Mg_4Al(OH)_2[Si_4O_{11}]^{\infty}_2$

*Ferroeckermannite $NaNa_2(Fe^{2+}_4Al)(OH)_2[Si_4O_{11}]^{\infty}_2$

Fluoreckermannite $Na_3(Mg,Fe^{2+})_4(Al,Fe^{3+})(F,OH)_2[Si_4O_{11}]^{\infty}_2$
Arfvedsonite series
*Arfvedsonite $NaNa_2(Fe^{2+}_4Fe^{3+})(OH)_2[Si_4O_{11}]^{\infty}_2$
Magnesioarfvedsonite $NaNa_2(Mg_4Fe^{3+})(OH)_2[Si_4O_{11}]^{\infty}_2$
*Potassic-arfvedsonite $KNa_2(Fe^{2+}_4Fe^{3+})(OH)_2[Si_4O_{11}]^{\infty}_2$
*Potassic-magnesio-arfvedsonite $KNa_2(Mg,Fe^{2+})_4Fe^{3+}(OH,F)_2[Si_4O_{11}]^{\infty}_2$
*Fluoro-magnesio-arfvedsonite $NaNa_2(Mg_4Fe^{3+})F_2[Si_4O_{11}]^{\infty}_2$
Hydroxylarfvedsonite = arfvedsonite $Na_3(Fe^{2+}, Fe^{3+})_4(OH)_2[Si_4O_{11}]^{\infty}_2$
Fluorarfvedsonite $Na_3(Fe^{2+},Mg)_4Fe^{3+}F_2[Si_4O_{11}]^{\infty}_2$
Kalifluorarfvedsonite $KNa_2(Fe^{2+}_4Fe^{3+})_2F_2[Si_4O_{11}]^{\infty}_2$
Kozulite = mangano-ferri-eckermanite $NaNa_2(Mn^{2+}_4Fe^{3+})(OH_2[Si_4O_{11}]^{\infty}_2$

Mono-disilicato-oxides with K = 1,5   Neutral
*Mangano-mangani-ungarettiite $NaNa_2(Mn^{2+}_2Mn^{3+}_3)O_2[Si_4O_{11}]^{\infty}_2$

Mono-disilicates with K = 1,(6)   Neutral
Pellyite $Ba_4Ca_2Fe_4[Si_{12}O_{34}]^{\infty}$
                                          Basic and hydrates

Xonotlite family
Xonotlite $Ca_6(OH)_2[Si_6O_{17}]^{\infty}$
Inesite $Ca_2Mn_7(OH)_2(H_2O)_5[Si_{10}O_{28}]^{\infty}$

Mono-disilicato-oxides with K = 1,6

                                          Basic
Hillebrandite $Ca_2(OH)_2[SiO_3]^{\infty}$
Deerite $Fe^{2+}_6Fe^{3+}_3(OH)_5[Si_6O_{20}]^{\infty}$
  Howieite series
  Taneyamalite $Na(Mn^{2+}, Mg, Fe^{3+},Al)_{12}[Si_6O_{17}]_2(O,OH)_{10}$
  Howieite $Na(Fe^{2+},Fe^{3+},Mn,Al,Mg)_{12}(O,OH)_{10}[Si_6O_{17}]^{\infty}_2$

*Mono-disilicato carbonato-chlorides with K = 1,75
                               *Hydrates
*Fencooperite $Ba_6Fe^{3+}_3[Si_8O_{23}][CO_3]_2Cl_3 \cdot H_2O$

Disilicates (K = 2)
                               Neutral
Imandrite $Na_{12}Ca_3Fe^{3+}_2[Si_6O_{18}]_2$
*Unnamed $(Na_{0.06}Ca_{0.02}Mg_{0.71}Fe_{0.20}Al_{0.11})_{\Sigma 1.1}[Si_{0.94}O_3]$
 Pyroxenes family
 Mg-Fe(Mn)- pyroxenes subfamily
 *Akimotoite $(Mg,Fe)[SiO_3]$
  Enstatite series (orthopyroxenes)
  Enstatite $Mg_2[Si_2O_6]^{\infty}$
  Hypersthene = Fe-энстатин $(Mg,Fe,Al)_2[(Si,Al)_2O_6]^{\infty}$
  Donpeacorite $Mg(Mn,Mg)[Si_2O_6]^{\infty}$
  Clinoenstatite series
  Clinoenstatite $Mg_2[Si_2O_6]^{\infty}$
  Pigeonite $(Mg,Fe,Ca)(Mg,Fe)[Si_2O_6]^{\infty}$
  Kanoite $(Mn^{2+},Mg)_2[Si_2O_6]^{\infty}$

Clinoferrosilite	$(Fe,Mg)_2[Si_2O_6]^{\infty}$
*Ferrosilite	$(Fe^{2+}Mg)_2[Si_2O_6]^{\infty}$
Ca-Na-pyroxenes subfamily	
Augite series	
*Kushiroite	$CaAl[AlSiO_6]^{\infty}$
Esseneite	$CaFe^{3+}[AlSiO_6]^{\infty}$
*Davisite	$CaSc[AlSiO_6]$
*Grossmanite	$CaTi^{3+}[AlSiO_6]$
Augite	$(Ca,Na)(Mg,Fe,Al)[(Si,Al)_2O_6]^{\infty}$
Omphacite	$(Ca,Na)(Mg,Fe^{2+},Fe^{3+},Al)[(Si,Al)_2O_6]^{\infty}$
Diopside series	
Diopside	$CaMg[Si_2O_6]^{\infty}$
Hedenbergite	$CaFe^{2+}[Si_2O_6]^{\infty}$
Johannsenite	$Ca(Mn,Fe)^{2+}[Si_2O_6]^{\infty}$
Jervisite	$(Na,Ca,Fe^{2+})(Sc,Mg,Fe^{2+})[Si_2O_6]^{\infty}$
*Na-Mg pyroxene	$(Na,Mg,Ca,Mn)(Mg,Al,Cr,Fe)[Si_2O_6]^{\infty}$
Aegirine series	
Natalyite	$NaV[Si_2O_6]^{\infty}$
Kosmochlor	$NaCr[Si_2O_6]^{\infty}$
Aegirine	$NaFe^{3+}[Si_2O_6]^{\infty}$
Jadeite	$Na(Al,Fe^{3+})[Si_2O_6]^{\infty}$
*Namansilite	$NaMn^{3+}[Si_2O_6]$
*Vladykinite	$Na_3Sr_4(Fe^{2+}Fe^{3+})[Si_2O_6]_4$
Pyroxenoids family	
	Neutral
Rhodonite subfamily	
***Pyroxmangite** series	
Pyroxmangite	$Mn^{2+}[SiO_3]^{\infty}$
*Pyroxferroite	$Fe_7[Si_7O_{21}]^{\infty}$
Rhodonite	$Mn_5[Si_5O_{15}]^{\infty}$
Bustamite	$CaMn^{2+}[Si_2O_6]^{\infty}$
*Mendigite	$Mn_2Mn_2MnCa[Si_3O_9]_2$
Ferrobustamite	$Ca_3(Fe^{2+},Ca)_3[Si_3O_9]^{\infty}_2$
Wollastonite subfamily	
Wollastonite-1T	$Ca_3[Si_3O_9]^{\infty}$
Parawollastonite or wollastonite-2M	$Ca_3[Si_3O_9]^{\infty}$
*Manganoparawollastonite	$Ca[SiO_3]$
Wollastonite-7T	$Ca_3[Si_3O_9]^{\infty}$
Pseudowollastonite (synthetic)	$3\beta s\text{-}CaSiO_3 \rightarrow Ca_3[Si_3O_9]$
Walstromite	$BaCa_2[Si_3O_9]$
Combeite	$Na_4Ca_4[Si_6O_{18}]$
*Unnamed (synthetic)	$K_{2.9}Rb_{0.1}Er[Si_3O_9]$
	Acid pyroxenoids $\rightarrow$ hydrates
*Neotocite	$(Mn^{2+},Fe^{2+})[SiO_3]\cdot H_2O$
*Nchwaningite	$Mn^{2+}_2(OH)_2[SiO_3]\cdot H_2O$
*Imogolite	$Al_2(OH)_4[SiO_3]$
Babingtonite series	
Babingtonite	$Ca_2Fe^{2+}Fe^{3+}[Si_5O_{14}(OH)]^{\infty}$

   *Scandiobabingtonite $Ca_2(Fe,Mn)Sc[Si_5O_{14}(OH)]^\infty$

   Manganbabingtonite $Ca_2(Mn,Fe)^{2+}Fe^{3+}[Si_5O_{14}(OH)]^\infty$

   Marsturite $NaCaMn^{2+}{}_3[Si_5O_{14}(OH)]^\infty$

   *Ruizite $Ca_2Mn^{3+}{}_2[Si_4O_{11}(OH)_4]\cdot2H_2O$

   Cascandite $CaSc[Si_3O_8(OH)]^\infty$

   Rosenhahnite $Ca_3[Si_3O_8(OH)_2]$

   *Trabzonite $Ca_4[Si_3O_9(OH)_2]$

  Tobermorite family

   Tobermorite-9A $Ca_5[HSi_3O_9]^\infty{}_2\,2H_2O$

   Tobermorite-11A = Clinotobermorite $Ca_5[Si_6O_{17}]\cdot5H_2O$

   Tobermorite-14A $\{Ca_4[Si_3O_8(OH)]_2\,2H_2O\}\cdot(Ca\cdot5H_2O)$

   *Clinotobermorite $Ca_5[Si_6O_{17}]\cdot5H_2O$

  Pectolite series

   Serandite $NaMn^{2+}{}_2[Si_3O_8(OH)]^\infty$

   Pectolite $NaCa_2[Si_3O_8(OH)]^\infty$

   *Pectolite M2abc $Na(Ca,Mn^{2+})_2[Si_3O_8(OH)]^\infty$

                                   Oxido-disilicates

  Krinovite series

   Dorrite $Ca_4(Mg_3Fe^{3+}{}_9)O_4[Si_3Al_8Fe^{3+}O_{36}]$

   Wilkinsonite $Na_4(Fe^{2+}{}_8Fe^{3+}{}_4)O_4[Si_6O_{18}]_2$

   Krinovite $Na_4(Mg_8Cr^{3+}{}_4O_4)[Si_6O_{18}]_2$

                                   Basic disilicates →hydrates

   *Bunnoite $Mn^{2+}{}_6Al(OH)_3[Si_6O_{18}]$

  Carpholite series

   Magnesiocarpholite $MgAl_2(OH)_4[Si_2O_6]^\infty$

   Ferrocarpholite $(Fe,Mg)Al_2(OH)_4[Si_2O_6]^\infty$

   Carpholite $Mn^{2+}Al_2(OH)_4[Si_2O_6]^\infty$

   *Potassiccarpholite $(K,\square)(Mn^{2+},Li)_2Al_4(OH,F)_8[Si_2O_6]_2$

   *Vanadiocarpholite $Mn^{2+}V^{3+}Al(OH)_4[Si_2O_6]^\infty$

  Gageite series

   Balangeroite $Mg_{21}O_3(OH)_{20}[Si_2O_6]^\infty{}_4$

   Gageite-*1Tc* $Mn^{2+}{}_{21}O_3(OH)_{20}[Si_2O_6]^\infty{}_4$

   Gagei -*2M* $Mn^{2+}{}_{21}O_3(OH)_{20}[Si_2O_6]^\infty{}_4$

  Foshagite $Ca_4(OH)_2[Si_3O_9]^\infty$

  Saneroite $Na_{1.15}(H,Mn^{2+},Mn^{3+})_5(OH)[Si_{5.5}V_{0.5})_{\Sigma6}O_{18}]$

  *Braccoite $NaMn^{2+}{}_5(OH)[Si_5As^{5+}O_{17}(OH)]$

  *Cerchiaraite-(Al) $Ba_4Al_4O_3(OH)_3[Si_4O_{12}][Si_2O_3(OH)_4]Cl$

  *Cerchiaraite-(Fe) $Ba_4Fe^{3+}{}_4O_3(OH)_3[Si_4O_{12}][Si_2O_3(OH)_4]Cl$

  *Cerchiaraite-(Mn) $Ba_4Mn^{3+}{}_4O_2(OH)_4{}^+[Si_4O_{12}][Si_2O_3(OH)_4]Cl_2$

  Verplanckite family

  Verplanckite $|(Mn^{3+},Ti,Fe^{3+})_6(OH,O)_2[Si_4O_{12}]_3|^{\infty3}\cdot Ba_{12}(OH,H_2O)_7Cl_9$

  Muirite $Ba_{10}Ca_2Mn^{2+}TiSi_{10}O_{30}(OH,Cl,F)_{10}$

  *Hubeite $Ca_2Mn^{2+}Fe^{3+}[Si_4O_{12}(OH)]\cdot(H_2O)_2$

  *Bavsiite $Ba_2V_2O_2[Si_4O_{12}]$

                             Hydrates; acid; acid-basic

  Santaclaraite $CaMn_4(H_2O)(OH)[Si_5O_{14}(OH)]^\infty$

  *Middendorfite $K_3Na_2Mn_5[Si_{12}(O,OH)_{36}]\cdot2H_2O$

  Krauskopfite $Ba_2(H_2O)_4[Si_4O_8(OH)_4]^\infty$

  *Jennite $Ca_9(OH)_6[Si_6O_{18}]\cdot8H_2O$

*Oxido-phosphato-carbonato-disilicates

Hydrates

*Devitoite $[Ba_6(PO_4)_2(CO_3)]\{Fe^{2+}_7Fe^{3+}_2O_2[Si_4O_{12}]_2(OH)_4\}$

*Disilicato-Trisilicates
*Strakhovite $NaBa_3(Mn^{2+},Mn^{3+})_4[Si_4O_{10}(OH)_2][Si_2O_7]O_2(F,OH)·H_2O$
*Varennesite $Na_8Mn^{2+}_2(OH,Cl)_2Si_{10}O_{25}·12H_2O$

Trisilicates (K = 3)
Proper trisilicates

Neutral

Melilite group
Gehlenite $Ca_2Al[AlSiO_7]$
Melilite $(Ca,Na)_2(Al,Mg)[(Si,Al)_2O_7]$
Akermanite $Ca_2Mg[Si_2O_7]$
*Alumoakermanite $(Ca,Na)_2(Al,Mg,Fe^{2+})[Si_2O_7]$
Hardystonite $Ca_2Zn[Si_2O_7]$
Rankinite family
Rankinite $Ca_3[Si_2O_7]$
Kilchoanite $Ca_3[Si_2O_7]$
Andremeyerite $BaFe^{2+}_2[Si_2O_7]$
Taikanite $(Sr,Ba)_4Mn_2[Si_2O_7]_2$

Basic

Ilvaite family
Ilvaite $CaFe^{2+}_2Fe^{3+}O(OH)[Si_2O_7]$
*Manganilvaite $CaFe^{2+}Mn^{2+}Fe^{3+}O(OH)[Si_2O_7]$
Orthoericssonite $BaMn_2Fe^{3+}O(OH)[Si_2O_7]$
Ericssonite $BaMn_2Fe^{3+}O(OH)[Si_2O_7]$
*Ferroericssonite $BaFe^{2+}_2Fe^{3+}O(OH)[Si_2O_7]$
Cuspidine family
Cuspidine $Ca_4(F,OH)_2[Si_2O_7]$
Jaffeite $Ca_6(OH)_6[Si_2O_7]$

Hydrates
Basic

Lawsonite family
Lawsonite $CaAl_2(OH)_2[Si_2O_7]H_2O$
*Cortesognoite $CaV_2(OH)_2[Si_2O_7]·H_2O$
*Itoigawaite $SrAl_2(OH)_2[Si_2O_7]·H_2O$
*Hennomartinite $SrMn^{3+}_2(OH)_2[Si_2O_7]·H_2O$
*Noélbelsonite $BaMn^{3+}_2(OH)_2[Si_2O_7]·H_2O$
Neutral
Killalaite $Ca_3[Si_2O_7]·H_2O$

*Trisilicato-chlorides
*Rusinovite $Ca_{10}[Si_2O_7]_3Cl_2$

Trisilicato-tetrasilicates
*Trisilicato-tetrasilicates whis K=3,2
*Medaite $(Mn^{2+},Ca)_6(V,As)^{5+}[Si_5O_{18}(OH)]$

*Pavlovskiite $Ca_8[Si_3O_{10}][SiO_4]_2$

Trisilicato-tetrasilicates with $Si_2O_7 : SiO_4 = 2 : 1$
                          Hydroxido-silicato-chlorides
Rustumite $Ca_{10}(OH)_2Cl_2[Si_2O_7]_2[SiO_4]$

Trisilicato-tetrasilicates with $Si_2O_7 : SiO_4 = 1 : 1$ Basic
 Epidote family (compare with allanite (series)
 Zoisite $Ca_2Al_3O(OH)[Si_2O_7][SiO_4]$
  Epidote series
  Clinozoisite $Ca_2Al_3O(OH)[Si_2O_7][SiO_4]$
  *Niigataite $CaSrAl_3O(OH)[Si_2O_7][SiO_4]$
  Epidote $Ca_2Fe^{3+}Al_2O(OH)[Si_2O_7][SiO_4]$
  *Epidote-(Pb) $(Ca,Pb)(Al_2Fe^{3+})O(OH)[Si_2O_7][SiO_4]$
  *Epidote-(Sr) $CaSrFe^{3+}Al_2O(OH)[Si_2O_7][SiO]$
  Piemontite $Ca_2Mn^{3+}Al_2O(OH)[Si_2O_7][SiO_4]]$
  *Piemontite-(Pb) $CaPbMn^{3+}Al_2O(OH)[Si_2O_7][SiO_4]$
  *Piemontite-(Sr) $SrCaMn^{3+}Al_2O(OH)[Si_2O_7][SiO_4]$
  *Манганипьемонтит-(Sr) = *Tweddillite
                                   $CaSrMn^{3+}_2AlO(OH)[Si_2O_7][SiO_4]$
  Mukhinite $Ca_2V^{3+}Al_2O(OH)[Si_2O_7][SiO_4]$
  *Cassagnaite $Ca_4Fe_4^{3+}V_2^{3+}(OH)_6O_2[Si_3O_{10}][SiO_4]_2$
  *Uedaite-(Ce) $(Mn^{2+}Ce)Al_2Fe^{2+}O(OH)[Si_2O_7][SiO_4]$
  Sursassite series
  Sursassite $Mn_2Al_3(OH)_3[Si_2O_7][SiO_4]$
  Macfallite $Ca_2Mn^{3+}_3(OH)_3[Si_2O_7][SiO_4]$
 Cebollite $Ca_4Al_2O(OH)_2[Si_2O_7][SiO_4]$ (?)
 Dellaite $Ca_6(OH)_2[Si_2O_7][SiO_4]$
                          Hydrates
  Pumpellyite series
  *Pumpellyite-(Al) $Ca_2AlAl_2O(OH)[Si_2O_7][SiO_4] \cdot H_2O$
  Pumpellyite-(Fe^{2+}) $Ca_2Fe^{2+}Al_2(OH)_2[Si_2O_7][SiO_4] \cdot H_2O$
  *Pumpellyite-(Fe^{3+}) $Ca_2Fe^{3+}Al_2O(OH)[Si_2O_7][SiO_4] \cdot H_2O$
  *Pumpellyite-(Mg) $Ca_2MgAl_2(OH)_2[Si_2O_7][SiO_4] \cdot H_2O$
  Pumpellyite-(Mn) $Ca_2MnAl_2(OH)_2[Si_2O_7][SiO_4] \cdot H_2O$
  Shuiskite $Ca_2(Mg,Al,Fe)(Cr,Al)_2(OH)_2[Si_2O_7][(Si,Al)O_4]H_2O$
  Julgoldite-(Fe^{2+}) $Ca_2Fe^{2+}(Fe^{3+},Al)_2(OH)_2[Si_2O_7][SiO_4]H_2O$
  *Julgoldite-(Fe^{3+}) $Ca_2Fe^{3+}(Fe^{3+})_2O(OH)[Si_2O_7][SiO_4] \cdot H_2O$
  *Julgoldite-(Mg) $Ca_2Mg(Fe^{3+})_2(OH)_2[Si_2O_7][SiO_4] \cdot H_2O$
  *Poppiite $Ca_2V^{3+}V^{3+}_2O(OH)[Si_2O_7][SiO_4] \cdot H_2O$
  *Okhotskite $Ca_2(Mn^{2+},Mg,Mn^{3+},Al,Fe^{3+})_3(O,OH)_3[Si_2O_7][SiO_4]$

Trisilicato-tetrasilicates with K=3,4 *Hydrates
*Vertumnite $Ca_8[Al_4Al_4Si_5O_{12}(OH)_{36}] \cdot 10H_2O$
*Strätlingite $Ca_8[Al_4Al_4Si_4O_8(OH)_{40}] \cdot 10H_2O$
*Aerinite $(Ca_{5,1}Na_{0,5})(Fe^{3+}AlFe^{2+}_{1,7}Mg_{0,3})(Al_{5,1}Mg_{0,7})[Si_{12}O_{36}(OH)_{12}H][(H_2O)_{12}(CO_3)_{1,2}]$ or
              *$Ca_4Al_{10}[Si_{12}O_{36}(OH)_{12}][CO_3](H_2O)$ [Krivovochev, 2008]

Trisilicato-tetrasilicates with $Si_2O_7 : SiO_4 = 0,4 : 1$ Basic
Vesuvianite $(Ca,Na)_{19}(Al,Mg,Fe)_{13}(OH,F,O)_{10}[SiO_4]_{10}[Si_2O_7]$

*Alumovesuvianite $Ca_{19}Al(Al_{10}Mg_2)O(OH)_9[Si_2O_7]_4[SiO_4]_{10}$
*Fluorvesuvianite $Ca_{19}(Al,Mg)_{13}O(F,OH)_9[Si_2O_7]_4[SiO_4]_{10}$
*Manganovesuvianite
$(Ca,Na,\square)_{19}(Al,Mg,Fe^{3+})_{13}(\square,B,Al,Fe^{3+})_5(OH,F,O)_{10}[Si_2O_7]_4[SiO_4]_{10}$
*Wiluite $Ca_{19}(Al,Mg,Fe,Ti)_{13}(B,Al,\square)_5(O,OH)_{10}[Si_2O_7]_4[SiO_4]_{10}$

*Трисиликато-тетрасиликато-фосфаты   (К = 3,(3))
*Lavoisierite $Mn^{2+}_8[Al_{10}(Mn^{3+}Mg)](OH)_{12}[(Si_{11}P)O_{44}]$

Tetrasilicates (orthosilicates) (K = 4) Neutral
   Garnet series
   Knorringite $Mg_3Cr_2[SiO_4]_3$
   Pyrope $Mg_3Al_2[SiO_4]_3$
   *Menzerite-(Y) $CaY_2Mg_2[SiO_4]_3$
   Almandine $Fe^{2+}_3Al_2[SiO_4]_3$
   Spessartine $Mn^{2+}_3Al_2[SiO_4]_3$
   Majorite $Mg_3(MgSi)[SiO_4]_3$
   *Eringaite $Ca_3Sc_2[SiO_4]_3$
   Grossular $Ca_3Al_2[SiO_4]_3$
   *Irinarassite $Ca_3Sn_2Al_2SiO_{12}$ или $Ca_3[Al_{2/3}Sn_{2/3}Si_{1/3}O_4]_3$
   Calderite $(Mn^{2+},Ca)_3(Fe^{3+},Al)_2[SiO_4]_3$
   Andradite $Ca_3Fe^{3+}_2[SiO_4]_3$
   *Ti-andradite $Ca_3Ti^{4+}_2[(Fe^{3+}_{0.66}Si_{0.33})O_4]_3$
   Schorlomite $Ca_3Ti^{4+}_2[(Fe^{3+}_2Si)O_{123}$
   *Morimotoite $Ca_3TiFe^{2+}[SiO_4]_3$
   *Hutcheonite $Ca_3Ti^{4+}_2[(Si_{0.33}Al_{0.66})O_4]_3$
   Kimzeyite $Ca_3Zr_2[Al_2Si]O_{12}]$
   *Kerimasite $Ca_3Zr_2[(Fe^{3+}_2Si)O_{12}]$
   Uvarovite $Ca_3Cr_2[SiO_4]_3$
   Goldmanite $Ca_3(V,Al,Fe)^{3+}_2[SiO_4]_3$
   *Momoite $Mn^{2+}_3V^{3+}_2[SiO_4]_3$
   Yamatoite $(Mn^{2+},Ca)_3(V^{3+},Al)_2[SiO_4]_3$
   *Toturite $Ca_3Sn_2[(Fe^{3+}_2Si)O_{12}]$
   Hydrogarnet family
   Hydrougrandite $(Ca,Mg,Fe^{2+})_3(Fe^{3+},Al)_2[SiO_4]_2(OH)_4$
   *Henritermierite $Ca_3(Mn^{3+},Al)_2[SiO_4]_2(OH)_4$
   *Holstamite $Ca_3Al_2[SiO_4]_2(OH)_4$
   Hibschite (plazolite) $Ca_3Al_2[SiO_4]_{3-x}(OH)_{4x}$ $(0,2 \leq x \leq 1,5)$
   Katoite $Ca_3Al_2[SiO_4]_{3-x}(OH)_{4x}$ $(1,5 < x \leq 3)$
   Olivine family
   *Calcio-olivine $Ca_2[SiO_4]$
   Wadsleyite $\beta\text{-}(Mg,Fe^{2+})_2[SiO_4]$
   *Wadsleyite II $(Mg,Fe)_{11}[(Al,Si)_6(OH)_2O_{22}]$
   Ringwoodite $(Mg,Fe^{2+})_2[SiO_4]$
      Forsterite series
      Forsterite $Mg_2[SiO_4]$
      Fayalite $Fe^{2+}_2[SiO_4]$
      Tephroite $Mn^{2+}_2[SiO_4]$
      Liebenbergite $Ni^{2+}[SiO_4]$
      Laihunite $Fe^{2+}Fe^{3+}_2[SiO_4]_2$

*Unnamed $(Na_{0.08}Ca_{0.03}Mg_{0.95}Fe_{0.26}Al_{0.15}Si_{0.25}\square_{0.28})_2[SiO_4]$

Monticellite family

Monticellite $CaMg[SiO_4]$
Kirschsteinite $CaFe^{2+}[SiO_4]$
Glaucochroite $CaMn^{2+}[SiO_4]$
Merwinite $Ca_3Mg[SiO_4]_2$
Bredigite $(Ca,Ba)Ca_{13}Mg_2[SiO_4]_8$
Larnite $\beta\text{-}Ca_2[SiO_4]$

Oxido-tetrasilicates (at that number sulfido-oxido tetrasilicates)
*Hatrurite $Ca_3O[SiO_4]$

Kyanite family

(Sillimanite) see monoalumosilicates, sillimanite (family)
Andalusite $^{(6)}Al^{(5)}AlO[SiO_4]$
Kanonaite $^{(6)}(Mn^{3+},Al)^{(5)}AlO[SiO_4]$
Kyanite $^{(6)}Al_2O[SiO_4]$

Staurolite family

Yoderite $Mg_2Al_6O_2(OH)_2[SiO_4]_4$
Staurolite $Fe_2^{2+}Al_9Si_4O_{23}(OH)$
*Magnesiostaurolite $\square_4Mg_4Al_{16}(Al_2\square_2)Si_8O_{40}[O_6(OH)_2]$
*Zincostaurolite $\square_4Zn_4Al_{16}(Al_2\square_2)Si_8O_{40}[O_6(OH)_2]$
Jasmundite $Ca_{11}SO_2[SiO_4]_4$

Basic oxido-tetrasilicates

Davreuxite $Mn^{2+}Al_6(OH)_2O[SiO_4]_4$

Chloritoid family

  Chloritoid series

  Magnesiochloritoid $\{Mg_2Al(OH)_4Al_3O_2[SiO_4]_2\}^{\infty 2}$
  Chloritoid $\{(Fe^{2+},Mg)_2Al(OH)_4Al_3O_2[SiO_4]_2\}^{\infty 2}$
Ottrelite $\{(Mn^{2+},Fe^{2+},Mg)_2Al(OH)_4Al_3O_2[SiO_4]_2\}^{\infty 2}$

Humite polysomatic series $M_{3+n}X_2[SiO_4]_{1+0,5n}$, where $M^{2+} = Mg^{2+}$, Fe^{2+}, Mn^{2+};
$X = F^-$, OH^-; $n = 0$; 2; 4; 6

Norbergite $Mg_3(F,OH)_2[SiO_4]$ (n = 0)

  Chondrodite group

  Chondrodite $Mg_5(F,OH)_2[SiO_4]_2$ (n = 2)
  *Hydroxychondrodite $Mg_5(OH)_2[SiO_4]_2$
  *Kumtyubeite $Ca_5F_2[SiO_4]_2$
  Alleghanyite $Mn^{2+}_5(OH)_2[SiO_4]_2$ (n = 2)
  Reinhardbraunsite $Ca_5(OH,F)_2[SiO_4]_2$
Ribbeite $Mn^{2+}_5(OH)_2[SiO_4]_2$ (n = 2)

  Humite series

  Humite $(Mg,Fe^{2+})_7(F,OH)_2[SiO_4]_3$ (n = 4)
  Manganhumite $(Mn^{2+},Mg)_7(OH)_2[SiO_4]_3$ (n = 4)
  *Chegemite $Ca_7(OH)_2[SiO_4]_3$
  *Fluorchegemite $Ca_7F_2[SiO_4]_3$
Leucophoenicite $Mn^{2+}_7(OH)_2[SiO_4]_3$ (n = 4)

  Clinohumite series

  Titanclinohumite $(Mg,Fe,Ti)_9F_2[SiO_4]_4$ (n = 6)
  Clinohumite $(Mg,Fe^{2+})_9(F,OH)_2[SiO_4]_4$ (n = 6)
  *Hydroxylclinohumite $Mg_9(OH,F)_2[SiO_4]_4$ (n = 6)
Jerrygibbsite $Mn^{2+}_9(OH)_2[SiO_4]_4$ (n = 6)

Sonolite $\qquad$ $Mn^{2+}_9(OH)_2[SiO_4]_4$ (n = 6)
*Poldervaartite $\qquad$ $Ca(Ca_{0.5}Mn_{0.5})(OH)[SiO_3OH]$
*Olmiite $\qquad$ $CaMn[SiO_3(OH)](OH)$
Welinite family
Welinite $\qquad$ $Mn^{2+}_3(Mn^{4+},W)(O,OH)_3[SiO_4]$
Franciscanite $\qquad$ $Mn_3(V^{5+}_x,\square_{1-x})(O,OH)_3[SiO_4]$ (x ~ 0.5)
*Vuagnatite $\qquad$ $CaAl(OH)[SiO_4]$
*Mozartite $\qquad$ $CaMn^{3+}(OH)[SiO_4]$
Chantalite $\qquad$ $CaAl_2(OH)_4[SiO_4]$

Hydrates (basic)
*Spadaite $\qquad$ $Mg[SiO_2(OH)_2]\cdot H_2O$
*Chesnokovite $\qquad$ $Na_2[SiO_2(OH)_2]\cdot 8H_2O$
Kittatinnyite(compare with wallkilldellite (gr.)) $Ca_4Mn^{2+}_2Mn^{3+}_4(OH)_8[SiO_4]_4\cdot 18H_2O$
*Orientite $\qquad$ $Ca_8Mn^{3+}_{10}(OH)_{10}[Si_3O_{10}]_3[SiO_4]_3\cdot 4H_2O$

*Oxido-tetrasilicato-halogenides
*Wadalite $\qquad$ $Ca_6Al_5O_8[SiO_4]_2Cl_3$
*Eltyubyuite $\qquad$ $Ca_{12}Fe^{3+}_{10}Si_4O_{32}Cl_6$ or
$\qquad$ $Ca_{12}Fe^{3+}_6[(Fe^{3+}_4Si_4)O_{32}]Cl_6$

Acid →tetrasilicato-fluorides
Bultfonteinite $\qquad$ $Ca_4[SiO_3(OH)]_2F_2\cdot 2H_2O$

*Tetrasilicato-phosphates
*Harrisonite $\qquad$ $Ca(Fe,Mg)_6[SiO_4]_2[PO_4]_2$

Silicato-halogenides
Di-trisilicato-halogenides

Basic
Zunyite   (K = 2,4) $\qquad$ $Al_{12}(OH)_{14}\{[AlO_4][Si_5O_{16}]\}F_4Cl$
Magbasite   (K = 2,(6)) $\qquad$ $KBa(Al,Sc)(Mg,Fe^{2+})_6[Si_6O_{20}]F_2$
*Jagoite   (K = 2,(6)) $\qquad$ $Pb_{18}Fe_4^{3+}[Si_4(Si,Fe^{3+})_6][Pb_4Si_{16}(Si,Fe)_4]O_{82}Cl_6$

Tetrasilicato-halogenides $\qquad$ Basic
Topaz $\qquad$ $Al_2[SiO_4](F,OH)_2$
*Topaz-(OH) $\qquad$ $Al_2[SiO_4](OH,F)_2$
*Podnoginite $\qquad$ $Ca_2[SiO_4]CaF_2$
*Rondorfite $\qquad$ $Ca_8Mg[SiO_4]_4Cl_2$
*Edgrewite $\qquad$ $Ca_9[SiO_4]_4F_{1.2}(OH)_{0.8}$
*Hydroxyledgrewite $\qquad$ $Ca_9[SiO_4]_4(OH)_2$

*Tetrasilicato-oxido-sulphates
*Gazeevite $\qquad$ $BaCa_6[SiO_4]_2[SO_4]_2O$
*Nabimusaite $\qquad$ $KCa_{12}[SiO_4]_4[SO_4]_2O_2F$

Silicato-borates
Borosilicates
Zero-borosilicates ($K_\Sigma = 0$) $\qquad$ Neutral
Danburite $\qquad$ $Ca[Si_2B_2O_8]^{\infty 3}\rightarrow Ca[(Si_2O_7)B_2O]^{\infty 3}$
*Maleevite $\qquad$ $Ba[Si_2B_2O_8]$
*Pekovite $\qquad$ $Sr[Si_2B_2O_8]$

Reedmergnerite $Na[Si_3BO_8]^{\infty 3} \rightarrow Na[SiO_3BO_5]^{\infty 3}$
*Malinkoite $Na[SiBO_4]$
*Pudrettite $KNa_2[Si_{12}B_3O_{30}]$
*Jadarite $LiNa[SiB_3O_7(OH)]$
*Lisitsynite $K[Si_2BO_6]$
*Kirchhoffite $Cs[Si_2BO_6]$

*Zero-monoborosilicates (K_Σ = 0,32) Acid
*Martinite $(Na,\square,Ca)_{12}Ca_4[(Si,S,B)_{14}B_2O_{38}](OH,Cl)_2F_2 \cdot 4H_2O$

*Zero-monoborosilicates (K = 0.5)   Basic (hydrates)
*Steedeite $NaMn_2[BSi_3O_9](OH)_2$
*Nolzeite $NaMn_2[Si_3BO_9](OH)_2 \cdot 2H_2O$

Zero-monoborosilicates (K_Σ = 0,(6))   Acid
Searlesite $Na[(Si_2O_5B(OH)_2]^{\infty 2}$
*Okayamalite $Ca_2[B_2SiO_7]$
*Itsiite $Ba_2Ca[BSi_2O_7]_2$

*Zero-monoborosilicates (K = 0.75)   Basic
*Odigitriaite $CsNa_5Ca_5[Si_{14}B_2O_{38}]F_2$

*Zero-monoborosilicates (K_Σ = 0,9)   Basic (hydrates)
*Kasatkinite $Ba_2Ca_8[B_5Si_8O_{32}](OH)_3 \cdot 6H_2O$

Monoborosilicates (K_Σ = 1)   Neutral
Homilite $Ca_2Fe^{2+}[Si_2B_2O_{10}]^{\infty 2} \rightarrow Ca_2Fe^{2+}[(SiO_4)_2B_2O_2]^{\infty 2}$
*Boromullite $Al_8O_9[AlBSi_2O_{10}]$
                                  Acid
Datolite family
Datolite $Ca[(SiO_4)BOH]^{\infty 2}$
Bakerite $Ca_4[Si_3B_5O_{15}(OH)_5]^{\infty 2} \rightarrow Ca_4[(SiO_4)_3(BO_3OH)(BOH)_4]^{\infty 2}$

*Zero-monoborosilicat-fluorides
*Kapitsaite-(Y) $(Ba,Pb)_4(Y,Ca)_2[Si_8B_4O_{28}]F$
*Khvorovite $Pb^{2+}_4Ca_2[Si_8B_2(S,B)O_{28}]F$

*Mono-diborosilicates (K_Σ = 1,7)
*Piergorite-(Ce) $Ca_8Ce_2(Al_{0.5}Fe^{3+}_{0.5})(\square Li,Be)_2[Si_6B_8O_{36}](OH)_2$
*Vistepite $Mn^{2+}_5Sn^{4+}[B_2Si_5O_{20}]$

*Mono-diborosilicates (K = 1,8)   Hydrates
*Oyelite $Ca_{10}[B_2Si_8O_{29}] \cdot 12H_2O$

Diborosilicates (K_Σ = 2)   Oxido-diborosilicates
Serendibite $Ca_4(Mg_6Al_6)O_4[Si_6B_3Al_3O_{36}]$
(compare with aenigmatite (family))
                                 Basic diborosilicates
   Axinite series

*Axinite-(Mg) = magnesioaxinite $\{Ca_2MgAl_2(OH)[(Si_2O_7)_2BO]\}^{\infty 2}$
*Axinite-(Fe) $\{Ca_2FeAl_2(OH)[(Si_2O_7)_2BO]\}^{\infty 2}$
*Axinite-(Mn) $\{Ca_2(Mn,Fe)Al_2(OH)[(Si_2O_7)_2BO]\}^{\infty 2}$
Tinzenite $*CaMn^{2+}_4Al_4(OH)_2[(Si_2O_7)_4B_2O_2]$
Neutral$\rightarrow$ acid diborosilicato-chlorides

Taramellite family
Taramellite $Ba_4(Fe^{3+},Ti,Fe^{2+},Mg,V^{3+})_4[(Si_8B_2O_{27})O_2Cl_x]$
Nagashimalite $Ba_4(V^{3+},Ti)_4[(Si_8B_2O_{27})(O,OH)_2Cl]$
Titantaramellite $Ba_4(,Ti,Fe^{3+},Fe^{2+},Mg)_4[(Si_8B_2O_{27})O_2Cl_{0-1}]$

Di-tri-borosilicates $(2 < K_\Sigma < 3)$ Basic
Kornerupine $Mg_3Al_6(OH)O_4[(Al,Si)_2(Si,B)O_{10}][Si_2O_7]$
*Prismatine $Mg_3Al_6(OH)BO_7[Si_2O_7]_2$

Silicato-(4)-borate
Zero-silicato-(4)-borates   Acid zero-silicato-(4)-borato-chlorides
Kalborsite $[Al_2Si_3O_{10}]^{\infty 3}_2[B(OH)_4]K_6Cl$

Silicato-(3)-borates
Disilicato-(3)-borates
Disilicato-(3)-borates $(K_{Si} = 2)$ Basic
   Tourmaline series (compare with elbaite (series))
Olenite $NaAl_3Al_6O_3F[Si_6O_{18}][BO_3]_3$
*Luinaite-(OH) $(Na,\square)(Fe^{3+},Mg)_3Al_6(OH)_3(OH)[Si_6O_{18}][BO_3]_3$
Buergerite = fluor-buergerite $NaFe^{3+}_3Al_6O_3F[Si_6O_{18}][BO_3]_3$
*Povondraite $NaFe^{3+}_3Fe^{3+}_6O(OH)_3[Si_6O_{18}][BO_3]_3$
*Chromo-alumino-povondraite $NaCr_3(Al_4Mg_2)O(OH)_3[Si_6O_{18}][BO_3]_3$
Dravite $NaMg_3Al_6(OH)_3(OH)[Si_6O_{18}][BO_3]_3$
*Oxy-dravite $Na(Al_2Mg)(Al_5Mg)O(OH)_3[Si_6O_{18}][BO_3]_3$
*Fluor-dravite $NaMg_3Al_6(OH,F)_{3+1}[Si_6O_{18}][BO_3]_3$
*Vanadiodravite $NaMg_3V_6(OH,F)_{3+1}[Si_6O_{18}][BO_3]_3$
*Vanadio-oxy-dravite $NaV_3(V_4Mg_2)(OH)_3O[Si_6O_{18}][BO_3]_3$
*Oxy-vanadium-dravite $NaV_3(V_4Mg_2)(OH)_3O[Si_6O_{18}][BO_3]_3$
*Vanadio-oxy-chromium dravite $NaV_3(Cr_4Mg_2)(OH)_3O[Si_6O_{18}][BO_3]_3$
Chromdravite $NaMg_3Cr_6(OH)_3(OH)[Si_6O_{18}][BO_3]_3$
*Oxy-chromium-dravite $NaCr_3(Cr_4Mg_2)O(OH)_3[Si_6O_{18}][BO_3]_3$
Schorl $NaFe^{2+}_3Al_6(OH)_3(OH)[Si_6O_{18}][BO_3]_3$
*Fluor-schorl $NaFe^{2+}_3Al_6(OH)_3F[Si_6O_{18}][BO_3]_3$
*Oxy-schorl $NaFe^{2+}_3Al_6(OH)_3O[Si_6O_{18}][BO_3]_3$
*Bosiite $NaFe^{3+}_3(Al_4Mg_2)(OH)_3O[Si_6O_{18}][BO_3]_3$
Ferridravite $NaFe_3^{3+}(Mg_2Fe_4^{3+})(OH)_3O[Si_6O_{18}][BO_3]_3$
*K-and O-dominate dravite $KFe^{3+}_3(Mg_2Fe^{3+}_4)O(OH)_3[Si_6O_{18}][BO_3]_3$
*Tsilaisite $Na(Mn^{2+})_3Al_6(OH)_3(OH)[Si_6O_{18}][BO_3]_3$
Uvite $Ca(Mg,Fe^{2+})_3Al_5Mg(OH)_3(OH)[Si_6O_{18}][BO_3]_3$
*Lucchesiite $CaFe^{2+}_3Al_6O(OH)_3[Si_6O_{18}][BO_3]_3$
*Feruvite $Ca(Fe,Mg)_3Al_6(OH)_3(OH)[Si_6O_{18}][BO_3]_3$
*Adachiite $CaFe^{2+}_3Al_6(OH)_3(OH)[Si_5AlO_{18}][BO_3]_3$
*Foitite $(\square,Na)(Fe_2^{2+}Al)Al_6(OH)_3(OH)[Si_6O_{18}][BO_3]_3$
*Magnesio-foitite $\square(Mg_2Al)Al_6(OH)_4[Si_6O_{18}][BO_3]_3$
*Capranicaite $(K,\square)CaNaAl_4[Si_2O_6][BO_3]_4$

Di-tri-alumosilicatj-(3)-borato-carbonates Asid
*Harkerite $Ca_{12}Mg_4Al[SiO_4]_4[BO_3]_3[CO_3]_5 \cdot H_2O$

Tetrasilicato-(3)-borates
Oxido-tetrasilicato-(3)-borates
Dumortierite $Al_7O_3[SiO_4]_3[BO_3]$
*Magnesiodumortierite $MgAl_6O_3[SiO_4]_3[BO_3]$
*Werdingite $(Mg,Fe^{2+})_2Al_{14}B_4Si_4O_{37} = (Mg,Fe^{2+})_2Al_{14}O_9[SiO_4]_4[BO_3]_4$
*Fe analog werdingite $(Fe^{2+},Mg)_2Al_{14}O_9[SiO_4]_4[BO_3]_4$
*Boralsilite $Al_{16}B_6Si_2O_{37} = Al_{16}O_{11}[SiO_4]_2[BO_3]_6$
*Holtite $(Ta,\square)Al_6O_3[SiO_4]_3[BO_3]$
*Nioboholtite $(Nb_{0.6}\square_{0.4})Al_6O_3[SiO_4]_3[BO_3]$
*Titanoholtite $(Ti_{0.75}\square_{0.25})Al_6O_3[SiO_4]_3[BO_3]$
Grandidierite $^{(5)}Mg^{(6)}Al_2^{(5)}AlO_2[SiO_4][BO_3]$
*Ominelite $(Fe,Mg)Al_2AlO_2[SiO_4][BO_3]$

Borosilicato-(4)-(3)-borates   Acid
Howlite $Ca_2(OH)_5SiB_5O_9 \rightarrow Ca_2\{[SiB_2O_5(OH)_3]^\infty B_3O_4(OH)_3\}^{\infty 3}$
Garrelsite $NaBa_3(OH)_4Si_2B_7O_{16} \rightarrow CaBa_3(OH)_2[(SiO_4)_2^{(4)}B_2O_2OH_2]^\infty B_5O_6\}^{\infty 2}$
*Hundholmenite-(Y) $Y_{15}AlCa_x(As^{3+})_{1-x}(Si,As^{5+})Si_6B_3(O,F)_{48}$

Silicato-phosphates
Disilicato-phosphates   Neutral
Phosinaite family
Clinophosinaite $Na_3Ca[SiO_3][PO_4]$
Phosinaite-(Ce) $Na_{13}Ca_2Ce[SiO_3]_4[PO_4]_4$

Tetrasilicato-phosphates   Neutral$Ca_3Al_{7.7}Si_3P_4O_{23.5}(OH)_{14.1} \cdot 8H_2O$
Nagelschmidtite $Ca_7[SiO_4]_2[PO_4]_2$
*Silicocarnotite $Ca_5[(SiO_4)(PO_4)][PO_4]$
*Flamite $(Ca,Na,K)_2[(Si,P)O_4]$
                                            Hydrates
Perhamite $Ca_3Al_{7.7}Si_3P_4O_{23.5}(OH)_{14.1} \cdot 8H_2O$
*Krásnoite $Ca_3Al_{7.7}Si_3P_4O_{23.5}(OH)_{12.1}F_2 \cdot 8H_2O$

*Tetrasilicato-phosphato-halogenides
*Zadovite $BaCa_6[(SiO_4)(PO_4)][PO_4]_2F$

*Silicato-arsenates
*Mono-disilicato-arsenates (K=1,(6))
                                            *Basic
*Johninnesite $Na_2Mn^{2+}_9(Mg,Mn^{2+})_7(OH)_8[Si_6O_{17}]_2[AsO_4]_2$

*Disilicato-arsenates   *Hydrates
*Tiragalloite $Mn^{2+}_4[Si_3O_8(OH)][AsO_4]$

Zero-monosilicato-carbonates   Hydrates
Carletonite $KNa_4Ca_4(F,OH)(H_2O)[Si_8O_{18}]^{\infty 2}[CO_3]_4$

*Monosilicato-carbonates   *Hydrates
*Niksergievite $Ba_2Al_3[(Si,Al)_4O_{10}][CO_3](OH)_6 \cdot nH_2O$

*Monosilicato-carbonato-halogenides
                                            *Основные
*Hanjiangite $Ba_2Ca(V^{3+}Al)(OH)_2[Si_3AlO_{10}]F[CO_3]_2$

Disilicato-carbonates   Basic
Fukalite $Ca_4(OH,F)_2[Si_2O_6]^{\infty}[CO_3]$
                                            Hydrates
Scawtite $Ca_7(H_2O)_2[Si_6O_{18}][CO_3]$

Trisilicato-carbonates Neutral
Tilleyite $Ca_5[Si_2O_7][CO_3]_2$

Tetrasilicato-carbonates   Neutral
 Spurrite family
 Spurrite $Ca_5[SiO_4]_2[CO_3]$
*Galuskinite $Ca_7[SiO_4]_3[CO_3]$

Silicato-sulfates
Tetrasilicato-sulfates   Neutral
*Ternesite $Ca_5[SiO_4]_2[SO_4]$
                                Basic
  Ellestadite series (compare with apatite (family))
  Ellestadite-(OH) $Ca_{10}(OH)_2[SiO_4]_3[SO_4]_3$
  Ellestadite-(F) $Ca_{10}F_2[SiO_4]_3[SO_4]_3$
  *Ellestadite-(Cl) $Ca_2Cl_2[SiO_4]_3[SO_4]_3$
                                Hydrates
Chessexite $(Na,K)_4Ca_2(Mg,Zn)_3Al_8(OH)_{10}[SiO_4]_2[SO_4]_{10} \cdot 40H_2O$

Silicates of Li
Proper silicates of Li
Zeroalumosilicates (K = 0)   Neutral
Eucryptite $^{(4)}Li[AlSiO_4]^{\infty 3}$
Petalite $^{(4)}Li[AlSi_4O_{10}]^{\infty 3}$
Virgilite $Li[AlSi_2O_6]^{\infty 3}$
Bikitaite $Li[AlSi_2O_6]^{\infty 3} \cdot H_2O$

Monoalumo-and monosilicates (K = 1)
                                Neutral *and hydrates
 Emeleusite family (compare with osumilite (family))
 Emeleusite $Na_4Li_2Fe^{3+}_2[Si_{12}O_{30}]^{\infty}$
 Sugilite $KNa_2(Fe^{3+},Mn^{2+},Al)_2Li_3[Si_{12}O_{30}]^{\infty}$
 *Silinaite $NaLi[Si_2O_5] \cdot 2H_2O$
                                Basic
 Lithium mica family
 Fragile lithium mica subfamily
 Ephesite $Na\{LiAl_2(OH)_2[AlSiO_5]^{\infty 2}_2\}^{\infty 2}$

Usual lithium mica subfamily
 Lepidolite series
Hydroxyllepidolite, Fluorlepidolite a Li-rich micas in, or close to, the so-called
                                            Polylithionite-Trilithionite join.
  *Voloshinite $Rb(LiAl_{1.5}\square_{1.5})[Al_{0.5}Si_{3.5}]O_{10}F_2$
  *Zinnwaldite is series by Fleischer's, 2014.
  *Zinnwaldite 1M, 2M, 3T polytyps
  Masutomilite $K\{LiMn^{2+}AlF_2[AlSi_3O_{10}]^{\infty 2}\}^{\infty 2}$
  Polylithionite $K\{Li_2AlF_2[Si_2O_5]^{\infty 2}_2\}^{\infty 2}$
  *Orlovite $KLi_2TiF[Si_4O_{11}]$
  *Cs-polylithionite
       $(Cs_{0.75}K_{0.23}Rb_{0.02})_{1.00}(Al_{1.33}Li_{1.27}Mn_{0.01})_{2.61}[(Si_{3.72}Al_{0.28})_{4.00}O_{10}]F_{1.46}(OH)_{0.54}$
  *Sokolovaite $CsLi_2AlF_2[Si_4O_{10}]$
 *Trilylithionite $KLi_{1.5}Al_{1.5}F_2[AlSi_3O_{10}]$
  Tainiolite $K\{LiMg_2F_2[Si_2O_5]^{\infty 2}_2\}^{\infty 2}$
  Norrishite $K\{LiMn^{3+}_2O_2[Si_2O_5]^{\infty 2}_2\}^{\infty 2}$

 *Saliotite family
 *Saliotite $Li_{0.5}Na_{0.5}Al_3(OH)_2[AlSi_3O_{10}]$

Lithium chlorite family (in what numbers borosilicate of Li)
Cookeite $LiAl_2(OH)_6\{Al_2(OH)_2[(Si,Al)_2O_5]^{\infty 2}_2\}^{\infty 2}$
*Borocookeite $LiAl_4(OH)_8[BSi_3O_{10}]$ = $LiAl_2(OH)_6Al_2(OH)_2[BSi_3O_{10}]$
Manandonite $Li_2Al_2(OH)_6\{Al_2(OH)_2[Si_2AlBO_{10}]^{\infty 2}\}^{\infty 2}$

Lithium smectite family
Hectorite $Na_{0.3}\{(Mg,Li)_3(F,OH)_2[Si_2O_5]^{\infty 2}_2\}^{\infty 2}\cdot nH_2O$

Mono-dialumosilicates (1 < K < 2)   Basic
Lithium amphibole family
 Holmquistite series
 Magnesioclinoholmquistite $\square Li_2(Mg,Fe^{2+})_3Al_2(OH)_2[Si_4O_{11}]^{\infty}_2$
 Ferroclinoholmquistite $\square Li_2(Fe^{2+},Mg)_3Al_2(OH)_2[Si_4O_{11}]^{\infty}_2$
 Clino-holmquistite series
 Magnesioclinoholmquistite $\square Li_2(Mg,Fe^{2+})_3Al_2(OH)_2[Si_4O_{11}]^{\infty}_2$
 Clino-holmquistite $\square Li_2(Mg,Fe^{2+})_3Al_2(OH)_2[Si_4O_{11}]^{\infty}_2$
 Ferroclinoholmquistite $\square Li_2(Fe^{2+},Mg)_3Al_2(OH)_2[Si_4O_{11}]^{\infty}_2$
 *Ferriclinoferroholmquistite $\square Li_2(Fe^{2+}_3Fe^{3+}_2)(OH)_2[Si_4O_{11}]^{\infty}_2$
 *Kornite $(Na,K)Na_2(Mg_2Mn^{3+}_2Li)(OH)_2[Si_4O_{11}]_2$
 *Ottoliniite $(Na,Li)(Mg_3Fe^{3+}Al)(OH)_2[Si_4O_{11}]$
 *Ferri-ottolinite $(Na,Li)_2(Mg_3Fe^{3+}_2)(OH)_2[Si_4O_{11}]_2$
 *Ferriwhittakerite $Na(Na_{1+x}Li_{1-x})(Mg_2Fe^{3+}_2Li)(OH)_2[Si_4O_{11}]_2$
 *Sodic-ferripedrizite $NaLi_2(Fe^{3+}_2Mg_2Li)(OH)_2[Si_4O_{11}]_2$
 *Sodic-ferri-ferropedrizite $NaLi_2(Fe^{3+}_2Fe_2Li)(OH)_2[Si_4O_{11}]_2$
 *Ferropedrizite $NaLi_2(Fe^{2+}_2Al_2Li)(OH)_2[Si_4O_{11}]_2$
 *Fluoro-sodic-ferropedrizite $NaLi_2(Fe^{2+}_2Al_2Li)F_2[Si_4O_{11}]_2$
 *Fluoro-sodic-pedrizite $NaLi_2(Mg_2Al_2Li)F_2[Si_4O_{11}]_2$
 *Leakeite $NaNa_2(Mg_2Al_2Li)(OH)_2[Si_4O_{11}]_2$

*Ferroleakeite $\qquad$ $NaNa_2(Fe_2Al_2Li)(OH)_2[Si_4O_{11}]_2$
*Potassic-leakeite $\qquad$ $KNa_2(Mg_2Al_2Li)(OH)_2[Si_4O_{11}]_2$
*Fluoroleakeite $\qquad$ $NaNa_2(Mg_2Al_2Li)F_2[Si_4O_{11}]_2$
*Fluoro-aluminoleakeite $\qquad$ $NaNa_2(Mg_2Al_2Li)F_2[Si_4O_{11}]_2$
*Oxo-mangani-leakeite $\qquad$ $NaNa_2(Mn^{3+}_4Li)O_2[Si_4O_{11}]_2$
*Lunijianlaite $\qquad$ $Li(OH)_{10}[Al_{3.5}Si_{3.5}O_{20}]$

Disilicates (K = 2) $\qquad$ Neutral
Spodumene (compare with pyroxene (family)) $\qquad$ $LiAl[Si_2O_6]^{\infty}$
*Watatsumiite $\qquad$ $KNa_2LiMn_2V^{4+}_2[Si_2O_6]_4$
*Balestraite $\qquad$ $KLi_2V^{5+}[Si_2O_6]_2$
$\qquad$ Acid
Nambulite series (compare with marsturite (group); pyroxenoids (family))
Nambulite $\qquad$ $NaLiMn_8[Si_5O_{14}(OH)]^{\infty}_2$
Natronambulite $\qquad$ $(Na,Li)(Mn,Ca)_4[Si_5O_{14}(OH)]^{\infty}$
*Tanohataite $\qquad$ $LiMn_2[Si_3O_8(OH)]$
Lithiomarsturite $\qquad$ $LiCa_2Mn_2[Si_5O_{14}(OH)]^{\infty}$
$\qquad$ Basic
Balipholite $\qquad$ $LiBaMg_2Al_3(OH)_4F_4[Si_2O_6]_2$
*Katayamalite $\qquad$ $KLi_3Ca_7Ti_2(OH)_2[Si_6O_{18}]_2$
*Aleksandrovite $\qquad$ $KLi_3Ca_7Sn_2F_2[Si_6O_{18}]_2$

Silicato-borates
Borosilicates
Monoborosilicates $K_\Sigma = 1$ $\qquad$ Basic
Manandonite $\qquad$ $Li_2Al_2(OH)_6\{Al_2(OH)_2[Si_2AlBO_{10}]^{\infty 2}\}^{\infty 2}$

Silicato-(3)-borates
Disilicato-(3)-borates $\qquad$ Basic
Elbaite series (compare with tourmaline (series))
Liddicoatite = fluor-liddicoatite
*Fluor-liddicoatite $\qquad$ $Ca(Li_3Al)Al_6(OH)_3F[Si_6O_{18}][BO_3]_3$
Elbaite $\qquad$ $Na(Li_{1.5}Al_{1.5})Al_6(OH)_3(OH)[Si_6O_{18}][BO_3]_3$
*Fluorelbaite $\qquad$ $Na(Li_{1.5}Al_{1.5})_{\Sigma 3}Al_6(OH)_3F[Si_6O_{18}][BO_3]_3$
*Rossmanite $\qquad$ $\square(LiAl_2)Al_6(OH)_3(OH)[Si_6O_{18}][BO_3]_3$
*Oxyrossmanite $\qquad$ $\square(Li_{0.5}Al_{2.5})Al_6O(OH)_3[Si_6O_{18}][BO_3]_3$
*Darrellhenryite $\qquad$ $Na(LiAl_2)Al_6O(OH)_3[Si_6O_{18}][BO_3]_3$

Beryllium silicates
Proper beryllium silicates $\qquad$ Hydrates
Beryllite $\qquad$ $Be_3(OH)_2[SiO_4]\cdot H_2O$

Beryllosilicates
Zero-beryllosilicates $(K_\Sigma = 0)$
Zero-beryllosilicates with $Be_\Sigma : Si > 1$
$\qquad$ Neutral
Phenakite $\qquad$ $Be[BeSiO_4]^{\infty 3}$
$\qquad$ Acid
Bertrandite $\qquad$ $Be_3[Be_5(OH)_4(Si_2O_7)_2]^{\infty 3}$

*Sphaerobertrandite $Be_3[Be_5(OH)_4(Si_2O_7)_2]^{\infty3}$

Zero-beryllosilicates with (Be,Al) : Si = 1   Neutral
Trimerite $CaMn^{2+}_2[BeSiO_4]_3$
Liberite $Li_2Be[SiO_4]$
  Helvine series
  Danalite $[BeSiO_4]_6\,Fe_8S_2$
  Helvine $[BeSiO_4]_6\,Mn_8S_2$
  Genthelvite $[BeSiO_4]_6\,Zn_8S_2$

*Zero-beryllosilicates with (Be,Al) : Si = 0,75 Hydrates
*Alflarsenite $NaCa_2[Be_3Si_4O_{13}(OH)]\cdot2H_2O$

Zero-beryllosilicates with (Be,Al) : Si = 0,5
                                  Neutral
  Beryl series
  Beryl $Al_2[Be_3(Si_6O_{18})]^{\infty3}$
  *Pezzottaite $CsAl_2[(Be_2Li)(Si_6O_{18})]$
  Bazzite $(Sc,Al)_2[Be_3(Si_6O_{18})]^{\infty3}$
  Chkalovite $Na_2[Be(Si_2O_6)]^{\infty3}$
  Tugtupite $[BeAlSi_4O_{12}]^{\infty3}\cdot Na_4Cl$
                              *Hydrates
  *Stoppanite $Fe^{3+}_3(Fe^{2+},Mg)Na[Be_3(Si_6O_{18})]^{\infty3}_2(H_2O)_2$

Zero-beryllosilicato-fluorides with (Be,Al) : Si = 0,7
  Meliphanite family
  Meliphanite $(Na,Ca,)_4Ca_4(F,O)_4[Be_4AlSi_7O_{24}]$

Zero-beryllosilicato-fluorides with (Be,Al) : Si = 0,5
Leucophanite $Ca_4Na_4F_4[Be_4Si_8O_{24}]$

Zero-beryllosilicates with (Be,Al) : Si < 0,5
                                  Neutral
*Oftedalite $K(Cs,Ca,Mn^{2+})_2[Be_3AlSi_{11}O_{30}]$
*Agakhanovite-(Y) $YCa\square_2K[Be_3Si_{12}O_{30}]$

Zero-beryllosilicates with (Be,Al) : Si < 0,5   Neutral $\rightarrow$ hydrates
  Epididymite family
  Epididymite $Be[Be_3(Si_6O_{15})_2]^{\infty3}\,Na_4(H_2O)_2$
  Eudidymite $Be[Be_3(Si_6O_{15})_2]^{\infty3}\,Na_4(H_2O)_2$
  Milarite $KCa_2[Be_2Al(Si_{12}O_{30})]^{\infty3}\,(H_2O)_{0,5}$
  *Almarudite $K(\square,Na)_2(Mn,Fe,Mg)_2[(Be,Al)_3(Si_{12}O_{30})]$
  *Eirikite $KNa_6[Be_2(Al_3Si_{15})O_{39}F_2]$
  Lovdarite $K_4Na_{12}[Be_8Si_{28}O_{72}]\cdot18H_2O$
  *Nabesite $Na_2[BeSi_4O_{10}]\cdot4H_2O$

Zero-monoberyllosilicates $(0 < K_\Sigma < 1)$
Zero-monoberyllosilicates with $K_\Sigma = 0,1$
                                  *Neutral

*Telyushenkoite $\qquad$ $CsNa_6[Be_2(Al_3Si_{15})O_{39}F_2]$
Leifite $\qquad$ $NaNa_6[Be_2Al_3Si_{15}O_{39}F_2]$

Zero-monoberyllosilicates with $K_\Sigma = 0,3$
Bavenite $\qquad$ $Ca_4[Be_2(OH)_2Al_2(Si_3O_{10})(Si_6O_{16})^x]^{\infty 3}$
$\qquad$ Hydrates (acid)
Chiavennite $\qquad$ $CaMn[Be_2Si_5O_{13}(OH)_2]\,2H_2O$
*Ferrochiavennite $\qquad$ $Ca_{1-2}Fe[Be_2Si_5O_{13}(OH)_2]\cdot 2H_2O$
*Roggianite $\qquad$ $Ca_2[BeAl_2Si_4O_{13}(OH)_2]\cdot 2,5H_2O$

Zero-monoberyllosilicates with $K_\Sigma = 0,(6)$ $\qquad$ Neutral
Gugiaite family
Gugiaite $\qquad$ $Ca_2[Be(Si_2O_7)]^{\infty 2}$
Jeffreyite $\qquad$ $(Ca,Na)_2[(Be,Al)Si_2(O,OH)_7]^{\infty 2}$
Barylite $\qquad$ $Ba[Be_2(Si_2O_7)]^{\infty 3.}$
*Clinobarylite $\qquad$ $Ba[Be_2Si_2O_7)]$

Zero-monoberyllosilicates with $K_\Sigma = 0,8$
$\qquad$ Acid
Aminoffite family
Harstigite $\qquad$ $Ca_6Mn[Be_2(OH)Si_3O_{11}]^{\infty 2}_2$
Aminoffite $\qquad$ $Ca_3[Be_2(OH)_2Si_3O_{10}]$

Monoberyllosilicates ($K_\Sigma = 1$) $\qquad$ Acid
Euclase $\qquad$ $Al[Be(OH)(SiO_4)]^\infty$
Bityite $\qquad$ $Ca\{LiAl_2(OH)_2[(BeAl)Si_2O_{10}]^{\infty 2}\}^{\infty 2}$

Diberyllosilicates ($K_\Sigma = 2$) $\qquad$ Beryllosilicato-fluorides
Hsianghualite $\qquad$ $Ca_3Li_2[Be_3Si_3O_{12}]^{\infty 3}F_2$
*Khmaralite $\qquad$ $Mg_4(Mg_3Al_9)O_4[Si_5Be_2Al_5O_{36}]$

*Di-triberillosilicates

Silicates and alumosilicates of Sn^{4+}
Proper silicates and alumosilicates
*Zero-monoberyllosilicates ($K = 1,25$)
*Eakerite $\qquad$ $Ca_2Sn^{4+}[Al_2Si_6O_{18}](OH)_2\cdot 2H_2O$

Monosilicates ($K = 1$) $\qquad$ Neutral
Brannockite (compare with osumilite (family)) $\qquad$ $KLi_3Sn^{4+}_2[Si_{12}O_{30}]^\infty$

Disilicates ($K = 2$) $\qquad$ Neutral
Pabstite (compare with benitoite (group) $\qquad$ $Ba(Sn,Ti)^{4+}[Si_3O_9]$
$\qquad$ Hydrates
Stokesite $\qquad$ $2CaSnSi_3O_9\,2H_2O \rightarrow Ca_2Sn^{4+}_2[Si_6O_{18}]^\infty\,4H_2O$

*Trisilicates ($K = 3$)
*Kristiansenite $\qquad$ $Ca_2ScSn[Si_2O_7][Si_2O_6OH]$

Tetrasilicates (K = 4)   Oxido-tetrasilicates
Malayaite (compare with titanite (group)) $Ca|Sn^{4+}O[SiO_4]|^{\infty 2}$

Beryllosilicates of Sn
Zero-monoberyllosilicates (K_Σ = 0,25)
                                          Hydrates
Sorensenite $Na_4Sn^{4+}[Be_2(Si_3O_9)_2]^{\infty 3}\cdot 2H_2O$

Mono-diberyllosilicates (K_Σ = 1,2)   Acid
Sverigeite $NaMn^{2+}_2Sn^{4+}[Be_2Si_3O_{12}(OH)]$

Silicates of Zn^{2+}, Pb^{2+}, As^{3+}, Sb^{3+} и Sb^{5+} paragenetic association of Franclin and Sterling Hill, New Jersey, USA, Langban and Jacobsberg, Sweden.

Minerals of Zn
Zero-mono(zinc)alumosilicates   Acid-neutral
Minehillite $(K,Na)_{2-3}Ca_{28}(OH)_{12}[(Zn_5Al_4Si_{40})O_{112}(OH)_4]$
(compare with reyerite, truscottite)

*Mono-disilicates   *Hydrates
*Gaultite $Na_4Zn_2[Si_7O_{18}]\cdot 5H_2O$

Disilicates   Neutral
Petedunnite $CaZn[Si_2O_6]^{\infty}$

Trisilicates   Neutral
Hardystonite (compare with melilite) $Ca_2|^{(4)}Zn[Si_2O_7]|^{\infty 2}$

*Tri-tetrasilicates
*Scheuchzerite (K = 3.8) $Na(Mn,Mg,Zn)_9[VSi_9O_{28}(OH)](OH)_3$

Tetrasilicates
Proper tetrasilicates   Neutral
Willemite $\{^{(4)}Zn_2[SiO_4]\}^{\infty 3}$
*Xingsaoite = Co-willemite) $(Zn,Co)_2[SiO_4]$
   Larsenite group (CN Zn^{2+} = 6)
   Esperite $Ca_2PbZn_3[SiO_4]_3$
   Larsenite $PbZn[SiO_4]$
                                       Basic
Hodgkinsonite family
   Gerstmannite $(Mn,Mg)(OH)_2\{^{(4)}Mg^{(4)}Zn[SiO_4]\}^{\infty 2}$
   Hodgkinsonite $Mn(OH)_2\{^{(4)}Zn_2[SiO_4]\}^{\infty 2}$
   *Franklinfurnaceite $Ca_2Fe^{3+}Mn^{3+}Mn^{2+}_3[Zn_2Si_2O_{10}](OH)_8$
                                       Hydrates
   Clinohedrite $Ca\{^{(4)}Zn(H_2O)[SiO_4]\}^{\infty 2}$

Tetrasilicato-arsenates   Basic
 Holdenite family
 Holdenite $(Mn,Mg)_6Zn_3(OH)_8[SiO_4][AsO_4]_2$

Kolicite $Mn_7Zn_4(OH)_8[SiO_4]_2[AsO_4]_2$

*Tetrasilicato -arsenito- arsenates   Основные
*Mcgovernite $Zn_3(Mn^{2+},Mg)_{42}(OH)_{40}[SiO_4]_8[As^{3+}O_3]_2[As^{5+}O_4]_4$

Minerals of Pb
*Zerosilucates c K = 0
*Plumbotsumite $Pb_5[Si_4O_8](OH)_{10}$

*Zero-monosilicates with K = 0,9
*Wickenburgite $Pb_3CaAl[AlSi_{10}O_{27}]\cdot(H_2O)_4$

*Monosilicates with K = 1   Hydrates
*Plumbophyllite $Pb_2[Si_4O_{10}]\cdot H_2O$

*Mono-disilicates with K = 1,6   Hydrates
*Luddenite $Pb_2Cu^{2+}_2[Si_5O_{14}]\cdot14H_2O$

*Mono-disilicato-(3)-borato-oxido-carbonates K = 1,6
*Britvinite $Pb_{14}Mg_9[Si_{10}O_{28}][BO_3]_4[CO_3]_2F_2(OH)_{12}$

Disilicates
Proter disilicates   Neutral
Margarosanite $Pb(Ca,Mn)_2[Si_3O_9]$

Disilicato-sulfates   Hydrates
Roeblingite $Pb_2Ca_6Mn(H_2O)_4(OH)_2[Si_3O_9]_2[SO_4]_2$

Trisilicates   Neutral
Barysilite $MnPb_8[Si_2O_7]_3$

Oxido-trisilicates
  Melanotekite series
  Melanotekite $Pb_2Fe^{3+}_2O_2[Si_2O_7]$
  Kentrolite $Pb_2Mn^{3+}_2O_2[Si_2O_7]$

*Trisilicato-halogenides
*Nasonite $Pb_6Ca_4[Si_2O_7]_3Cl_2$

Tri-tetrasilicates   Basic
Hancockite $CaPbFe^{3+}Al_2O(OH)[Si_2O_7][SiO_4]$
(compare with epidote (family))

*Tri-tetrasilicato-oxido-carbonates   Hydrates
Molybdophyllite   (K = 3.6) $Pb_8Mg_9[Si_{10}O_{30}(OH)_8][CO_3]_3\cdot H_2O$

Beryllosilicates
Mono-diberyllosilicates (K_Σ = 1,5)   Basic
Joesmithite (structure of amphibole) $PbCa_2(Mg_3Fe_2^{3+}(OH)_2[BeSi_3O_{11}]^\infty{}_2$

*Tri-tetrasilites (K = 3,4) $Si_2O_7 : SiO_4 = 2 : 3$ *Основные
*Samfowlerite $Ca_{28}Mn_6Zn_4(Zn_{1,5}Be_{2,5})_{\Sigma 4}Be_{12}[SiO_4]_{12}[Si_2O_7]_8(OH)_{12}$

Berylloborosilicates
Zero-monoberylloborosilicates $(0 < K_\Sigma < 1)$ Neutral
Zero-monoberylloborosilicato-fluorides
Hyalotekite $Pb_2Ba_2Ca_2[(Be_{0,5}Si_{9,5})_{\Sigma 10}B_2O_{28}]^{\infty 3}F$

*Zero-monoberylloborosilicato-halogenides Basic

*Wawayandaite $Ca_6Mn_2Be_5[BBe_4Si_6O_{23}](OH,Cl)_{15}$

Minerals of As and Sb
Proper silicates
Disilicates   Basic
 Schallerite family
 Schallerite $(Mn,Fe)_{16}As^{3+}_3(OH)_{17}[Si_{12}O_{36}]$
 Nelenite $(Mn,Fe)_{16}As^{3+}_3(OH)_{17}[Si_{12}O_{36}]$
 *Långbanite $Mn^{2+}_4Mn^{3+}_9Sb^{5+}[Si_2O_{24}]$

*Di-triberyllosilicates $(K_\Sigma = 2.2)$ *Oxido-di-triberyllosilicates
Welshite $Ca_4(Mg_9Sb^{5+}_3)O_4[Be_3Si_6AlO_{36}]$

Silicato-arsenates
*Tri-tetrasilicato-arsenates
*Ardennite-(As) $Mn^{2+}_4(Al,Fe^{3+})_5Mg[SiO_4]_2[Si_3O_{10}][(As,V)O_4](OH)_6$
*Ardennite-(V) $Mn^{2+}_4(Al,Fe^{3+})_5Mg[SiO_4]_2[Si_3O_{10}][(V,Si,As)O_4](OH)_6$

Tetrasilicato-arsenates   Basic
 Dixenite family
 Dixenite $CuMn^{2+}_{14}Fe^{3+}(As^{3+}O_3)_5(OH)_6[SiO_4]_2[As^{5+}O_4]$
 Kraisslite $Zn_3(Mn,Mg)_{25}(Fe^{3+},Al)(As^{3+}O_3)_2(OH)_{16}[(Si,As^{5+})O_4]_{10}$
                             Oxido-tetrasilicato-arsenates
 Parwelite $Mn^{2+}_{10}(Sb^{5+}O_4)_2[SiO_4]_2[AsO_4]_2$

*Tetrasilicato-antimonates
*Tegengrenite $(Mg,Mn)_4(Sb^{5+}O_4)[(Mn^{3+},Si,Ti)O_4]$

Silicates of f-elements
Proper silicates
*Zero-monosilicates with K = 0,(6)
*Chiappinoite-(Y) $Y_2Mn[Si_3O_7]_4$

*Zero-monosilicates with K = 0,75   Hydrates
*Thornasite $Na_{12}Th_3[Si_8O_{19}]_4 \cdot 18H_2O$
*Monteregianite –(Y) $Na_4K_2Y_2[Si_{16}O_{38}] \cdot 10H_2O$

Monosilicates (K = 1)   Neutral
 Ekanite family
 Ekanite $Ca_2Th[Si_4O_{10}]^{\infty 2}_2$

Iraqite-(La) $(K_{1-x}\square_x)Ca_2(La,Ce,Th)[Si_4O_{10}]^{\infty2}{}_2$
Steacyite $(K_{1-x}\square_x)(Na,Ca)_2Th[Si_4O_{10}]^{\infty2}{}_2$ $(x = 0,1\text{-}0,4)$
*Moskvinite-(Y) $Na_2K(Y,REE)[Si_6O_{15}]$
                                Hydrates
*Mendeleevite-(Ce) $Cs_6(Ce_{22}Ca_6)[Si_{70}O_{175}](OH,F)_{14}(H_2O)_{21}$
*Mendeleevite-(Nd) $Cs_6(Nd,REE)_{23}Ca_7[Si_{70}O_{175}](OH,F)_{19}(H_2O)_{16}$
*K-mendeleevite-(Ce) $K_6Cs_6(REE_{22}Ca_6)[Si_{70}O_{175}](OH,F)_{20}(H_2O)$
*Yakovenchukite-(Y) $K_3NaCaY_2[Si_{12}O_{30}](H_2O)_4$
*Turkestanite $(K_{1-x},\square_x)(Ca,Na)_2Th[Si_8O_{20}]\cdot nH_2O$
*Arapovite $(K_{1-x},\square_x)(Ca,Na)_2(U,Th)[Si_8O_{20}]\cdot H_2O$ $(x\sim0.5)$
Sazhinite-(Ce) $Na_3Ce[Si_6O_{15}]^{\infty2}6H_2O$
*Sazhinite-(La) $Na_3La[Si_6O_{15}](H_2O)_2$
Ashcroftine-(Y) $K_{10}Na_{10}(Y,Ca)_{24}(OH)_4(CO_3)_{16}[Si_{56}O_{140}]\cdot16H_2O$

Monosilicato-trisilicates (mixed)   Basic
Miserite-(Y) $K_2Ca_{10}Y_2(OH)_2F_2[Si_{12}O_{30}]^{\infty}[Si_2O_7]_2$
Miserite $K_2(Ca,Ce)_{12}(OH,F)_4[Si_{12}O_{30}]^{\infty}[Si_2O_7]_2$

Mono-disilicates (1 < K < 2) Neutral
Nordite-(La) $Na_6(Sr,Ca)_2(La,Ce)_2(Mn,Zn,Mg)_2[Si_{12}O_{34}]^{\infty}$
*Nordite-(Ce) $Na_6(Sr,Ca)_2(Ce,La)_2(Mn,Zn,Mg)_2[Si_{12}O_{34}]^{\infty}$
*Manganonordite-(Ce) $Na_6(Sr,Ba)_2(Ce,La)_2(Mn,Zn,Fe,Mg)_2[Si_{12}O_{34}]^{\infty}$
*Ferronordite-(La) $Na_3Sr(La,Ce)Fe^{2+}[Si_6O_{17}]^{\infty}$
*Ferronordite-(Ce) $Na_3SrCeFe^{2+}[Si_6O_{17}]^{\infty}$
                                Basic
*Atelisite-(Y) $Y_4(OH)_8[Si_3O_8]$

Disilicates (K = 2)   Hydrates
*Gerenite-(Y) $(Ca,Na)_2(Y,Th)_3[Si_6O_{18}]^{\infty}\cdot2H_2O$
                                *Asid
Thorosteenstrupine $CaThMn^{2+}[Si_4O_{11}(O,F)]\,6H_2O$

Disilicato-trisilicates (K = 2,(6))   Basic
Thalénite-(Y) $Y_3(OH)[Si_3O_{10}]$
*Fluorthalénite-(Y) $Y_3F[Si_3O_{10}]$

Trisilicates (K = 3)
Proper trisilicates   Neutral
 Thortveitite family
  Thortveitite group
  Keiviite-(Yb) $Yb_2[Si_2O_7]$
  *Keiviite-(Y) $(Y,Yb)_2[Si_2O_7]$
  Thortveitite-(Sc) $(Sc,Y)_2[Si_2O_7]$
  Yttrialite-(Y) $Y_2[Si_2O_7]$
  *Percleveite-(Ce) $(Ce,La,Nd)_2[Si_2O_7]$
Oxido-trisilicates
 Perrierite family
Perrierite-(Ce) $(Ce,La,Ca,Sr)_4(Fe^{2+},Mg,Mn^{2+})(Ti^{4+},Fe^{3+})_4O_8[Si_2O_7]_2$
*Perrierite-(La) $(La,Ce,Ca)_4Fe^{2+}(Ti^{4+},Fe^{3+})_4O_8[Si_2O_7]_2$
*Matsubaraite $Sr_4Ti_5O_8[Si_2O_7]_2$

Chevkinite-(Ce) $(REE,Ca)_4Fe^{2+}_2(Ti^{3+},Fe^{3+})_3O_8[Si_2O_7]_2$
*Dingdaohengite-(Ce) $(Ce,La)_4Fe^{2+}(Ti,Fe^{2+},Mg,Fe^{3+})_2Ti_2O_8[Si_2O_7]_2$
*Maoniupingite-(Ce) $(Ca,Ce)_4(Fe^{3+}TiFe^{2+}\square)(TiFe^{3+}Nb)_4O_8[Si_2O_7]_2$
*Cr-chevkinite (Ce,La,Nd,Pr,Th)$_4(Mg,Fe,Ca)_2Cr^{3+}_2(Ti,Al,Nb)_2O_8[Si_2O_7]_2$
*Polyakovite-(Ce) $(REE,Ca)_4(Mg,Fe^{2+})(Cr,Fe^{3+})_2(Ti,Nb)_2O_8[Si_2O_7]_2$
Strontiochevkinite $(Sr,La,Ce,Ca)_4(Fe^{2+},Fe^{3+})(Ti^{3+},Zr)_4O_8[Si_2O_7]_2$
*Christofschäferite-(Ce) $(Ce,La,Ca)_4Mn(Ti,Fe)_3(Fe,Ti)O_8[Si_2O_7]_2$
*Fogoite-(Y) $Na_3Ca_2Y_2TiOF_3[Si_2O_7]_2$
*Stavelotite-(La) $La_3Mn^{2+}_3Cu(Mn^{3+}_{24}Fe^{3+}Mn^{4+})O_{30}[Si_2O_7]_6$

                                    Trisilicato-fluorides
Rowlandite-(Y) $Fe^{2+}Y_4[Si_2O_7]_2F_2$

Trisilicato-tetrasilicates   Neutral
*Perböite-(Ce) $(CaCe_3)(Al_3Fe^{2+})O(OH)_2[Si_2O_7][SiO_4]_3$
*Alnaperböite-(Ce) $(CaCe_{2.5}Na_{0.5})Al_4O(OH)_2[Si_2O_7][SiO_4]_3$

Oxido(fluorido)-hydroxido-trisilicato-tetrasilicates
  Allanite series (compare with epidote (family))(K = 3,3)
    *Allanite-(La) $LaCaAl_2Fe^{2+}_3O(OH)[Si_2O_7][SiO_4]$
    Allanite-(Ce) (orthite) $CaCeAl_2Fe^{2+}O(OH)[Si_2O_7][SiO_4]$
    *Allanite-(Nd) $CaNdAl_2Fe^{2+}O(OH)[Si_2O_7][SiO_4]$
    Allanite-(Y) (yttriumorthite) $CaYAl_2Fe^{2+}O(OH)[Si_2O_7][SiO_4]$
    *Ferriallanite-(Ce) $CaCeFe^{3+}AlFe^{2+}O(OH)[Si_2O_7][SiO_4]$
    *Uedaite-(Ce) $(Mn^{2+}Ce)(Al_2Fe^{2+})O(OH)[Si_2O_7][SiO_4]$
    *Dissakisite-(Ce) $Ca(Ce,La)MgAl_2O(OH)[Si_2O_7][SiO_4]$
    *Androsite-(La) $Mn^{2+}LaMn^{2+}Mn^{3+}AlO(OH)[Si_2O_7][SiO_4]$
    *Manganiandrosite-(Ce) $Mn^{2+}CeMn^{3+}AlMn^{2+}O(OH)[Si_2O_7][SiO_4]$
    *Vanadoandrosite-(Ce) $Mn^{2+}CeV^{3+}AlMn^{2+}O(OH)[Si_2O_7][SiO_4]$
    *Áskagenite-(Nd) $Mn^{2+}NdAl_2Fe^{3+}O_2[Si_2O_7][SiO_4]$
    Dollaseite-(Ce) $Ca(Ce,La,Nd)Mg_2AlF(OH)[Si_2O_7][SiO_4]$
    *Västmanlandite-(Ce) $(Ce,La)_3CaAl_2Mg_2F(OH)_2[Si_2O_7][SiO_4]_3$
    *Khristovite-(Ce) $CaCeMgMn^{2+}AlF(OH)[Si_2O_7][SiO_4]$
    *Gatelite-(Ce) $(CaCe_3)(Al_3Mg)O(OH)_2[Si_2O_7][SiO_4]_3$

Tetrasilicates (K = 4)   Neutral→ acid
*Unnamed $CaCe_2[SiO_4]_2$
Thorite family
Thorite $Th[SiO_4]$
*Stetindite-(Ce) $Ce^{4+}[SiO_4]$
Thorogummite $Th[(SiO_4)_{1-x}(OH)_{4x}]$
Coffinite $U[SiO_4]\cdot nH_2O$
Huttonite $Th[SiO_4]$
Tombarthite-(Y) $Y_4(Si,H_4)_4O_{12-x}(OH)_{4+2x}$
Vyuntspakhkite-(Y) $Y(Al,Si)[SiO_4](OH,O)_2$
                                    Acid-basic
Törnebohmite-(Ce) $(Ce,La)_2Al(OH)[SiO_4]_2$
*Törnebohmite-(La) $(La,Ce)_2Al(OH)[SiO_4]_2$
Cerite-(Ce) $(Ce,La,Ca)_9(Mg,Fe^{3+})(OH)_3[SiO_4]_3[SiO_3(OH)]_4$
*Aluminocerite-(Ce) $(Ce,La)_9(Al,Fe^{3+})(OH)_3[SiO_4]_3[SiO_3(OH)]_4$
*Cerite-(La) $(La,Ce,Ca)_9(Fe^{3+},Ca,Mg)(OH)_3[SiO_4]_3[SiO_3(OH)]$

Britholite series
Britholite-(Ce) $(Ce,Ca)_5(OH)[SiO_4]_3$
Britholite-(Y) $(Y,Ca)_5(OH)[SiO_4]_3$
Tetrasilicato-halogenides
 *Fluorbritholite-(Ce) $(Ce,Ca)_5F[SiO_4]_3$
 Fluorbritholite-(Y) $(Y,Ca)_5F[SiO_4)]_3$
 *Fluorcalciobritholite $(Ca,REE)_5F[(Si,P)O_4]_3$
                                        Basic
Kuliokite-(Y) $(Y,Yb)_4Al(OH)_2[SiO_4]_2F_5$

Silicates of f-cations with unknown structure
Umbozerite $Na_3Sr_4Th[Si_8O_{23}OH]$

Beryllosilicates
*Zero-monoberyylosilicates $(K_\Sigma = 0,3)$
*Bussyite-(Ce) $(Ce,REE)_3(Na,H_2O)_6Mn[Be_5Si_9(O,OH)_{30}]F_4$

Zero-monoberyylosilicates $(K_\Sigma = 0,8)$ Acid
Semenovite-(Ce) $Na_8Ca_2Ce_2Fe^{2+}[Be_6Si_{14}O_{40}]F_4(OH)_4$

Monoberillosilicates   Neutral $\rightarrow$Acid
 Gadolinite series
 Gadolinite-(Ce) $Ce_2Fe^{2+}[BeO(SiO_4)]^{\infty2}_2$
 Gadolinite-(Y) $Y_2Fe^{2+}[BeO(SiO_4)]^{\infty2}_2$
 Minasgeraisite-(Y) $CaY_2[BeO(SiO_4)]^{\infty2}_2$
 Hingganite-(Ce) $(Ce,Y,Yb)[Be(OH)(SiO_4)]^{\infty2}$
 Hingganite-(Y) $(Y,Yb,Er)[Be(OH)(SiO_4)]^{\infty2}$
 Hingganite-(Yb) $(Yb,Y)[Be(OH)(SiO_4)]^{\infty2}$
 *Calcybeborosilite-(Y) $(Y,Ca)_2(\square,Fe^{2+})[(B,Be)_2(OH,O)(SiO_4)]_2$

*Tetraberillosilicatesr $(K = 4)$
*Makarochkinite $Ca_2Fe^{2+}_4Fe^{3+}TiO_2[BeAlSi_4O_{18}]$

Borosilicates
Zero-borosilicates   Basic
Tritomite-(Ce) $Ce_5(OH,O)[(SiO_4,BO_4)_3]$
Tritomite-(Y) $Y_5(OH,O,F)[(SiO_4,BO_4)_3]$

*Zero-monoborosilicates $(K = 0.8)$
*Perettiite-(Y) $Y^{3+}_2Mn^{2+}_4Fe^{2+}[Si_2B_8O_{24}]$

Monoborosilicates   Neutral
Stillwellite-(Ce) $(Ce,La,Ca)BSiO_5 \rightarrow (Ce,La,Ca)_3[(SiO_3)_3(B_3O_6)^{\infty}]^{\infty}$

Mono-diborosilicates   Basic
Tadzhikite-(Ce) $Ca_2(Ca,Y)_2(Ti,Al,Fe^{3+})(Ce,Y,\square)(OH)_2[Si_4B_4O_{16}(O,OH)_6]$
Tadzhikite-(Y) $(Ca,Ce)_4(Ti^{4+},Fe^{3+},Al)(Y,Ce)_2(OH)_2[B_4Si_4O_{22}]$
                                        Acid
*Hellandite-(Ce) $(Ca,REE)_4Ce_2Al\square_2(OH)_2[Si_2B_2O_{11}]_2$

Hellandite-(Y) $(Ca,REE)_4Y_2Al\square_2(OH)_2[B_2Si_2O_{11}]_2$
*Mottanaite-(Ce) $Ca_4(Ce,Ca)_2AlBe_2O_2[Si_2B_2O_{11}]_2$
*Ciprianite $Ca_4[(Th,U)REE]_2Al\square_2(OH)_2[Si_2B_2O_{11}]_2$

*Tri-tetraborosilicates K = 3,(1)
*Proshchenkoite-(Y) $Ca(Y,REE,CaNa,Mn)_{15}(Fe^{2+},Mn)(P,Si)[Si_6B_3O_{34}]F_{14}$

*Tri-tetraborosilicates K = 3,(1)
*Laptevite-(Ce) $NaFe^{2+}(REE_7Ca_5Y_3)[SiO_4]_4[Si_3B_2PO_{18}][BO_3]F_{11}$

*Tri-tetraborosilicates K = 3,6
Okanoganite-(Y) $(Ca,Na,REE,Th)_{16}Fe^{3+}[Si_7B_3O_{34}(OH)_4]F_{10}$

*Tetraborosilicates
*Vicanite-(Ce) $(Ca,REE,Th)_{15}As^{5+}(As^{3+}_{0,5},Na_{0,5})Fe^{3+}[Si_6B_4O_{40}]F_7$

Borosilicates of *f*-cations with unknown structure
Melanocerite-(Ce) $Ce_5(OH,O)[(SiO_4,BO_4)]_3$
Cappelenite-(Y) $Ba(Y,Ce)_6[Si_3B_6O_{24}]F_2$

Silicato-phosphates
Disilicato-phosphates   Neutral
Phosinaite-(Ce) $Na_{13}Ca_2Ce[SiO_3]_4[PO_4]_4$

Tetrasilicato-phosphates-sulfates   Hydrates
Saryarkite-(Y) $Ca(Y,Th)Al_5(OH)_7[SiO_4]_2[PO_4][SO_4]\cdot6H_2O$

*Silicato-phosphato-carbonates
*Abenakiite-(Ce) $Na_{26}REE_6[SiO_3]_6[PO_4]_6[CO_3]_6(S^{4+}O_2)O$

Silicato-carbonates
Monosilicato-carbonates   Hydrates
Caysichite-(Y) $Y_4Ca_3Gd(OH)[Si_8O_{20}][CO_3]_6\cdot7H_2O$

Disilicato-carbonates   Hydrates
Kainosite-(Y) $Ca_2(Y,Ce)_2[SiO_3]_4[CO_3]H_2O$

*Trisilicato-carbonates   Средние
*Biraite-(Ce) $Ce_2Fe^{2+}[Si_2O_7][CO_3]$

Tetrasilicato-carbonates   Neutral
Iimoriite-(Y) $Y_2[SiO_4][CO_3]$

Subclass: Silicates of cations with middle FC
Silicates of V^{4+}
Monosilicates   Hydrates
 Cavansite family
 Cavansite $Ca(V^{4+}O)[Si_4O_{10}]^{\infty2}\cdot4H_2O$
 Pentagonite $Ca(V^{4+}O)[Si_4O_{10}]^{\infty2}\cdot4H_2O$

Trisilicates   Neutral $BaV^{4+}Si_2O_7$

   Haradaite group

   Haradaite $SrV^{4+}[Si_2O_7]$

   Suzukiite $BaV^{4+}[Si_2O_7]$

*Tetrasilicates

                           *Oxido-tetrasilicates

*Vanadomalayaite $Ca\{V^{4+}O[SiO_4]\}^{\infty 2}$

Silicates of $Zr \rightarrow$ zirconium silicates

Zirconium silicates of s-, d_s- and p_s-cations

Zirconium silicates of s-, d_s- and p_s-cations without Li^+ and Be^{2+}

Proper zirconium silicates

Zirconomonosilicates (K = 1)   Neutral

Dalyite $K_2|Zr[Si_6O_{15}]^{\infty 2}|^{\infty 3}$

                           Hydrates

Elpidite family

   Armstrongite $Ca|Zr[Si_6O_{15}]^{\infty 2}|^{\infty 3}(H_2O)_3$

   Elpidite $Na_2|Zr[Si_6O_{15}]^{\infty 2}|^{\infty 3}(H_2O)_3$

   *Yusupovite $Na_2Zr[Si_6O_{15}]\cdot 2.5H_2O$

   *Zeravshanite $Cs_4Na_2Zr_3[Si_6O_{15}]_3(H_2O)_2$

Zirconomono-disilicates ($1 < K < 2$)

Zirconomono-disilicates with K = 1,2 Neutral

Lemoynite $|CaZr_2[Si_{10}O_{26}]^{\infty 2}|^{\infty 3}(Na,K)_2(H_2O)_{5-6}$

*Natrolemoynite $|Na_4Zr_2[Si_{10}O_{26}]^{\infty 2}|^{\infty 3}\cdot 9H_2O$

Zirconomono-disilicates with K = 1,(3)   Hydrates

*Kapustinite $Na_{5,5}Mn_{0,25}Zr[Si_6O_{16}](OH)_2$

Terskite $Na_4Zr[Si_6O_{16}]\cdot 2H_2O$

*Hydroterskite $Na_2Zr[Si_6O_{12}(OH)_4](OH)_2$

Zirconomono-disilicates with K = 1,5   Neutral

Vlasovite $Na_2|Zr[Si_4O_{11}]^{\infty}|^{\infty 3}$

                          *Hydrates

*Tumchaite $Na_2(Zr,Sn)[Si_4O_{11}]\cdot 2H_2O$

*Zirconomono-disilicates with K = 1,54

*Rastsvetaevaite $Na_{27}K_8Ca_{12}Fe_3Zr_6[Si_{52}O_{144}](O,OH)_6Cl_2$

*Zirconomono-disilicates with K = 1,77

*Aqualite $(H_3O)_8(Na,R,Sr)_5Ca_6Zr_3[Si_{26}O_{66}(OH)_9]Cl$

Zirconomono-disilicates with K = 1,84

                           Neutral

*Khomyakovite $Na_{12}Sr_3Ca_6Fe_3WZr_3[Si_{25}O_{73}](O,OH,\cdot H_2O)_3(Cl,OH)_2$

*Manganokhomyakovite $Na_{12}Sr_3Ca_6Mn_3WZr_3[Si_{25}O_{73}](O,OH,\cdot H_2O)_3(Cl,OH)_2$

*Raslakite $Na_{15}Ca_3Fe_3(Na,Zr)_3Zr_3(Si,Nb)[Si_{25}O_{73}](OH,\cdot H_2O)_3(Cl,OH)_2$

*Mn analog raslakite $[Na,H_3O]_{15}[Ca_3Mn]_3Na_3Zr_3(Si,Ti)[Si_{25}O_{72}OH](OH)_2\cdot 2H_2O$

Zirconodisilicates with (K = 2)   Neutral
Wadeite family
Bazirite $Ba|Zr[Si_3O_9]|^{\infty 3}$
Wadeite $K_2|Zr[Si_3O_9]|^{\infty 3}$

Neutral → acid → hydrates

*__Eudialyte__
Eudialyte $Na_{15}Ca_7Fe^{2+}_3Zr_3SiO[Si_3O_9]_2[Si_9O_{27}]_2\}^{\infty 3}$ $(OH)_2Cl$
*Davinciite $Na_{12}K_3Ca_6Fe^{2+}_3Zr_3[Si_{26}O_{72}](O,OH)_2Cl_2$
*Ikranite $Na_{15}Ca_6Fe^{3+}_2Zr_3(\square,Zr)(\square,Si)[Si_{24}O_{66}(O,OH)]\cdot nH_2O$
*Georgbarsanovite $Na_{12}(Mn,Sr,REE)_3Ca_6Fe^{2+}_3Zr_3Nb[Si_{25}O_{76}]Cl_2\cdot H_2O$
*Zirsilite-(Ce) $(Na,\square)_{12}(Ce,Na)_3Ca_6Mn_3Zr_3Nb[Si_{25}O_{73}](OH)_3(CO_3)\cdot H_2O$
*Kentbrooksite $(Na,REE)_{15}(Ca,REE)_6Mn^{2+}_3Zr_3Nb[Si_{25}O_{74}]F_2\cdot 2H_2O$
*Carbokentbrooksite $(Na,\square)_{12}(Na,Ce)_3Ca_6Mn_3Zr_3Nb[Si_{25}O_{73}](OH)_3(CO_3)\cdot H_2O$
*Ferrokentbrooksite $Na_{15}Ca_6Fe_3Zr_3Nb[Si_{25}O_{73}](O,OH,\cdot H_2O)_3(F,Cl)_2$
*Andrianovite $Na_{12}(Ca,Sr,Ce)_3Ca_6Mn_3Zr_3Nb[Si_{25}O_{73}](O,OH,\cdot H_2O)_5$
*Voronkovite $Na_{15}(Na,Ca,Ce)_3(Mn,Ca)_3Fe_3Zr_3[Si_{26}O_{72}](OH,O)_4Cl\cdot H_2O$
*Manganoeudialyte $Na_{14}Ca_6Mn_3Zr_3[Si_{26}O_{72}](OH)_2]Cl_2\cdot 4H_2O$
*Ilyukhinite $(H_3O,Na)_{14}Ca_6Mn_2Zr_3[Si_{26}O_{72}(OH)_2]\cdot 3H_2O$
*Golyshevite $(Na,Ca)_{10}Ca_9(Fe^{3+},Fe^{2+})_2Zr_3Nb[Si_{25}O_{72}](OH)_3(CO_3)\cdot H_2O$
*Mogovidite $Na_9(Ca,Na)_6Ca_6(Fe^{3+},Fe^{2+})_2Zr_3\square[Si_{25}O_{72}](CO_3)(OH,\cdot H_2O)_4$
*Taseqite $Na_{12}Sr_3Ca_6Fe_3Zr_3Nb[Si_{25}O_{73}](O,OH,\cdot H_2O)_3Cl_2$
*Johnsenite-(Ce) $Na_{12}(Ce,La,Sr,Ca,\square)_3Ca_6Mn_3Zr_3W[Si_{25}O_{73}][CO_3](OH,Cl)_2$
*Oneillite $Na_{15}Ca_3Mn_3Fe_3Zr_3Nb[Si_{25}O_{73}](O,OH,\cdot H_2O)_3(OH)_2$
*Feklichevite $Na_{11}Ca_9(Fe^{3+},Fe^{2+})_2Zr_3Nb[Si_{25}O_{73}](OH,\cdot H_2O,Cl,O)_5$
*Bobtreilite $(Na,Ca)_{13}Sr_{11}(Zr,Y,Nb)_{14}[Si_{42}B_6O_{132}(OH)_{12}]\cdot 12H_2O$
*Labyrinthite $(Na,K,Sr)_{35}Ca_{12}Fe_3Zr_6Ti[Si_{51}O_{144}(O,OH,\cdot H_2O)_9]Cl_3$
*Dualite
          $Na_{30}(Ca,Na,Ce,Sr)_{12}(Na,Mn,Fe,Ti)_6Zr_3Ti_3Mn[Si_{51}O_{144}](OH,\cdot H_2O,Cl)_9$
Lovozerite family
*Townendite $Na_8Zr[Si_6O_{18}]$
Koashvite $Na_6Ca)|Ti^{4+}[Si_6O_{18}]|^{\infty 3}$
Zirsinalite $Na_6(Ca,Mn,Fe^{2+})|Zr[Si_6O_{18}]|^{\infty 3}$
Lovozerite $Na_3Ca\,|\,Zr[Si_6O_{15}(OH,O)_3]\,|^{\infty}$
Petarasite $Na_5Zr\,|\,Zr[Si_6O_{18}]\,|^{\infty 3}(Cl,OH)\cdot 2H_2O$
*Litvinskite $Na_3Zr[Si_6O_{13},(OH)_5]^{\infty 3}$
Catapleiite family
  Catapleiite series
  Catapleiite $|Zr[Si_3O_9]|^{\infty 3}Na_2\cdot 2H_2O$
  Gaidonnayite $Zr[Si_3O_9]|^{\infty 3}Na_2(H_2O)_2$
  *Calcigaidonnayite $|\,Zr[Si_3O_9]\,|^{\infty 3}(Ca,Na,K)_{2-x}\cdot n(H_2O)$
  Georgechaoite $|Zr[Si_3O_9]|^{\infty}NaK(H_2O)_2$
  Calciohilairite $|Zr[Si_3O_9]|^{\infty}Ca(H_2O)_3$
  Komkovite $|Zr[Si_3O_9]|^{\infty 3}Ba(H_2O)_3$
  *Rogermitchellite $Na_6(Sr,Na)_{12}Ba_2Zr_{13}[Si_{39}(B,Si)_6O_{123}](OH)_{12}\cdot 9H_2O$
  Hilairite $|Zr[Si_3O_9]|^{\infty 3}Na_2(H_2O)_3$
  *REE analog hilairite $Na_{4.34}K_{0.57}(Y_{0.69}REE_{0.17})(Zr_{0.65}Ti_{0.20}Nb_{0.11})[Si_6O_{18}]\cdot 6H_2O$
Kostylevite family
Kostylevite $|Zr[Si_3O_9]|^{\infty 3}K_2(H_2O)$
Umbite $|Zr[Si_3O_9]|^{\infty 3}K_2(H_2O)$

Paraumbite $\qquad$ $|Zr_2[Si_3O_9]|^{\infty3}{}_2K_3H\cdot3H_2O$

Zirconotrisilicates $(K = 3)$

Neutral

Keldyshite family

Gittinsite $\qquad$ $Ca|Zr[Si_2O_7]|^{\infty3}$
Parakeldyshite $\qquad$ $Na_2|Zr[Si_2O_7]|^{\infty3}$
Keldyshite $\qquad$ $Na_{2-x}H_x|Zr[Si_2O_7]|^{\infty3}(H_2O)_n$
Khibinskite $\qquad$ $K_2|Zr[Si_2O_7]|^{\infty3}$

Basic

*Dovyrenite $\qquad$ $Ca_6|Zr[Si_2O_7]_2(OH)_4$

Oxido-hydroxido-zirconotrisilicato-fluorides $(Na,Ca)_2Ca_4Zr(Mn,Ti,Fe)(F,O)_4[Si_2O_7]_2$

Baghdadite family

Baghdadite $\qquad$ $Ca_3|ZrO_2[Si_2O_7]|$
Burpalite $\qquad$ $Na_2Ca|ZrF_2[Si_2O_7]|$
Hiortdahlite $\qquad$ $(Na,Ca)_2Ca_4Zr(Mn,Ti,Fe)(F,O)_4[Si_2O_7]_2$
*Rengeite $\qquad$ $Sr_4ZrTi_4[Si_4O_{22}] \rightarrow Sr_4ZrTiTi_3O_8[Si_2O_7]_2$

Tetrasilicates of Zr $\qquad$ Neutral

Zircon series

Zircon $\qquad$ $Zr[SiO_4]$
*Reidite $\qquad$ $Zr[SiO_4]$
Hafnon $\qquad$ $(Hf,Zr)[SiO_4]$
Kimzeyite (compare with garnet (series)) $\qquad$ $Ca_3Zr_2[(SiAl_2)O_{12}]$

Zirconosilicates with unknown structure
Loudounite $\qquad$ $NaCa_5Zr_4Si_{16}O_{40}(OH)_{11}\cdot8H_2O$

Zirconodisilicato-phosphates $\qquad$ Hydrates
Steenstrupine-(Ce) $\qquad$ $Na_{14}|Ce^{3+}{}_6Mn^{2+}Mn^{3+}Fe^{2+}{}_2(Zr,Th)[Si_6O_{18}]_2[PO_4]_7|^{\infty2}\cdot3H_2O$

Zirconosilicates of Li
Zirconomonosilicates $(K = 1)$ $\qquad$ Neutral

Sogdianite family (compare with osumilite (family))

Sogdianite $\qquad$ $KNa|Li_2Zr_2[Si_{12}O_{30}]|^{\infty3}$
Darapiosite $\qquad$ $KNa_2(Mn, Zr)_2|Li_2 Zn[Si_{12}O_{30}]|^{\infty3}$
Zektzerite $\qquad$ $Na_2|Li_2ZrZr[Si_{12}O_{30}]|^{\infty3}$
*Dusmatovite $\qquad$ $K(K,Na,\square)(Zn,Li)_3(Mn^{2+},Y,Zr)_2[Si_{12}O_{30}]$

Zirconosilicates of *f*-cations
Tranquillityite $\qquad$ $Fe_8(Zr,Y)_2Ti_3Si_3O_{24}$
*Sazykinaite-(Y) $\qquad$ $Na_5YZr[Si_6O_{18}]\cdot6H_2O$

Titanosilicates (with niobo- and tantalosilicates)
Titanosilicates of *s*-, *d$_s$*- and *p$_s$*-cations
Titanosilicates of *s*-, *d$_s$*- and *p$_s$*-cations without Li^+ and Be^{2+}
Proper titanosilicates
*Titano-zero-monosilicates

*Lourenswalsite   (K = 0.66) $(K,Ba)_2Ti_4[(Si,Al,Fe)_6O_{14}](OH)_{12}$

*Oxido-titano-zero-monosilicates (K = 0,86)
                              *Hydrates
*Nafertisite $Na_3Fe^{2+}_{10}O_2(OH)_6[Ti_2Si_{12}O_{34}]F \cdot 2H_2O$

Titano-monosilicates (K = 1)   Neutral
Davanite $K_2|Ti[Si_6O_{15}]^{\infty2}|^{\infty3}$
                              *Hydrates
*Ershovite $Na_4K_3(Fe^{2+},Mn^{2+},Ti)_2(OH)_4[Si_8O_{20}] \cdot 4H_2O$
*Paraershovite $Na_3K_3Fe^{3+}_2(OH)_2[Si_8O_{20}(OH)_2] \cdot (H_2O)_4$

Oxido-titanosilicates →   Hydrates
Narsarsukite family
Narsarsukite $Na_2|TiO[Si_4O_{10}]^{\infty}|^{\infty3}$
Penkvilksite $Na_2|TiO[Si_4O_{10}]^{\infty}|^{\infty3} \cdot 2H_2O$
*Intersilite $Na_6Mn^{2+}Ti[Si_{10}O_{24}(OH)](OH)_3 \cdot 4H_2O$

Titanomono-disilicates with mixed silicooxygen anions   Hydrates
Vinogradovite $Na_4|(TiO)_4[Si_4O_{10}]^{\infty}[Si_2O_6]^{\infty}_2|^{\infty3 \cdot} (H_2O,Na,K)_3$
*Paravinogradovite $(Na,\square)_2(Ti^{4+},Fe^{3+})_4[Si_3AlO_{10}][Si_2O_6]_2(OH)_4 \cdot H_2O$

*Titanomono-disilicates with K = 1.14   Basic
*Senkevichite $CsNaKCa_2TiO[Si_7O_{18}](OH)$

*Titanomono-disilicates with K = 1.28
*Caryochroite $(Na,Sr)_3(Fe^{3+},Mg)_{10}[Ti_2Si_{12}O_{37}](O,OH)_9 \cdot 8H_2O$

Titanomono-disilicates (K = 1,4)   Basic
Tinaksite $NaK_2Ca_2|Ti(OH)[Si_7O_{19}]^{\infty}|^{\infty2}$

*Titano(niobo)mono-disilicates   *Hydrates
*Haineaultite (K = 1,(6)) $(Na,Ca)_5Ca(Ti,Nb)_5[(Si,S)_{12}O_{34}](OH,F)_8 \cdot 5H_2O$
*Chivruaiite $Ca_4(Ti,Nb)_5[Si_6O_{17}]_2(OH,O)_5 \cdot 13-14H_2O$

Титано(ниобо)моно-дисиликато-хлориды   (K = 1,7)
*Alluaivite $Na_{19}(Ca,Mn^{2+})_6(Ti,Nb)_3[Si_{26}O_{74}]Cl \cdot 2H_2O$

Titanodisilicates (K = 2)   Neutral
Benitoite $Ba|Ti[Si_3O_9]|^{\infty3}$
                              *Hydrates
*Zorite $Na_6Ti_5[Si_{12}O_{36}](O,OH)_3 \cdot 11H_2O$

Oxido-titanodisilicates
Baotite $Ba_4Ti_4(Ti,Nb,Fe)_4O_{16}[Si_4O_{12}]Cl$
*Niobobaotite $Ba_4(Nb,Ti,Fe)_8O_{16}[Si_4O_{12}]Cl$
Lorenzenite $|Na_2Ti_2O_3[Si_2O_6]^{\infty}|^{\infty2}$
Shcherbakovite $K_2NaTi_2^{4+}[Si_4O_{12}]O(OH)$
Batisite family
Batisite $Na_2BaO_2|Ti_2[Si_4O_{12}]|^{\infty3}$

*Noonkanbahite $\qquad$ $KNaBaO_2Ti_2[Si_4O_{12}]$

Aenigmatite series (compare with krinovite (series))

Rhönite $\qquad$ $|Ca_4,Mg_8Ti_2O_4[(SiAl)_6O_{36}]^{\infty}|^{\infty 2}$

Aenigmatite $\qquad$ $|Na_2Fe^{2+}_5TiO_2[Si_6O_{18}]^{\infty}|^{\infty 2}$

*Høgtuvaite $\qquad$ $Ca_4(Fe_6^{2+}Fe_6^{3+})O_4[Si_8Be_2Al_2O_{36}]$

*Hydrates

Kukisvumite $\qquad$ $Na_6ZnO_4\,|\,Ti_2[Si_4O_{12}]\,|^{\infty 3}_2\cdot 4H_2O$

*Manganokukisvumite $\qquad$ $Na_6MnO_4\,|\,Ti_2[Si_4O_{12}]\,|^{\infty 3}_2\cdot 4H_2O$

Titanodisilicato-fluorides $\qquad$ Basic

Yuksporite

$K_4(Ca,Na)_{14}(Sr,Ba)_2(\square,Mn,Fe)(Ti,Nb)_4(O,OH)_4[Si_6O_{17}]_2[Si_2O_7]_3(H_2O,OH)_3$

*Eveslogite $\qquad$ $(Ca_{25}K_{24})Ti_{12}[Si_4O_{12}]_{12}(OH)_{12}F_{14}$

Labuntsovite family

Nenadkevichite group

Nenadkevichite $\qquad$ $Na_{8-x}Nb_4[Si_4O_{12}]_2(O,OH)_4\cdot 8H_2O$

*Korobitsynite $\qquad$ $Na_{8-x}Ti_4[Si_4O_{12}]_2(O,OH)_4\cdot 8H_2O$

Vuoriyarvite group

*Vuoriyarvite-K $\qquad$ $(K,Na,\square)_{12}Nb_8[Si_4O_{12}]_4O_8\cdot 12\text{-}16H_2O$

*Tsepinite-Ca $\qquad$ $(Ca,K,Na)_{2-x}(Ti,Nb)_2[Si_4O_{12}](OH,O)_2\cdot 4H_2O$

*Tsepinite-K $\qquad$ $(K,Ba,Na)_2(Ti,Nb)_2[Si_4O_{12}](OH,O)_2\cdot 3H_2O$

*Tsepinite-Na $\qquad$ $(Na,H_3O,K,Sr,Ba,\square)_{12}Ti_8[Si_4O_{12}]_4(OH,O)_8\cdot 12\text{-}16H_2O$

*Tsepinite-Sr $\qquad$ $(Sr,Ba,K)(Ti,Nb)_2[Si_4O_{12}](OH,O)_2\cdot 3H_2O$

Paratsepinite group

*Paratsepinite-Ba $\qquad$ $(Ba,Na,K)_{2-x}(Ti,Nb)_2[Si_4O_{12}](OH,O)_2\cdot 4H_2O$

*Paratsepinite-Na $\qquad$ $(Na,Sr,K,Ca)_2(Ti,Nb)_2[Si_4O_{12}](O,OH)_2\cdot 4H_2O$

Kuzmenkoite group

*Kuzmenkoite-Mn $\qquad$ $K_4Mn_2Ti_8[Si_4O_{12}]_4(OH,O)_8\cdot 10\text{-}12H_2O$

*Kuzmenkoite-Zn $\qquad$ $K_2ZnTi_4[Si_4O_{12}]_2(OH)_4\cdot 6\text{-}8H_2O$

*Burovaite-Ca $\qquad$ $(Na,K)_4Ca_2(Ti,Nb)_8[Si_4O_{12}]_4(OH,O)_8\cdot 12H_2O$

*Lepkhenelmite-Zn $\qquad$ $Ba_2Zn(Ti,Nb)_4[Si_4O_{12}]_2(OH,O)_4\cdot 7H_2O$

*Gjerdingenite-Ca $\qquad$ $K_2Ca(Nb,Ti)_4[Si_4O_{12}]_2(O,OH)_4\cdot 6H_2O$

*Gjerdingenite-Fe $\qquad$ $K_2Fe(Nb,Ti)_4[Si_4O_{12}]_2(O,OH)_4\cdot 6H_2O$

*Gjerdingenite-Mn $\qquad$ $K_2Mn(Nb,Ti)_4[Si_4O_{12}]_2(O,OH)_4\cdot 6H_2O$

*Gjerdingenite-Na $\qquad$ $(K,Na)_2Na(Nb,Ti)_4[Si_4O_{12}]_2(OH,O)_4\cdot 5H_2O$

*Karupmøllerite-Ca $\qquad$ $(Na,Ca,K)_2Ca(Nb,Ti)_4[Si_4O_{12}]_2(O,OH)_4\cdot 7H_2O$

Lemmleinite group

*Lemmleinite-K $\qquad$ $Na_4K_8Ti_8[Si_4O_{12}]_4(O,OH)_8\cdot 8H_2O$

*Lemmleinite-Ba $\qquad$ $Na_4K_4Ba_{2+x}Ti_8[Si_4O_{12}]_4(O,OH)_8\cdot 8H_2O$

Labuntsovite group

*Labuntsovite-Fe $\qquad$ $Na_4K_4Fe^{2+}_2Ti_8[Si_4O_{12}]_4(O,OH)_8\cdot 10\text{-}12H_2O$

*Labuntsovite-Mg $\qquad$ $Na_4K_4Mg_2Ti_8O_4[Si_4O_{12}](OH)_4\cdot 10\text{-}12H_2O$

*Labuntsovite-Mn $\qquad$ $Na_4K_4Mn^{2+}_2Ti_8O_4[Si_4O_{12}]_4(OH)_4\cdot 10\text{-}12H_2O$

Paralabuntsovite group

*Paralabuntsovite-Mg $\qquad$ $Na_8K_8Mg_4Ti_{16}[Si_4O_{12}](O,OH)_{16}\cdot 20\text{-}24H_2O$

Organovaite group

*Organovaite-Mn $\qquad$ $K_2MnNb_4[Si_4O_{12}]_2O_4\cdot 5\text{-}7H_2O$

*Organovaite-Zn $\qquad$ $K_2Zn(Nb,Ti)_4[Si_4O_{12}]_2(O,OH)_4\cdot 6H_2O$

*Parakuzmenkoite-Fe $\qquad$ $(K,Ba)_8Fe_4Ti_{16}[Si_4O_{12}]_8(OH,O)_{16}\cdot 20\text{-}28H_2O$

Gutkovaite group

*Gutkovaite-Mn $CaK_2Mn(Ti,Nb)_4[Si_4O_{12}]_2(O,OH)_4 \cdot 5H_2O$

*Alsakharovite-Zn $NaSrKZn(Ti,Nb)_4[Si_4O_{12}]_2(O,OH)_4 \cdot 7H_2O$

*Neskevaraite-Fe $NaKK_2Fe(Ti,Nb)_4[Si_4O_{12}]_2(O,OH)_4 \cdot 6H_2O$

Astrophyllite family (that is "titanosilicates micas " by N.V.Belov's opinion)

Niobophyllite $*K_2NaFe_7^{2+}(Nb,Ti)O_2[Si_4O_{12}]_2(OH)_4(O,F)$

Kupletskite $K_2NaMn_7^{2+}Ti_2(OH)_4O_2[Si_4O_{12}]_2F_4$

Astrophyllite $K_2NaFe_7^{2+}Ti_2(OH)_4O_2[Si_4O_{12}]_2F$

Magnesioastrophyllite = *Lobanovite

*Lobanovite $K_2Na(Fe_4^{2+}Mg_2Na)Ti_2(OH)_4O_2[Si_4O_{12}]_2$

*Tarbagataite $(K\square)Ca(Fe^{2+},Mn)_7Ti_2(OH)_5O_2[Si_4O_{12}]_2$

*Sveinbergeite $Ca(Fe_6^{2+}Fe^{3+})Ti_2(OH)_5O_2[Si_4O_{12}]_2 (H_2O)_4$

Kupletskite-(Cs) $Cs_2NaMn_7^{2+}Ti_2O_2[Si_4O_{12}]_2(OH)_4F$

Zircophyllite $K_2NaMn_7^{2+}Zr_2O_2[Si_4O_{12}]_2(OH)_4F$

*Ferrozircophyllite $K_2Na(Fe^{2+},Mn)_7Zr_2O_2[Si_4O_{12}]_2(OH)_4F$

*Niobokupletskite $K_2NaMn_7^{2+}(Nb,Zr,Ti)O_2[Si_4O_{12}]_2(OH)_4(O,F)$

Acid → Hydrates

Kazakovite family

Kazakovite $Na_6Mn|Ti[Si_6O_{18}]|^{\infty 3}$

Tisinalite $Na_3Mn^{2+}Ti[Si_6O_{15}OH)_3]$

Ohmilite $Sr_3|(Ti,Fe^{3+})(O,OH)[Si_4O_{12}]|^{\infty}|^{\infty}(H_2O)_{2-3}$

Strontio-orthojoaquinite series (compare with joaquinite (family))

Strontio-orthojoaquinite $Sr_2Ba_2(Na,Fe^{2+})_2|Ti(O,OH)[Si_4O_{12}]|^{\infty 2}_2 \cdot H_2O$

*Bario-orthojoaquinite $(Ba,Sr)_4Fe^{2+}_2Ti_2[Si_8O_{26}] \cdot H_2O$

*Strontiojoaquinite mon. $Sr_2Ba_2(Na,Fe^{2+})_2|Ti(O,OH)[Si_4O_{12}]|^{\infty 2}_2 \cdot H_2O$

*Titanodi-trisilicates (K = 2.5)

Traskite $Ba_{21}Ca(Fe^{2+},Mn,Ti)_4(Ti,Fe,Mg)_{12}[Si_{12}O_{36}][Si_2O_7]_6(O,OH)_{30}Cl_6$

Titanotrisilicates (K = 3)

Oxido-titanotrisilicates

Belkovite $Ba_3|[NbO_2]_6[Si_2O_7]_2|^{\infty 3}$

Fresnoite $Ba_2|(^{(5)}TiO)[Si_2O_7]|^{\infty 2}$

*Unnamed $(Ca,Fe)_3TiO_2[Si_2O_7]$

*Batisivite $BaV^{3+}_8Ti_6O_{22}[Si_2O_7]$

*Greenwoodite $Ba_{2-x}(V^{3+}OH)_xV^{3+}_9(Fe^{3+},Fe^{2+})_2O_{15}[Si_2O_7]$

*Kolskyite $(Ca,\square)Na_2Ti_4O_4[Si_2O_7]_2 \cdot 7H_2O$

*Laurentianite $[Na(H_2O)_2]_3[NbO(H_2O)]_3[Si_2O_7]_2$

*Kazanskyite $BaNa_3Ti_2NbO_2(OH)_2[Si_2O_7]_2(H_2O)_4$

*Titanosilicato-halogenides (fluorides)

*Altisite $K_6Na_3Al_2Ti_2Si_8O_{26}Cl_3$

Hydrates

*Bulgakite $Li_2(Ca,Na)Fe^{2+}_7Ti_2[Si_8O_{24}]O_2(OH)_4(F,O)(H_2O)_2$

*Nalivkinite $Li_2NaFe^{2+}_7Ti_2[Si_8O_{24}]O_2(OH)_4F(H_2O)_2$

*Titano-trisilicato-fluorides (K = 3)

*Roumaite $(Ca,Na,Ce,\square)_7(Nb,Ti)[Si_2O_7]_2(OH)F_3$

Hydrates

*Saamite $Ba\square Na_3Ti_2NbO_2(OH)_2[Si_2O_7]_2(H_2O)_2$

Oxido-titano (niobo) trisilicato-fluorides

Nacareniobsite-(Ce)　　　　$Na_3Ca_3(Ce,La)NbO[Si_2O_7]_2F_3$

*Lileyite　　　　$Ba_2(Na,Fe,Ca)_3MgTi_2O_2[Si_2O_7]_2F$

Oxido-hydroxido (fluorido)-titanosilicates

Rosenbuschite　　　　$Ca_6Zr_2Na_6ZrTi[Si_2O_7]_4(OF)_2F_4$

*Hainite　　　　$Na_2Ca_4(Y,REE)TiO[Si_2O_7]_2F_3$

*Kochite　　　　$Ca_2MnZrNa_3Ti[Si_2O_7]_2(OF)F_2$

Lamprophyllite family

Janhaugite　　　　$Na_3Mn_3|(Ti,Zr,Nb)O(OH,F)[Si_2O_7]|^{\infty2}_2$

Låvenite　　　　$Na_2Ca_2Mn_2|(Zr,Ti)O(F,OH)[Si_2O_7]|^{\infty2}_2$

*Normandite　　　　$NaCa(Mn^{2+},Fe^{2+})|(Ti,Nb,Zr)OF[Si_2O_7]|^{\infty2}$

*Perraultite　　　　$(Na,Ca)_2(Ba,K)_2(Mn,Fe)_8(Ti,Nb)_4O_4[Si_2O_7]_4(OH)_2(OH,F)_4$

*Jinshajiangite　　　　$NaBaFe_4^{2+}Ti_2O_2[Si_2O_7]_2(OH)_2F$

Seidozerite　　　　$(Na,Ca)_4Mn_2|Zr_{0,5}Ti_{0,5}O(F,OH)[Si_2O_7]|^{\infty2}_2$

*Grenmarit　　$(Na,Ca)_4(Mn,Na)(Zr,Mn)_2(Zr,Ti)(O,F)_4[Si_2O_7]_2$

Wöhlerite　　　　$Na_2Ca_4|Zr_{0,5}(Nb,Ti)_{0,5}(O,F)_2[Si_2O_7]|^{\infty2}_2$

*Marianoite　　　　$Na_2Ca_4(Nb,Zr)_2(O,F)_4[Si_2O_7]_2$

*Schüllerite　　　　$Na_2Ba_2Mg_2Ti_2O_2F_2[Si_2O_7]_2$

*Surkhobite　　　　$KBa_3Ca_2Na_2Mn_{16}Ti_8O_8[Si_2O_7]_8(OH)_4(F,O,OH)_8$

Bafertisite　　　　$*Ba_2Fe_4^2Ti_2O_2(OH)_2F_2[Si_2O_7]_2$

*Camaraite　　　　$Ba_3Na(Fe^{2+},Mn)_8(OH,F)_7Ti_4O_4[Si_2O_7]_4$

*Ba-Mn titanosilicate　　　　$Ba[Mn^{2+}_2(OH)]\{TiO(OH)[Si_2O_7]\}$

*Hejtmanite　　　　$Ba(Mn^{2+},Fe^{2+})_2\{TiO(OH,F)_2[Si_2O_7]\}$

*Vigrishinite　　　　$Zn_2Ti_{4-x}O_2(OH,F,O)_2[Si_2O_7]_2(H_2O,OH,▢)_4$　　(x<1)

　Lamprophyllite series

Lamprophyllite　　　　$(SrNa)Ti_2Na_3TiO_2[Si_2O_7]_2(OH)_2$

*Fluorlamprophyllite　　　　$Na_3(SrNa)Ti_3O_2F_2[Si_2O_7]_2$

Barytolamprophyllite　　　　$(BaK)Ti_2Na_3TiO_2[Si_2O_7]_2(OH)_2$

*Emmerichite　　　　$Ba_2Na(Na,Fe^{2+})_2(Fe^{3+},Mg)Ti_2O_2F_2[Si_2O_7]_2$

*Nabalamprophyllite　　　　$(BaNa)Ti_2Na_3TiO_2[Si_2O_7]_2(OH)_2$

*Lamprophyllite orth.　　　　$(SrNa)Ti_2Na_3TiO_2[Si_2O_7]_2(OH)_2$

　Götzenite group

Götzenite　　　　$Ca_4NaCa_2Ti[Si_2O_7]_2(OF)F_2$

Niocalite　　　　$Ca_7|Nb[Si_2O_7]_2O_3F$

*Fersmanite　$Ca_5Na_3Ti_3Nb[Si_2O_7]_2O_8F_2$ or $Ca_4(Na,Ca)_4(Ti,Nb)_4[Si_2O_7]_2O_8F_3$

　　　　　　　　Hydrates

*Delindeite　　　　$Ba_2Ti_2(Na_2▢)Ti[Si_2O_7]_2(OH)_2(H_2O)_2O_2$

*Chirvinskyite　　$(Na,Ca)_{13}(Fe,Mn,□)_2(Ti,Zr)_5(OH,O)_{12}[Si_2O_7]_4·2H_2O$

*Oxido-fluorido-phosphato-titanosilicates

*Sobolevite　　$Na_6(Na_2Ca)(NaCaMn)Na_2Ti_2Na_2(TiMn)[Si_2O_7]_2[PO_4]_4O_2(OF)F_2$

*Polyphite　　　　$Na_6(Na_4Ca_2)_2Na_2Ti_2Na_2Ti_2[Si_2O_7]_2[PO_4]_6O_4F_4$

*Quadruphite　　　　$Na_{14}Ca_2Ti_4O_4F_2[Si_2O_7]_2[PO_4]_4$

*Hydroxido (fluorido)　　　　Hydrates

*Shkatulkalite　　　　$Na_{10}MnTi_3Nb_3(OH)_2F[Si_2O_7]_6·12H_2O$

Komarovite series

Komarovite　　　　$(Ca,Mn,H)_2|Nb_2O_3(OH,F)_2[Si_2O_7]|^{\infty3}(H_2O)_{3,5}$

Na-komarovite　　　　$(Na,Ca,H)_2|Nb_2O_3(OH,F)_2[Si_2O_7]|^{\infty3}·H_2O$ or

$$Na_6CaNb_6O_{14}[Si_4O_{12}]\cdot 4H_2O$$

Epistolite $Na_2(Nb,Ti)_2O_2[Si_2O_7]\cdot nH_2O$
*Zvyaginite $(Na\square)Nb_2Zn\square_2Ti[O(OH)][Si_2O_7]_2O_2[(OH)F](H_2O)_5$
Murmanite-lomonosovite family
Murmanite $Na_2Ti_2Na_2Ti_2[Si_2O_7]_2O_4(H_2O)_4$
*Calciomurmanite $(Na,\square)_2Ca(Ti,Mg,Nb)_4O_2(OH,O)_2[Si_2O_7]_2(H_2O)_4$
*Bykovaite $(Ba,Na,K)_2(Na,Ti,Mn)_4(Ti,Nb)_2O_2[Si_2O_7]_2(H_2O,F,OH)_2\cdot 3.5H_2O$
*Nechelustovite $(Ba,Sr,K)_2(Na,Ti,Mn)_4(Ti,Nb)_2O_2[Si_2O_7]_2(O,H_2O,F)_2\cdot 4.5H_2O$
Bornemanite $Na_6BaTi_2NbO_2[Si_2O_7]_2[PO_4](OH)F$
Vuonnemite $Na_6Na_2Nb_2Na_3Ti[Si_2O_7]_2[PO_4]_2O_2(OF)$
Lomonosovite $Na_6Na_2Ti_2Na_2Ti_2[Si_2O_7]_2[PO_4]_2O_4$
Betalomonosovite $Na_2\square_4Na_2Ti_2Na_2Ti_2[Si_2O_7]_2[PO_3(OH)][(PO_2(OH)_2]O_2(OF)$
Yoshimuraite $Ba_2Mn_2TiO[Si_2O_7][PO_4](OH)$
Innelite and polytypes 1T and 2M
                   $Ba_4Ti_2Na(NaCa)Ti[O(OH)][Si_2O_7]_2[(SO_4)(PO_4)]O_2$
*Phosphoinnelite $Na_3Ba_4Ti_3[Si_2O_7]_2[PO_4]_2O_2F$
                                   Basic
Ellenbergerite $Mg_6Al_6Ti(OH)_{10}[Si_2O_7]_4$

* Titanotri-tetrasilicates (K = 3,2)   *Hydrates
*Hogarthite $(Na,K)_2CaTi_2[Si_{10}O_{26}]\cdot 8H_2O$

*Titanotri-tetrasilicates (K = 3,4)   *Hydrates
*Tiettaite $(Na,K)_{17}FeTiSi_{16}O_{29}(OH)_{30}\cdot 2H_2O$

Titanotetrasilicates (K = 4)
Oxido-titanotetrasilicates
Titanite $Ca|TiO[SiO_4]|^{\infty 2}$
*Natrotitanite $(Na_{0.5}Y_{0.5})|TiO[SiO_4]|^{\infty 2}$
Natisite $Na_2|^{(5)}TiO)[SiO_4]|^{\infty 2}$
*Paranatisite $Na_2\{TiO[SiO_4]\}$
*Sitinakite $Na_2\{Ti_4O_5(OH)[SiO_4]_2\}K\cdot 4H_2O$
                                          Hydrates
Mongolite $Ca_4(OH)_8|Nb_6O_4(OH)_2[SiO_4]_5|^{\infty 2}\cdot 5\text{-}6H_2O$
*Ivanyukite-Na $Na_2\{Ti_4O_2(OH)_2[SiO_4]_3\}\cdot 6H_2O$
*Ivanyukite-K $K_2\{Ti_4O_2(OH)_2[SiO_4]_3\}\cdot 9H_2O$
*Ivanyukite-Cu $Cu\{Ti_4O_2(OH)_2[SiO_4]_3\}\cdot 7H_2O$

Titanoborosilicates
Oxido-and oxido-hydroxido-titanoborosilicates
Leucosphenite (K_Σ = 0,(6)) $Ba|Na_4Ti_2O_2[Si_{10}B_2O_{28}]^{\infty 2}|^{\infty 2}$
Tienshanite (K_Σ = 1,2) $KNa_9Ba_6Ca_2|(Mn,Fe)_6(Ti,Nb,Ta)_6(O,OH,F)_{11}[Si_{36}B_{12}O_{114}]^{\infty 2}|^{\infty 2}$

*Titanoborosilicato-phosphates   Hydrates
*Byzantievite
   $Ba_5(Ca,REE,Y)_{22}(Ti,Nb)_{18}(SiO_4)_4(PO_4,SiO_4)_4(BO_3)_9O_{22}[(OH),F]_{43}\cdot 1.5H_2O$

Titanosilicates of Li   Neutral
*Titanomonosilicates (K = 1)
*Berezanskite $KLi_3Ti_2[Si_{12}O_{30}]$

*Unnamed $KLi_3Zn_2[Si_{12}O_{30}]$

*Titanomono-disilicates (K = 1,4) Hydrates
*Punkaruaivite $Li\{Ti_2(OH)_2[Si_4O_{11}(OH)]\}\cdot 2H_2O$

Oxido-titanomono-disilicates
Titanomono-disilicates (K = 1,5)
 Neptunite series
 Neptunite $KNa_2|Li(Fe^{2+},Mn)_2Ti_2O_2[Si_4O_{11}]^{\infty3}_{2}|^{\infty3}$
 Manganneptunite $KNa_2|Li(Mn,Fe^{2+})_2Ti_2O_2[Si_4O_{11}]^{\infty3}_{2}|^{\infty3}$
 *Magnesioneptunite $KNa_2|Li(Mg,Fe^{2+})_2Ti_2O_2[Si_4O_{11}]^{\infty3}_{2}|^{\infty3}$

*Titanomono-disilicato-fluorides
*Faizievite $K_2Na(Ca_6Na)Ti_4Li_6F_2[Si_4O_{11}]_6$

Titanodisilicates (K = 2) Basic
Baratovite $(K,Na)Ca_7Li_3Ti_2F_2[SiO_3]_{12}$
*Katayamalite $KLi_3Ca_7Ti_2(OH)_2[Si_6O_{18}]_2$

Oxido-titanodisilicates *Hydrates
*Lintisite $Na_3LiTi_2O_2[Si_4O_{12}]\cdot 2H_2O$

*Titanotrisilicates (K = 3) *Hydrates
*Eliseevite $Na_{1.5}Li[Ti_2Si_4O_{12.5}(OH)_{1.5}]\cdot 2H_2O$

*Titanotrisilicato-carbonates -halogenides
*Bussenite $Ba_4(Na,\square)_2(Fe^{2+},Na)_2Ti_2[Si_2O_7]_2(CO_3)_2O_2(OH)_2(H_2O)_2F_2$

*Titanotrisilicates Be
*Odintsovite $K_2Na_4Ca_3Ti_2Be_4[Si_{12}O_{38}]$

Titanosilicates of f- elements
Proper titanosilicates
*Titanomono-disilicates (K = 1,5) *Hydrates
*Seidite-(Ce) $Na_4SrCeTi[Si_8O_{22}]F\cdot 5H_2O$

Titanodisilicates (K = 2)
Oxido-hydroxido-titanodsilicates Hydrates
 Joaquinite family (ср. стронциоджоакинита (с.))
  Byelorussite group
  Orthojoaquinite-(Ce) $Ba_2NaFe^{2+}Ce_2(OH,F)Ti_2O_2[Si_4O_{12}]_2\cdot H_2O$
  *Orthojoaquinite-(La) $Ba_2NaFe^{2+}La_2(O,OH)Ti_2O_2[Si_4O_{12}]_2\cdot H_2O$
  Byelorussite-(Ce) $Ba_2NaMnCe_2(F,OH)Ti_2O_2[Si_4O_{12}]_2\cdot H_2O$
  Joaquinite-(Ce) $Ba_2NaFe^{2+}Ce_2(OH)Ti_2O_2[Si_4O_{12}]_2\cdot H_2O$
 *K-Ti analog ilímaussite-(Ce)
  $(Ba,Na,K,Ca)_{11\text{-}12}(REE,Fe,Th)_4(Ti,Nb)_6[Si_6O_{18}]_4(OH)_{12}\cdot 4,5H_2O$
 *Pyatenkoite-(Y) $Na_5(Y,Dy,Gd)Ti[Si_6O_{18}]\cdot 6H_2O$

*Titanodo-trisilicates with mixed silicooxygenous anions
*Diversilite-(Ce) (K = 2,5)

$$Na_2(Ba,K)_6Ce_2Fe^{2+}Ti_3[Si_3O_9]_3[SiO_3OH]_3(OH)_7 \cdot nH_2O$$

Titanitrisilicates (K = 3)   Basic
Rinkite family (compare götzenite)
Rinkite $(Ca_3REE)Na(Na,Ca)Ti[Si_2O_7]_2(O,F)F_2$
*Mosandrite = Lovchorrite $NaCaCeTi[Si_2O_7]O_2$ (?)
                                                          Hydrates
*Batievaite-(Y) $Y_2Ca_2Ti(OH)_2[Si_2O_7]_2(H_2O)_4$

*Titanitrisilicato-oxido-arsenates
*Cervandonite-(Ce) $(Ce,Nd,La)(Fe^{3+}, Fe^{2+},Ti^{4+},Al)_3O_2[Si_2O_7]_{1-x+y}[AsO_3]_{1+x-y}(OH)_{3x-3y},$
$$x = 0,47, y = 0,31$$

*Titanotri-tetrasilicates with mixed silicooxygenous anions
*Oxido-titanotri-tetrasikicates   *Basic
Ilimaussite-(Ce) (K = 3,4) $(Ba,Na)_{10}K_3Na_{4,5}Ce_5(Nb,Ti)_6[Si_{12}O_{36}][Si_9O_{18}(O,OH)_{24}]O_6$

Titanotetrasilicates (K = 4)
Oxido-titanotetrasilicates
Trimounsite-(Y) $Y_2|Ti_2O_5[SiO_4]|^{\infty 3}$

Titanosilicates with unknown structure
Ilmajokite $(Na,Ce,Ba)_{10}Ti_5Si_{14}O_{22}(OH)_{44} \cdot nH_2O$

Titanosilicato-phosphates with unknown structure
Laplandite-(Ce) $Na_4CeTiSi_7O_{18}(PO_4) \cdot 5H_2O$
*Karnasurtite-(Ce) $(Ce,La,Th)(Ti,Nb)(Al,Fe^{3+})[(Si,P)_2O_7](OH)_4 \cdot 3H_2O$

Titanosilicato-carbonates
Titanotetrasilicato-carbonates   Hydrates
  Tundrite series
  Tundrite-(Ce) $Na_2Ce_2TiO_2[SiO_4][CO_3]_2$
  Tundrite-(Nd) $Na_2Nd_2TiO_2[SiO_4][CO_3]_2$

*Titanotetrasilicato-hydrocarbonates   *Hydrates
*Kihlmanite-(Ce) $Ce_2TiO_2[SiO_4][HCO_3]_2(H_2O)$

Subclass: Silicates of chalcophylic elements
Silicates of **I**b-cations (Cu^{2+})
*Zero-monoalumosilicates (K = 0,(6))
*Kurumsakite $(Zn,Ni,Cu^{2+})_8Al_8V^{5+}_2[Si_5O_{35}] \cdot 27(H_2O)$

*Zero-monoalumosilicates (K = 0,75) *Hydrates
*Ajoite $(K,Na)_3Cu_{20}[Al_3Si_{29}O_{76}](OH)_{16} \cdot 8H_2O$

Monosilicates (K = 1)   Neutral
  Cuprorivaite family
  Cuprorivaite $CaCu^{2+}[Si_4O_{10}]^{\infty 2}$
  *Effenbergerite $BaCu^{2+}[Si_4O_{10}]$
  *Wesselsite $SrCu^{2+}[Si_4O_{10}]$

Litidionite $\qquad$ $KNaCu^{2+}[Si_4O_{10}]^{\infty 2}$

Acid-basic

Chrysocolla $\qquad$ $\{Cu^{2+}{}_2H_2(OH)_4[Si_2O_5]^{\infty 2}\}^{\infty 2}$

with $(Al,Fe)^{3+}$ in place of $Cu^{2+}H^+$: $\{Cu_{2-x}H_{2-x}(Al,Fe)_x(OH)_4[Si_2O_5]^{\infty 2}\}^{\infty 2}$

if $x = 2$ $\qquad$ $(Al,Fe)_2(OH)_4[Si_2O_5]^{\infty 2}\}^{\infty 2}$ (kaolinite)

Mono-disilicates $\qquad$ *Neutral

*Lavinskyite $(K = 1,5)$ $\qquad$ $K(LiCu)Cu^{2+}{}_6[Si_4O_{11}]_2(OH)_4$

*Liebauite $(K = 1,8)$ $\qquad$ $Ca_3Cu_5[Si_9O_{26}]$

Hydrates

Plancheite $(K = 1,5)$ $\qquad$ $Cu^{2+}{}_8(OH)_4[Si_4O_{11}]^{\infty 2}{}_2 H_2O$

*Gilalite $(K = 1,6)$ $\qquad$ $Cu^{2+}{}_5[Si_6O_{17}]\cdot 7H_2O$

*Apachite $(K = 1,8)$ $\qquad$ $Cu^{2+}{}_9[Si_{10}O_{29}]\cdot 11H_2O$

Disilicates $(K = 2)$ $\qquad$ Basic

Papagoite $\qquad$ $Ca_2Cu^{2+}{}_2Al_2(OH)_6[Si_4O_{12}]$

Shattuckite $\qquad$ $Cu^{2+}{}_5(OH)_2[Si_2O_6]^{\infty 2}{}_2$

Hydrates

Dioptase $\qquad$ $Cu_6[Si_6O_{18}](H_2O)_6$

Di-trisilicates $(K = 2,(6))$ $\qquad$ Hydrates

Kinoite $\qquad$ $Ca_2\{Cu^{2+}{}_2(H_2O)_2[Si_3O_{10}]\}^{\infty 3}$

*Trisilicates

*Scottyite $\qquad$ $BaCu_2[Si_2O_7]$

*Tetrasilicates $\qquad$ *Hydrates

*Stringhamite $\qquad$ $CaCu^{2+}[SiO_4]\cdot H_2O$

*Unnamed $\qquad$ $Cu^{2+}{}_8(OH)_{12}[SiO_4]\cdot 8H_2O$

*Silicato-carbonates

*Disilicato-carbonates $(K = 2)$

*Whelanite $\qquad$ $Cu^{2+}{}_2Ca_6[Si_6O_{17}(OH)][CO_3](OH)_3(H_2O)_2$

*Silicato-hydrocarbonaro-chlorides $\qquad$ *Basic

*Ashburtonite $\qquad$ $HPb_4Cu^{2+}{}_4[Si_4O_{12}](HCO_3)_4(OH)_4Cl$

Silicates of IIb-cations

Silcates of Hg^+

Trisilicates $\qquad$ Neutral

Edgarbaileyite $\qquad$ $Hg^+{}_6[Si_2O_7]$

Silicates of $Zn \rightarrow$ zincosilicates

Alumosilicates and proper silicates

Zero-monozincoalumosilicates $(K = 0,8)$ $\qquad$ Acid-basic

Minehillite $\qquad$ $(K,Na)_2Ca_{28}(OH)_{12}[(Zn_5Al_4Si_{40})O_{112}(OH)_4]$

(compare with reyerite and truscottite)

Monoalumosilicates $(K = 1)$ $\qquad$ Basic

Hendrickite $K\{(Zn,Mg,Mn)_3(OH)_2[AlSi_3O_{10}]^{\infty 2}$
(compare with subfamily of common mica)
Baileychlore $(Fe^{2+},Mg)_3(OH)_6\{(Zn,Al)_3(OH)_2[AlSi_3O_{10}]^{\infty 2}\}^{\infty 2}$
(compare with chlorites (family))
Fraipontite = zinalsite $\{(Zn,Al)_3(OH)_4[(Si,Al)_2O_5]^{\infty 2}{}_2\}^{\infty 2}$
(compare with kaolinite (family))
*Klöchite $KNaFe^{2+}{}_2Zn_3[Si_{12}O_{30}]$
  Sauconite series (compare with smectite (family))
  Sauconite $Na_{0.33}\{Zn_3(OH)_2[(Si,Al)_4O_{10}]^{\infty 2}\}^{\infty 2}(H_2O)_4$
  Zincsilite $Zn_3(OH)_2[Si_4O_{10}]\cdot 4H_2O$
Trisilicates (K = 3)   Hydrates
Hemimorphite $\{^{(4)}Zn_4(OH)_2[Si_2O_7]\}^{\infty 3}\cdot H_2O$
Junitoite $CaZn_2[Si_2O_7]\cdot H_2O$

Silicates of Zn with unknown structure   Hydrates
Silicates of **IV**a-cations
Silicates of Pb^{2+}
Proper silicates
*Zero-monoalumosilicates (K = 0,4)
*Rongibbsite $Pb_2[(AlSi_4)O_{11}](OH)$

*Zero-monoalumosilicates (K = 0,9)
*Wickenburgite $Pb_3CaAl[AlSi_{10}O_{27}](H_2O)_3$

*Mono-disilicates (K = 1,3)   Hydrates
*Yangite $PbMn[Si_3O_8]\cdot H_2O$

*Mono-disilicates (K = 1,5)   Hydrates
*Mathewrogersite $Pb_7Fe^{2+}GeAl_3[Si_3O_9]_4(OH)_4\cdot 2H_2O$

Disilicates (K = 2)   Neutral
Plumalsite $Pb_4Al_2[SiO_3]_7$
Alamosite $PbSiO_3 \rightarrow Pb_{12}[Si_{12}O_{36}]$

*Disilicato-halogenides   Hydrates
*Hyttsjöite $Pb_{18}Ba_2Ca_5Mn^{2+}{}_2Fe^{3+}{}_2[Si_{30}O_{90}]Cl\cdot H_2O$

*Disilicato-trisilicates (K = 2,8)   *Hydrates
*Creaseyite $Pb_2Cu^{2+}{}_2Fe^{3+}{}_2[Si_5O_{17}]\cdot 6H_2O$

*Trisilicato-tetrasilicares (K = 3,(3))
*Ganomalite $Pb_9Ca_5Mn[Si_9O_{33}]$

*Silicato-carbonates
*Mono-silicatocarbonates   Basic
*Surite $(Pb,Ca)_3Al_2(OH)_3[(Si,Al)_4O_{10}][CO_3]_2\cdot 0.3H_2O$
                                         *Hydrates
*Ferrisurite $(Pb,Ca)_{2.4}Fe^{3+}{}_2(OH)_3[Si_4O_{10}][CO_3]_{1.7}\cdot nH_2O$

Silicato-sulfates

Trisilicato-tetrasilicato-sulfates   Neutral
Quéitite $Pb_4Zn_2[Si_2O_7][SiO_4][SO_4]$

*Silicato-sulfato-carbonates   *Basic
*Kegelite $Pb_8Al_4[Si_4O_{10}]_2[SO_4]_2[CO_3]_4(OH)_8$

Silicato-chromates
Tetrasilicato-chromates and tetrasilicato-chromato-fluorides
                                             Neutral
  Hemihedrite series
  Iranite $CuPb_{10}(OH)_2[SiO_4]_2[CrO_4]_6$
  Hemihedrite $ZnPb_{10}F_2[SiO_4]_2[CrO_4]_6$

Silicato-sulfato-chlorides   Neutral
Mattheddleite $Pb_{10}[SiO_4]_3[SO_4]_3Cl_2$
(compare with wulfenite (series); ellestadite (group); apatite (group))

Silicates of **Va**-elements
Silicates of nonfull-valence **Va**-cations (As^{3+}, Sb^{3+} and Bi^{3+})
Proper silicates
Monosilicates
                                             Basic
Chapmanite $Sb^{3+}\{Fe^{3+}_2(OH)O_3[Si_2O_5]^{\infty 2}\}^{\infty 2}$

Tetrasilicates   Neutral
Eulytine $Bi_4[SiO_4]_3$
Bismutoferrite $Fe^{3+}_2Bi[SiO_4]_2(OH)$

Titanoberyllosilicates
Titanotriberyllosilicates ($K_\Sigma = 3$)
Oxido-titanotriberyllosilicates
Asbecasite $Ca_3(Ti,Sn)As^{3+}_6O_6[BeSiO_7]_2$

3.2.1a. *Class:* Germanates (zone of oxidization of Tsumeb and at France ?)

Tetragermanates (orthogermanates)   Basic
Carboirite (compare with chloritoid (family)
          $2[Fe^{2+}Al_2GeO_5(OH)_2] \rightarrow \{Fe^{2+}_2Al(OH)_4Al_3O_2[GeO_4]_2\}^{\infty 2}$

Germanates (?) with unknown structure
Bartelkeite $PbFeGe[Ge_2O_7](OH)_2 \cdot H_2O$
Otjisumeite $PbGe_4O_9$
(see also sulfates of Ge – itoite, schaurteite, fleischerite)

3.2.2. *Class:* Borates
3.2.2.1. Quasiclass: (4)-Borates of cations with low FC
(4)-Borates of **s**-, **d$_s$**- and **p$_s$**- cations
(4)-Borates of **s**-, **d$_s$**- and **p$_s$**- cations without Li and Be
Proper (4)-Borates
*Zero-monoborates (K = 0,28)
                                             Asid

*Jarandolite (Serbianite) $Ca[B_3O_4(OH)_3]$

Zero-monoborates $(K = 0,(6))$
                                                  Neutral
Johachidolite $Ca\{^{(6)}Al[B_3O_7]^{\infty 2}\}^{\infty 3}$

Monoborates $(K = 1)$ Acid
*Shimazakiite $Ca_2B_2O_5$
Korzhinskite $CaB_2O_4 \cdot 0.5H_2O$

Diborates $(K = 2)$ Acid
 Vimsite family
 Vimsite $Ca[B_2O_2(OH)_4]^{\infty}$
 Uralborite $Ca[B_2O_2(OH)_4]^{\infty.}$

Triborates $(K = 3)$ Acid
Pinnoite $Mg[B_2O(OH)_6]$
                                                  Hydrates
Pentahydroborite $\{Ca[B_2O(OH)_6]\}^{\infty 2} \cdot 2H_2O)_2$

Tetraborates $(K = 4)$ Neutral
Sinhalite $MgAl[BO_4]$
                                                  Acid
*Pseudosinhalite $(Mg,Fe)_2Al_3B_2O_9(OH)$
Frolovite $\{Ca[B(OH)_4]_2\}^{\infty 2}$
                                                  Hydrates
Hexahydroborite $Ca(H_2O)_2[B(OH)_4]_2$

Tetraborates and tetraborato-halogenides Basic- acid
Teepleite $Na_2Cl[B(OH)_4]$

(4)-Borato-phosphates   Basic-acid
Seamanite $Mn_3(OH)_2[B(OH)_4][PO_4]$

(4)-Borato-arsenates   Acid
Cahnite $Ca_2[B(OH)_4][AsO_4]$

(4)-Borato-carbonates   Hydrates
Carboborite $Ca_2Mg(H_2O)_4[B(OH)_4]_2[CO_3]_2$
*Imayoshiite $Ca_3Al(OH)_6[B(OH)_4][CO_3] \cdot 12H_2O$

(4)-Borato-sulfates   Basic-acid
Sulfoborite $Mg_3(OH,F)_2[B(OH)_4]_2[SO_4]$
                                                  Hydrates
 Charlesite family
 Charlesite $Ca_6(Al,Si)_2(OH,O)_{12}[B(OH)_4][SO_4]_2 \cdot 26H_2O$
 Sturmanite $Ca_6Fe^{3+}_2(OH)_{12}[B(OH)_4][SO_4]_{2.5} \cdot 25H_2O$
 *Buryatite $Ca_3(Si,Fe^{3+},Al)(OH)_5O[B(OH)_4][SO_4] \cdot 12H_2O$

(4)-Borates Be $\rightarrow$ beryllo borates

Zero-monoberyllo-(4)-borates (K = 0,5)

                               Acid

Rhodizite $KBe_4Al_4(B_{11}Be)O_{28}$

*Zero-monoberyllo-(4)-borates (K = 0,6(6))

*Londonite $CsBe_5Al_4B_{11}O_{28}$

 (4)-Borates of f-cations

(4)-Borato-carbonates   Acid

Moydite-(Y) $Y[B(OH)_4][CO_3]$

(4)-Borates of cations with middle FC

(4)-Borates of Nb and Ta   Neutral

Behierite $Ta[BO_4]$

*Schiavinatoite $(Nb_{0,52}Ta_{0,48})[BO_4]$

(4)-Borates of chalcophylic elements

(4)-Borates of Cu

(4)-Borates and (4)-borato-halogenides   Acid

Henmilite $Ca_2Cu^{2+}[B(OH)_4]_2(OH)_4$

Bandylite $CuCl[B(OH)_4]$

*Jacquesdietrichite $Cu_2[BO(OH)_2](OH)_3$

3.2.2.2. Quasiclass: (3)-Borates

(3)-Borates of cations with low FC

(3)-Borates of s-, d_s- and p_s-cations

(3)- Borates of s-, d_s- and p_s-cations without Li and Be

Proper (3)-Borates

(3)-Diborates   Neutral

Suanite family

Suanite $Mg_2[B_2O_5]$

Kurchatovite $CaMg[B_2O_5]$

Clinokurchatovite $CaMg[B_2O_5]$

                               Basic

Wiserite $(Mn,Mg)_{14}(OH)_8(Si,Mg)(O,OH)_4Cl[B_2O_5]_4$

                               Acid

Szaibelyite family

Szaibelyite $Mg_2(OH)[B_2O_4(OH)]$

Sussexite $Mn_2(OH)[B_2O_4(OH)]$

                               Hydrates

Satimolite $KNa_2Al_4Cl_3[B_2O_5]_3 \cdot 13H_2O$

(3)-Monoborates   Neutral

Kotoite family

*Takedaite $Ca_3[BO_3]_2$

Kotoite $Mg_3[BO_3]_2$

Jimboite $Mn_3[BO_3]_2$

                    Oxido-(3)-monoborates

Warwickite $Mg(Ti,Fe,Al)_2O[BO_3]$

*Yuanfuliite $Mg(Fe^{3+},Al)O[BO_3]$

Ludwigite series
Azoproite $(Mg,Fe^{2+})_2(Fe^{3+},Ti,Mg)O_2[BO_3]$
Ludwigite $(Mg,Fe^{2+})_2Fe^{3+}O_2[BO_3]$
Bonaccordite $Ni_2Fe^{3+}O_2[BO_3]$
Vonsenite $(Fe^{2+},Mg)_2Fe^{3+}O_2[BO_3]$
Fredrikssonite $Mg_2Mn^{3+}O_2[BO_3]$
Orthopinakiolite series
Chestermanite $Mg_2(Fe^{3+},Mg,Al,Sb^{5+})O_2[BO_3]$
Takeuchiite $(Mg,Mn^{2+})_2(Mn,Fe)^{3+}O_2[BO_3]$
Orthopinakiolite $(Mg,Mn^{2+})_2Mn^{3+}O_2[BO_3]$
Pinakiolite $Mg_2Mn^{3+}O_2[BO_3]$
(compare with hulsite (group))   Basic
Jeremejevite $Al_6F_3[BO_3]_5$
Fluoborite (nocerite) $Mg_3(F,OH)_3[BO_3]$
*Pertsevite-(F) $Mg_2F[BO_3]$
*Pertsevite-(OH) $Mg_2(OH)[BO_3]$
Karlite $(Mg,Al_x)_7(OH)_4Cl_{1-x}[BO_3]_3$
                                       Acid
Sibirskite $CaH[BO_3]$
                                       Hydrates
*Parasibirskite $Ca_2B_2O_5 \cdot H_2O$
Wightmanite family
Wightmanite $Mg_5(OH)_5O[BO_3] \cdot 2H_2O$
Shabynite $Mg_5(OH)_5(Cl,OH)_2[BO_3] \cdot 4H_2O$
Nifontovite $Ca_3[BO(OH)_2]_6 \cdot 2H_2O$
Olshanskyite $Ca_2(OH)[BO(OH)_2]_3 \cdot 3H_2O$

*(3)-Borato-halogenides
*Hydroxylborite $Mg_3[BO_3(OH)_2]F$

(3)-Borato-carbonates
(3)-Monoborato-carbonates
(3)- Monoborato-carbonates with $BO_3 : CO_3 = 3 : 1$
                                       Basic
Gaudefroyite $Ca_4Mn^{3+}_3O_3[BO_3]_3[CO_3]$

(3)-Monoborato-carbonates with $BO_3 : CO_3 = 2 : 1$
                                  Hydrates (neutral)
Sakhaite $Ca_{48}Mg_{16}Al[SiO_3OH]_4[CO_3]_{16}[BO_3]_{28} (H_2O)_3(HCl)_3$

(3)- Monoborato-carbonates with $BO_3 : CO_3 = 1 : 1$
                                  Hydrates (acid)
Canavesite $Mg_2[BO_2(OH)][CO_3] \cdot 5H_2O$

(3)-Borato-phosphates
(3)- Monoborato-phosphates   Hydrates
Lüneburgite $Mg_3[B(OH)_3]_2[PO_4]_2 \cdot 6H_2O$

(3)-Borates of Be
(3)-Monoborates of Be   Basic

Hambergite $Be_2(OH)[BO_3]$
                                                    Hydrates
Berborite $Be_2(OH,F)[BO_3] \cdot H_2O$

*(3)-Borates f-cations
*Peprossite-(Ce) $(Ce,La)(Al_3O)_{2/3}[B_4O_{10}]$

(3)-Borates of cations with middle FC
(3)-Monoborates of Zr
Oxido-(3)-monoborates
Painite   *$CaZrAl_9O_{15}[BO_3]$

(3)-Borates of chalcophylic p-elements
(3)-Monoborates of Sn^{4+} Neutral
Nordenskiöldine family
Tusionite $MnSn[BO_3]_2$
Nordenskiöldine $CaSn[BO_3]_2$

Oxido-(3)-monoborates
  Hulsite series (compare with pinakiolite (group))
  Magnesiohulsite $(Mg,Fe)_2(Fe,Sn,Mg)O_2[BO_3]$
  Hulsite (paigeite) $(Fe^{2+},Mg)_2(Fe^{3+},Sn)O_2[BO_3]$
  *Aluminomagnesiohulsite $Mg_2(Al_{1-2x}Mg_xSn_x)O_2[BO_3]$ x = 0,18

(3)-Monoborates of Sb^{5+}
Oxido-(3)-monoborates
Blatterite $Sb^{5+}_3(Mn^{3+},Fe^{3+})_9(Mn^{2+},Mg)_{35}O_{32}[BO_3]_{16}$
(compare with ludwigite (family))

3.2.2.3. Quasiclass: (4)-(3)-Borates
(4)-(3)-Borates of s-, d_s- and p_s- cations
(4)-(3)- Borates of s-, d_s- and p_s- cations without Li and Be
Proper (4)-(3)-borates (x =MO : B_2O_3)
(4)-(3)-Borates with x = 1   Neutral
Calciborite $Ca[B_2O_4]^\infty$
                                              Acid
  Fedorovskite series
  Fedorovskite $Ca_2Mg_2(OH)_4[B_4O_7(OH)_2]$
  Roweit $Ca_2Mn^{2+}_2(OH)_4[B_4O_7(OH)_2]$
  *Mg-roweit $Ca_2(Mn^{2+},Mg)_2(OH)_4[B_4O_7(OH)_2]$

(4)-(3)-Borates with x = 0,8
Priceite family
 Priceite (pandermite) $Ca_2(H_2O)[B_5O_7(OH)_5]$
 Tertschite $Ca_2(H_2O)_7[B_5O_6(OH)_7]$ (?)

(4)-(3)-Borates with x = 0,(6)   Acid
Fabianite $Ca[B_3O_5(OH)]^{\infty 2}$
                                              Hydrates
 Hydroboracite family

Colemanite $Ca[B_3O_4(OH)_3]^\infty \cdot H_2O$
Meyerhofferite $Ca_2[B_3O_4(OH)_3]^\infty_2 \cdot 2H_2O$
Hydroboracite $CaMg[B_3O_4(OH)_3]^\infty_2 \cdot 3H_2O$
Inderborite $CaMg[B_3O_3(OH)_5]^\infty_2 \cdot 6H_2O$
Inyoite family
Inyoite $Ca[B_3O_3(OH)_5] \cdot 4H_2O$
Inderite $Mg[B_3O_3(OH)_5] \cdot 5H_2O$
Kurnakovite $Mg[B_3O_3(OH)_5] \cdot 5H_2O$
Veatchite family
Veatchite $Sr_2[B_5O_8(OH)]^\infty_2[B(OH)_3] \cdot H_2O$
Veatchite-A $Sr_2[B_5O_8(OH)]^\infty_2[B(OH)_3] \cdot H_2O$
Veatchite-P $Sr_2[B_5O_8(OH)]^\infty_2[B(OH)_3] \cdot H_2O]$
Ulexite family
*Tuzlaite $NaCa_2[B_5O_8(OH)_2]^\infty \cdot 3H_2O]$
Probertite $NaCa_2[B_5O_7(OH)_4]^\infty \cdot 3H_2O]$
Ulexite $NaCa[B_5O_6(OH)_6]^\infty \cdot 5H_2O]$

*(4)-(3)-Borates with x = 0,56
*Studenitsite $NaCa_2[B_9O_{14}(OH)_4]^\infty \cdot 2H_2O$

(4)-(3)-Borates with x = 0,54   Acid
Preobrazhenskite $Mg_3[B_{11}O_{15}(OH)_9]^{\infty 2}$

(4)-(3)-Borates with x = 0,5   Hydrates
Halurgite family
Halurgite $Mg_2(H_2O)[B_4O_5(OH)_4]_2$
*Hungchaoite $Mg(H_2O)_5[B_4O_5(OH)_4] \cdot 2H_2O$
Wardsmithite $Ca_5Mg[B_4O_5(OH)_4]_6 \cdot 18H_2O$
Borax family
Kernite $Na_2(H_2O)_3[B_4O_6(OH)_2]$
Tincalconite $Na_2(H_2O)_3[B_4O_5(OH)_4]$
Borax $Na_2B_4O_5(OH)_4 \cdot 8H_2O \rightarrow [Na(H_2O)_4]^{\infty 2}[B_4O_5(OH)_4]$
Aristarainite $Na_2(H_2O)_2\{Mg(H_2O)_2[B_6O_8(OH)_4]^\infty_2\}^{\infty 2}$
Kaliborite (paternoite) $HKMg_2(H_2O)_4[B_6O_8(OH)_5]^\infty_2$
*Alfredstelznerite $Ca_4(H_2O)_4[B_4O_4(OH)_6]_4 \cdot (H_2O)_{15}$

(4)-(3)-Borates with x = 0,4   Hydrates
Biringuccite family
Biringuccite $Na_2(H_2O)[B_5O_8(OH)]^{\infty 2}$
Nasinite $Na_2(H_2O)_2[B_5O_8(OH)]^{\infty 2}$
Ezcurrite $Na_2(H_2O)_2[B_5O_7(OH)_3]$

(4)-(3)-Borates with x = 0,(3)   Acid
Ameghinite $Na[B_3O_3(OH)_4]$
                                    Hydrates
Tunellite family
Nobleite $Ca(H_2O)_3[B_6O_9(OH)_2]^{\infty 2}$
Tunellite $Sr(H_2O)_3[B_6O_9(OH)_2]^{\infty 2}$
Gowerite $Ca(H_2O)_3[B_5O_8(OH)B(OH)_3]^{\infty 2} \rightarrow Ca(H_2O)[B_6O_8(OH)_4]^{\infty 2} \cdot 3H_2O$

Aksaite family

Aksaite	$Mg(H_2O)_2[B_6O_7(OH)_6]$
Mcallisterite	$Mg_2(H_2O)_3[B_6O_7(OH)_6]_2 \cdot 6H_2O$
Admontite	$Mg_2(H_2O)_3[B_6O_7(OH)_6]_2 \cdot 1.5H_2O$
Rivadavite	$Na_6Mg[B_6O_7(OH)_6]_4 \cdot 10H_2O$

(4)-(3)-Borates with $x = 0,28$ Hydrates

Ginorite family

Ginorite	$Ca_2[B_{14}O_{20}(OH)_6]^{\infty 2} \cdot 5H_2O$
Strontioginorite	$SrCa[B_{14}O_{20}(OH)_6]^{\infty 2} \cdot 5H_2O$

(4)-(3)-Borates with $x = 0,25$ Acid

Strontioborite $Sr[B_8O_{11}(OH)_4]^{\infty 2}$

(4)-(3)-Borates with $x = 0,2$ Hydrates

Sborgite family

Sborgite	$Na(H_2O)_3[B_5O_6(OH)_4]$
Santite	$K(H_2O)_2[B_5O_6(OH)_4]$
*Ramanite-(Cs)	$Cs(H_2O)_3[B_5O_6(OH)_4]$
*Ramanite-(Rb)	$Rb(H_2O)_3[B_5O_6(OH)_4]$
Larderellite	$NH_4(H_2O)[B_5O_7(OH)_2]$
Ammonioborite	$(NH_4)_3(H_2O)_4[B_{15}O_{20}(OH)_8]$

(4)-(3)-Borato-arsenates   Hydrates

Teruggite $Ca_4Mg[B_6O_7(OH)_6]_2[AsO_4]_2 \cdot 12H_2O$

(4)-(3)-Borato-carbonates   Acid

Borcarite	$Ca_4Mg[B_4O_6(OH)_6][CO_3]_2$
*Numanoite	$Ca_4Cu[B_4O_6(OH)_6][CO_3]_2$

*(4)-(3)-Borato-hydrocarbonates   Hydrates

*Qilianshanite $NaH_4[BO_3][CO_3] \cdot 2H_2O$

(4)-(3)-Borato-sulfates

*Vitimite $Ca_6[B_{14}O_{19}(OH)_{14}][SO_4] \cdot 5H_2O$

(4)-(3)- Borato-sulfato-chlorides

Heidornite $Ca_3Na_2Cl[B_5O_8(OH)_2][SO_4]_2$

(4)-(3)-Borato-chlorides
*(4)-(3)- Borato-chlorides $x = 2$ Hydrates

*Chelkarite $CaMg[B_2O_4]Cl_2 \cdot 7\ H_2O$

(4)-(3)-Borato-chlorides with $x = 1,(3)$ Acid

Solongoite $Ca_2[B_3O_4(OH)_4]Cl$

*(4)-(3)-Borato-chlorides with x = 1,2

*Brianroulstonite $Ca_3[B_5O_6(OH)_6](OH)Cl_2 \cdot 8H_2O$

(4)-(3)-Borato-chlorides with $x = 1$ Hydrates (acid)

Ekaterinite $Ca_2[B_4O_7(Cl,OH)_2]\cdot 2H_2O$
Hydrochlorborite $Ca_2[B_4O_4(OH)_7]Cl\cdot 7H_2O$

(4)-(3)-Borato-chlorides with c x = 0,86   Neutral
Boracite family
Boracite $Mg_3[B_7O_{13}]Cl \rightarrow [B_7O_{12}]^{\infty 3}\cdot Mg_3OCl$
*Trembathite $Mg_3[B_7O_{13}]Cl \rightarrow [B_7O_{12}]^{\infty 3}\cdot Mg_3OCl$
Chambersite $Mn_3[B_7O_{13}]Cl \rightarrow [B_7O_{12}]^{\infty 3}\cdot Mn_3OCl$
Ericaite $(Fe,Mg,Mn)_3[B_7O_{13}]Cl \rightarrow [B_7O_{12}]^{\infty 3} (Fe,Mg,Mn)_3OCl$
Congolite $(Fe,Mg,Mn)_3[B_7O_{13}]Cl \rightarrow [B_7O_{12}]^{\infty 3} (Fe,Mg,Mn)_3OCl$

(4)-(3)-Borato-chlorides with x = 0,8   Acid
Hilgardite family
Hilgardite $Ca_2[B_5O_8(OH)_2]Cl$
*Hilgardite-1TC $Ca_2[B_5O_8(OH)_2]Cl$
Parahilgardite $Ca_2[B_5O_8(OH)_2]Cl$
Cl-tyretskite $(Ca,Sr)_2[B_5O_8(OH)_2]Cl$
Tyretskite-1Tc $(Ca,Sr)_2[B_5O_9(OH)]\cdot H_2O$
                        *Hydrates
*Kurgantaite $CaSr[B_5O_9]Cl\cdot H_2O$

*(4)-(3)-Borato-chlorides with x = 0,7
*Pringleite $Ca_9B_{26}O_{34}(OH)_{24}Cl_4\cdot 13H_2O$ (tric.)
*Ruitenbergite $Ca_9B_{26}O_{34}(OH)_{24}Cl_4\cdot 13H_2O$ (mon.)
*Walkerite $Ca_{16}(Mg,Li,\square)_2[B_{13}O_{17}(OH)_{12}]_4Cl_6\cdot 28H_2O$

*(4)-(3)-Borato-chlorides with x = 0,4   *Hydrates
Volkovskite $KCa_4[B_5O_8(OH)]_4[B(OH)_3]_2Cl\cdot 4H_2O$

*(4)-(3)Borato-chlorides with x = 0,2   *Hydrates
*Penobsquisite $Ca_2FeCl[B_9O_{13}(OH)_6]\cdot 4H_2O$

*(4)-(3)-Borato-chromates
*Iquiqueite $K_3Na_4Mg[B_{24}O_{39}(OH)][CrO_4]\cdot 12H_2O$

*(4)-(3)-Borato-chlorides of f - cations
*(4)-(3)-Borato-chlorides with x = 1   *Hydrates
*Braitschite-(Ce) $Ca_6NaCe_2[B_6O_7(OH)_3(O,OH)_3]_4\cdot H_2O$

*(4)-(3)-Borates chalcophylic elements
*(4)-(3)-Borates Cu
*Santarosaite CuB_2O_4

*(4)-(3)-Borates Zn
*(4)-(3)-Borato-chlorides Zn
*Chubarovite $KZn_2[BO_3]Cl_2$

*(4)-(3)-Borates Pb   Hydrates
*Leucostaurite $Pb^{2+}_2[B_5O_9]Cl\cdot 0.5H_2O$

3.2.3. **Class:** Carbonates
3.2.3.1. Subclass: Carbonates of cations with low FC
3.2.3.1.1. Carbonates of *s-*, *d_s-* and *p_s-* cations
3.2.3.1.1.1 Carbonates of *s-*, *d_s-* and *p_s-* cations without Li^+ and Be^{2+}
3.2.3.1.1.1.1. Proper carbonates $x = M^{2+}/[CO_3]$
                                      3.2.3.1.1.1.1.1. Neutral ($x = 1$)

Calcite group (compare with smithsonite (group))	
Magnesite	$Mg[CO_3]$
Gaspeite	$(Ni,Mg,Fe)[CO_3]$
Sphaerocobaltite	$Co[CO_3]$
Siderite	$Fe[CO_3]$
Rhodochrosite	$Mn[CO_3]$
Calcite	$Ca[CO_3]$
Aragonite group (compare with cerussite (group))	
Aragonite	$Ca[CO_3]$
Strontianite	$Sr[CO_3]$
Witherite	$Ba[CO_3]$
*Unnamed mon.	$Ca[CO_3]$
Dolomite group (compare with minrecordite (group))	
Dolomite	$CaMg[CO_3]_2$
Ankerite	$Ca(Fe,Mg)[CO_3]_2$
Kutnohorite	$Ca(Mn,Mg,Fe)[CO_3]_2$
Benstonite	$(Ba,Sr)_6Ca_6Mg[CO_3]_{13}$
Eitelite	$Na_2Mg[CO_3]_2$
Huntite group	
Huntite	$CaMg_3[CO_3]_4$
Norsethite	$BaMg[CO_3]_2$
Fairchildite group	
Vaterite	$Ca[CO_3]$
Fairchildite	$K_2Ca[CO_3]_2$
Gregoryite	$Na_2[CO_3]$
Alstonite family	
Paralstonite	$(Ba,Sr)Ca[CO_3]_2$
*Olekminskite	$Sr(Sr,Ca,Ba)[CO_3]_2$
Barytocalcite	$BaCa[CO_3]_2$
Alstonite	$BaCa[CO_3]_2$
Shortite family	
Shortite	$Na_2Ca_2[CO_3]_3$
Nyerereite	$Na_2Ca[CO_3]_2$
Natrofairchildite	$Na_2Ca[CO_3]_2$
Zemkorite	$(Na,K)_2Ca[CO_3]_2$
Bütschliite	$K_2Ca[CO_3]_2$
Natrite	$Na_2[CO_3]$

3.2.3.1.1.1.1.2. Basic and carbonato-halogenides

Rouvilleite (M8) ($x = 1,1(6)$)	$Na_3Ca(Mn,Ca)F[CO_3]_3$
Northupite ($x = 1,25$)	$Na_3MgCl[CO_3]_2$
Dawsonite ($x = 2$)	$NaAl(OH)_2[CO_3]$
Nullaginite ($x = 2$)	$Ni_2(OH)_2[CO_3]$
Brenkite ($x = 2$)	$Ca_2F_2[CO_3]$
Tunisite ($x = 2,125$)	$NaCa_2Al_4(OH)_8Cl[CO_3]_4$

Holdawayite (x = 3) $Mn_6(OH)_7(Cl,OH)[CO_3]_2$
                                          3.2.3.1.1.1.3. Hydrates
*Alexkhomyakovite $K_6(Ca_2Na)Cl[CO_3]_5·6H_2O$
Kambaldaite (x = 1,1(6)) $NaNi_4(OH)_3[CO_3]_3·3H_2O$
Hydromagnesite family (x = 1,25)
Hydromagnesite $Mg_5(OH)_2[CO_3]_4·4H_2O$
Dypingite $Mg_5(OH)_2[CO_3]_4·5H_2O$
Giorgiosite $Mg_5(OH)_2[CO_3]_4·5H_2O$
Indigirite $Mg_2Al_2(OH)_2[CO_3]_4·15H_2O$
*Widgiemoolthalite $(Ni,Mg)_5(OH)_2[CO_3]_4·4-5H_2O$
Dresserite family (x = 2)
Strontiodresserite $(Sr,Ca)Al_2(OH)_4[CO_3]_2·H_2O$
Dresserite $BaAl_2(OH)_4[CO_3]_2·H_2O$
Hydrodresserite $BaAl_2(OH)_4[CO_3]_2·3H_2O$
Alumohydrocalcite family (x = 2)
Alumohydrocalcite $CaAl_2(OH)_4[CO_3]_2·3H_2O$
Para-alumohydrocalcite $CaAl_2(OH)_4[CO_3]_2·6H_2O$
*Kochsandorite $CaAl_2(OH)_4[CO_3]_2·H_2O$
Artinite family (x = 2)
Pokrovskite $Mg_2(OH)_2[CO_3]$
Artinite $Mg_2(OH)_2[CO_3]·3H_2O$
*Chlorartinite $Mg_2(OH)Cl[CO_3]·3H_2O$
Otwayite $(Ni,Mg)_2(OH)_2[CO_3]·H_2O$
Zaratite family (x = 3)
Zaratite $Ni_3(OH)_4[CO_3]·4H_2O$
Defernite $Ca_6(OH)_7(Cl,OH)_{1-2x}[CO_3]_{2-x}[SiO_4]_x$, where x≤0,5
Brugnatellite family (x = 7-7,5)
Hydroscarbroite $Al_{14}(OH)_{36}[CO_3]_3·nH_2O$
Scarbroite $Al_5(OH)_{13}[CO_3]·5H_2O$
Brugnatellite $Fe^{3+}Mg_6(OH)_{13}[CO_3]·4H_2O$
*Quintinite-3T $Mg_4Al_2(OH)_{12}[CO_3]·4H_2O$
*Quintinite $Mg_4Al_2(OH)_{12}[CO_3]·3H_2O$
Hydrotalcite family(x = 9)
*Caresite-3T $Fe_4Al_2(OH)_{12}[CO_3]·3H_2O$
Hydrotalcite $Mg_6Al_2(OH)_{16}[CO_3]·4H_2O$
*Charmarite-3T $Mn_4Al_2(OH)_{12}[CO_3]·3H_2O$
Pyroaurite $Mg_6Fe^{3+}_2(OH)_{16}[CO_3]·4H_2O$
Desautelsite $Mg_6Mn^{3+}_2(OH)_{16}[CO_3]·4H_2O$
Stichtite-3R $Mg_6Cr^{3+}_2(OH)_{16}[CO_3]·4H_2O$
*Stichtite-2H = Barbertonite $Mg_6Cr^{3+}_2(OH)_{16}[CO_3]·4H_2O$
Takovite $Ni_6Al_2(OH)_{16}[CO_3]·4H_2O$
Reevesite $Ni_6Fe^{3+}_2(OH)_{16}[CO_3]·4H_2O$
*Fe-reevesite $Fe^{2+}_4Ni_2(Fe^{3+}_{1,96}Al_{0,03}Cr_{0,01})_2(OH)_{16}[CO_3]·4H_2O$
Comblainite $Ni^{2+}_4Co^{3+}_2(OH)_{12}[CO_3]·3H_2O$
  Manasseite group (x = 9)
  Manasseite = hydrotalcite $Mg_6Al_2(OH)_{16}[CO_3]·4H_2O$
  Sjögrenite = pyroaurite $Mg_6Fe^{3+}_2(OH)_{16}[CO_3]·4H_2O$
Coalingite (x = 13) $Mg_{10}Fe^{3+}_2(OH)_{24}[CO_3]·H_2O$
                                          3.2.3.1.1.1.3.2. Neutral (x = 1)

Lansfordite family

Lansfordite	$Mg[CO_3] \cdot 5H_2O$
Baylissite	$K_2Mg[CO_3]_2 \cdot 4H_2O$
Monohydrocalcite	$Ca[CO_3] \cdot H_2O$
Gaylussite family	
Pirssonite	$Na_2Ca[CO_3]_2 \cdot 2H_2O$
Gaylussite	$Na_2Ca[CO_3]_2 \cdot 5H_2O$
Ikaite group	
Hellyerite	$Ni[CO_3] \cdot 6H_2O$
Ikaite	$Ca[CO_3] \cdot 6H_2O$
Natron family	
Thermonatrite	$Na_2[CO_3] \cdot H_2O$
Natron	$Na_2[CO_3] \cdot 10H_2O$

*3.2.3.1.1.1.2. Carbonato-borates   *3.2.3.1.1.1.2.1. Hydrates
                                                     (asid)

*Qilianshanite $NaH_4[CO_3][BO_3] \cdot 2H_2O$

3.2.3.1.1.1.3. Carbonato-phosphates
*3.2.3.1.1.1.3.2. Carbonato-phosphates with $CO_3 : PO_4 \sim 18$
                              *3.2.3.1.1.1.3.2.1. Hydrates
*Karchevskyite $Mg_{18}Al_9(OH)_{54}Sr_2[CO_3]_9(H_2O)_6(H_3O)_5$

3.2.3.1.1.1.2.1. Carbonato-phosphates with $CO_3 : PO_4 = 1$
                              3.2.3.1.1.1.2.1.1. Neutral

Bradleyite group	
Bradleyite	$Na_3Mg[CO_3][PO_4]$
*Crawfordite	$Na_3Sr[CO_3][PO_4]$
Bonshtedtite	$Na_3Fe^{2+}[CO_3][PO_4]$
Sidorenkite	$Na_3Mn^{2+}[CO_3][PO_4]$

3.2.3.1.1.1.2.2. Carbonato-phosphates with $CO_3 : PO_4 = 0,(3)$
Heneuite $CaMg_5(OH)[CO_3][PO_4]$

3.2.3.1.1.1.4. Carbonato-dihydrophosphato-phosphates
                            3.2.3.1.1.1.4.1. Hydrates (basic)
Girvasite $NaCa_2Mg_3(OH)_2[CO_3][PO_2(OH)_2][PO_4]_2 \cdot 4H_2O$

3.2.3.1.1.1.5. Carbonato-sulfates
*3.2.3.1.1.1.5.1. Carbonato-sulfates with $CO_3 : SO_4 = 8$
*3.2.3.1.1.1.5.1.1. Hydrates
*3.2.3.1.1.1.5.1.1.1. Basic
*Putnisite $SrCa_4Cr^{3+}_8[CO_3]_8[SO_4](OH)_{16} \cdot 23H_2O$

3.2.3.1.1.1.5.1. Carbonato-sulfates with $CO_3 : SO_4 = 4$
                            3.2.3.1.1.1.5.1.1. Neutral

Tychite group (x = 1)	
Tychite	$Na_6Mg_2[CO_3]_4[SO_4]$
Ferrotychite	$Na_6Fe^{2+}_2[CO_3]_4[SO_4]$
*Manganotychite	$Na_6Mn^{2+}_2[CO_3]_4[SO_4]$

3.2.3.1.1.1.5.2. Carbonato-sulfates with $CO_3 : SO_4 = 1$
                                        3.2.3.1.1.1.5.2.1. Hydrates
                                        3.2.3.1.1.1.5.2.1.1. Basic
Tatarskite (x = 2) $Ca_6Mg_2[CO_3]_2(SO_4)_2(OH)_4Cl_4 \cdot 7H_2O$

3.2.3.1.1.1.5.3. Carbonato-sulfates with $CO_3 : SO_4 = 1$
                                        3.2.3.1.1.1.5.3.1. Neutral
Burkeite (x = 1) $Na_6[CO_3][SO_4]_2$

                                        3.2.3.1.1.1.5.3.2. Hydrates
Rapidcreekite $Ca_2[CO_3][SO_4] \cdot 4H_2O$

3.2.3.1.1.1.5.4. Carbonato-sulfates with variable ratio $CO_3 : SO_4$
                with the proviso that $CO_3 < SO_4$
                                        3.2.3.1.1.1.5.4.1. Basic
Paraotwayite (x = 3) $Ni(OH)_{2-x}[(SO_4)(CO_3)]_{0.5x}$

3.2.3.1.1.1.5. Carbonato-sulfates with $CO_3 : SO_4 = 1$
3.2.3.1.1.1.5.1. Carbonato-sulfato-chlorides
Hanksite (x = 1,04) $KNa_{22}[CO_3]_2[SO_4]_9Cl$

*3.2.3.1.1.1.6. Carbonato-fluorides
*Podlesnoite $BaCa_2[CO_3]_2F_2$
                                        *3.2.3.1.1.1.6.1. Hydrates
*Sheldrikite $NaCa_3[CO_3]_2F_3 \cdot H_2O$

3.2.3.1.1.1.7. Carbonato-fluoraluminates
                                        3.2.3.1.1.1.7.1. Neutral
Stenonite $Sr_2AlF_5[CO_3] \rightarrow Sr_2[CO_3][AlF_5]^{\infty}$
                                        3.2.3.1.1.1.7.2. Hydrates
Montroyalite $Sr_4Al_8(OH,F)_{26}[CO_3]_3 \cdot 10H_2O \rightarrow Sr_4[CO_3]_3[Al_4(OH,F)_{13}]_2 \cdot 10H_2O$

3.2.3.1.1.1.8. Carbonato-fluoraluminato-hydrocarbonates
                                        3.2.3.1.1.1.8.1. Neutral
Barentsite $Na_7AlF_4[HCO_3]_2[CO_3]_2 \rightarrow Na_7[HCO_3]_2[CO_3]_2[AlF_4]^{\infty}$

3.2.3.1.1.1.9. Hydrocarbonato-carbonates
3.2.3.1.1.1.9.1. Hydrocarbonato-carbonates with $HCO_3 : CO_3 = 0,(4)$
                                        3.2.3.1.1.1.9.1.1. Hydrates (basic)
Sergeevite (x = 1) $Ca_2Mg_{11}(OH)_4[HCO_3]_4[CO_3]_9 \cdot 6H_2O$

3.2.3.1.1.1.9.2. Hydrocarbonato-carbonates with $HCO_3 : CO_3 = 1$
                                        3.2.3.1.1.1.9.2.1. Hydrates (neutral)
Trona (x = 0,75) $Na_3[HCO_3][CO_3] \cdot 2H_2O$

3.2.3.1.1.1.9.3. Hydrocarbonato-carbonates with $HCO_3 : CO_3 = 3$
                                        3.2.3.1.1.1.9.3.1. Neutral
Wegscheiderite (x = 0,625) $Na_5[HCO_3]_3[CO_3]$

3.2.3.1.1.1.10. Hydrocarbonates

3.2.3.1.1.1.10.1. Neutral

Nahcolite family (x = 0,5)

Nahcolite	$Na[HCO_3]$
Kalicinite	$K[HCO_3]$
Teschemacherite	$NH_4[HCO_3]$

3.2.3.1.1.1.9.2. Hydrates (basic)

Nesquehonite (x = 1) $MgOH[HCO_3]\cdot2H_2O$

3.2.3.1.1.2. Carbonates of Li
3.2.3.1.1.2.1. Neutral

Zabuyelite $Li_2[CO_3]$

*3.2.3.1.1.3. Carbonates of Be

*Niveolanite $NaBe[CO_3](OH)\cdot2H_2O$

3.2.3.1.2. Carbonates of *f*-elements

3.2.3.1.2.1. Neutral

Burbankite family (x = 1)

Sahamalite-(Ce)	$(Mg,Fe^{2+})Ce_2[CO_3]_4$
Carbocernaite	$(Ca,Na)(Sr,Ce,Ba)[CO_3]_2$
Khanneshite	$(Na,Ca)_3(Ba,Sr,Ce,Ca)_3[CO_3]_5$
Ewaldite	$Ba(Ca,Y,Na,K)[CO_3]_2\cdot2.6H_2O$
Burbankite	$(Na,Ca)_3(Sr,Ba,Ce)_3[CO_3]_5$
*Calcioburbankite	$Na_3(Ca,REE,Sr)_3[CO_3]_5$
Remondite-(Ce)	$Na_3(Ce,La,Ca,Na,Sr)_3[CO_3]_5$
*Remondite-(La)	$Na_3(La,Ce,Ca)_3[CO_3]_5$
*Petersenite-(Ce)	$Na_4REE_2[CO_3]_5$
*Paratooite-(La)	$(La,REE,Ca,Na,Sr)_6Cu^{2+}[CO_3]_8$

*3.2.3.1.2.1.1. Hydrates

*Shomiokite-(Y)	$Na_3Y[CO_3]_3\cdot3H_2O$
*Lecoqite-(Y)	$Na_3Y[CO_3]_3\cdot6H_2O$

3.2.3.1.2.2. Carbonato-fluorides and basic carbonares

Baiyuneboite-(Ce) $NaBaCe_2F[CO_3]_4$

Polysomatic series of bastnäsite $mM^{2+}[CO_3]\cdot nTR(F,OH)[CO_3]\cdot pH_2O$ or $M_{2+m}TR_n(F,OH)_n[CO_3]_{m+n}\cdot pH_2O$, where M^{2+} = Ca, Sr, Ba; TR = Ce, La, Nd, Y...Th with the proviso that $0 < p < 3$ (ratio m:n is reported at the name of groups, series, famileis ib round brackets)

  Parisite subseries (m : n = 2)

*Kukharenkoite-(Ce) = Zhonghuacerite-(Ce) (x = 1,1(6))	$Ba_2(Ce,REE)F[CO_3]_3$	
*Kukharenkoite-(La)	$Ba_2(La,REE)F[CO_3]_3$	
Cebaite group (x=1,2)		(m : n = 1,5)
Cebaite-(Ce)	$Ba_3Ce_2F_2[CO_3]_5$	
Synchysite family (x = 1,25)		(m : n = 1)
Synchysite-(Nd)	$CaNdF[CO_3]_2$	
Synchysite-(Ce)	$CaCeF[CO_3]_2$	
*Synchysite-(Ce) tet	$CaCeF[CO_3]_2$	
*Synchysite-(Ce) trig	$CaCeF[CO_3]_2$	
Synchysite-(Y)	$CaYF[CO_3]_2$	

Huanghoite-(Ce) $BaCeF[CO_3]_2$
*Qaqarssukite-(Ce) $BaCeF[CO_3]_2$
*Horvathite-(Y) $NaYF_2[CO_3]$
Röntgenite-(Ce) (x = 1,3) $Ca_2Ce_3F_3[CO_3]_5$ (m : n = 0,(6))
Parisite family (x = 1,(3))   (m : n = 0,5)
Parisite-(Ce)[*] $CaCe_2F_2[CO_3]_3$
*Lukechangite-(Ce) $Na_3Ce_2F[CO_3]_4$
Parisite-(Nd) $CaNd_2F_2[CO_3]_3$
Cordylite-(Ce) $NaBaCe_2F_2[CO_3]_4$
*Cordylite-(La) $NaCaBa_2La_3SrF_2[CO_3]_8$
  Bastnäsite group (x = 1,5)   (only n)
  Bastnäsite-(Ce) $(Ce,La)F[CO_3]$
  Bastnäsite-(Y) $(Y,REE)F[CO_3]$
  *Bastnäsite-(La) $(La,Ce)F[CO_3]$
  Hydroxylbastnäsite-(Ce) $(Ce,La,Nd)(OH,F)[CO_3]$
  *Hydroxylbastnäsite-(La) $(La,Nd)(OH,F)[CO_3]$
  *Hydroxylbastnäsite-(Nd) $(Nd,La)(OH,F)[CO_3]$
Ancylite subseries (hydrates basic)
Ancylite group
  Calcio-ancylite-(Ce) (x = 1,25) $(Ca,Sr)Ce(OH)[CO_3]_2 \cdot H_2O$ (m : n = 1)
  *Calcio-ancylite-(Nd) $Ca(Nd,Ce,Gd,Y)_3(OH)_3[CO_3]_4 \cdot H_2O$
  Ancylite-(Ce) $SrCe(OH)[CO_3]_2 \cdot H_2O$
  *Ancylite-(La) $SrLa(OH)[CO_3]_2 \cdot H_2O$
*Unnamed-(Nd) $Nd[CO_3][(OH), \cdot H_2O]$
*Kamphagite-(Y) $Ca_2(Y,REE)_2(OH)_2[CO_3]_4 \cdot 3H_2O$
*Unnamed (x = 1,42) $(Ca,Sr)((Nd,La,Pr,Sm)_5(OH)_5[CO_3]_6 \cdot H_2O$ (m:n = 0,2)
*Kozoite-(La) $(La,Nd,Ca)(OH)[CO_3]$
*Kozoite-(Nd) (x = 1,5) $(Nd,La,Ca)(OH)[CO_3]$
*Decrespignyite-(Y) (x = 1,75) (m:n = 0,17) $(Y,REE)^{3+}_4Cu^{2+}[CO_3]_4Cl(OH)_5 \cdot 2H_2O$
*Arisite-(Ce) (x = 1.75) (m:n = 0,33) $NaCe_2F[CO_3]_2[F_{2x}(CO_3)_{1-x}]$
*Arisite-(La) $NaLa_2F[CO_3]_2[F_{2x}(CO_3)_{1-x}]$
Thorbastnäsite-(Ce) (x = 1,5) $ThCeF_2[CO_3]_2 \cdot 3H_2O$
*Lusernaite-(Y) $Y_4Al[CO_3]_2(OH)_{10}F \cdot 6H_2O$
3.2.3.1.2.2.1. Hydrates

                               3.2.3.1.2.2.1.1. Neutral

Lanthanite family (x = 1)
Calkinsite-(Ce) $(Ce,La)_2[CO_3]_3 \cdot 4H_2O$
Lanthanite-(Ce) $(Ce,La,Nd)_2[CO_3]_3 \cdot 8H_2O$
Lanthanite-(La) $(La,Nd)_2[CO_3]_3 \cdot 8H_2O$
Lanthanite-(Nd) $(Nd,La)_2[CO_3]_3 \cdot 8H_2O$
Tengerite-(Y) $Y_2[CO_3]_3 \cdot 2\text{-}3H_2O$
Lokkaite family
Lokkaite-(Y) $CaY_4[CO_3]_7 \cdot 9H_2O$
Kimuraite-(Y) $CaY_2[CO_3]_4 \cdot 6H_2O$
*Adamsite-(Y) $NaY[CO_3]_2 \cdot 6H_2O$
Mckelveyite family

---

[*] According to Can. Min., 2001, v. 39, p.1713 the polytypes of parisite-(Ce): 4H, 8H, 10H, 14H, 16H, $6R_1$, $6R_2$, 18R, 25R, 30R, 36R, 42R be exist.

*Galgenbergite-(Ce)	$Ca(Ce,REE)_2[CO_3]_4 \cdot H_2O$
Donnayite-(Y)	$NaCaSr_3Y[CO_3]_6 \cdot 3H_2O$
Donnayite-(Y) trig.	$NaCaSr_3Y[CO_3]_6 \cdot 3H_2O$
Mckelveyite-(Y)	$Ba_3Na(Ca,U)Y[CO_3]_6 \cdot 3H_2O$
Tuliokite	$BaNa_6Th[CO_3]_6 \cdot 6H_2O$

*3.2.3.1.2.3. Carbonato -sulfato-halogenides
*Reederite-(Y) $(Na,Mn)_{15}Y_2[CO_3]_9[SO_3F]Cl$

*3.2.3.1.2.4. Carbonato-hydrocarbonato-sulfato-halogenides
                                        *3.2.3.1.2.4.1. Basic
*Mineevite-(Y) $Na_{25}Ba(Y,Gd,Dy)_2[CO_3]_{11}(HCO_3)_4(SO_4)_2F_2Cl$

3.2.3.1.2.5. Carbonato-phosphates   3.2.3.1.2.5.1. Basic
Daqingshanite-(Ce) $Sr_3Ce[CO_3]_3[PO_4]$

*3.2.3.1.2.6. Carbonato-hydrophosphates   *3.2.3.1.2.6.1. Hydrates
*Micheelsenite $(Ca,Y)_3Al[HPO_4,CO_3][CO_3](OH)_6 \cdot 12H_2O$
*Micheelsenite-(Y) $Ca_4Y_2(Al,Y,Dy)_2[(P,Al)O_4]_2[CO_3](OH)_{12} \cdot 25H_2O$

*3.2.3.1.2.7. Hydrocarbonates   *3.2.3.1.2.7.1. Basic
*Tomasclarkeite-(Y) $Na(Y,REE)(HCO_3)(OH)_3 \cdot 4H_2O$

3.2.3.2. Subclass: Carbonates of cations with middle FC
3.2.3.2.1. Carbonates of Zr   3.2.3.2.1.1. Hydrates (neutral)
Weloganite (x = 1) $Sr_3Na_2Zr[CO_3]_6 \cdot 3H_2O$
(compare with mckelveyite)

3.2.3.2.2. Carbonates of Ti   3.2.3.2.2.1. Oxido-carbonates
Sabinaite (x= 2) $Na_4Zr_2TiO_4[CO_3]_4$

3.2.3.2.3. Carbonates of Mn^{4+} 3.2.3.2.3.1. Hydrates (neutral)
Jouravskite $Ca_3Mn^{4+}[CO_3][SO_4](OH)_6 \cdot 12H_2O$

3.2.3.3. Subklass: Carbonates of chalcophylic cations
*3.2.3.3a. Carbonates VIIб – VIIIб cations   Основные
*Chukanovite $Fe^{2+}_2(OH)_2[CO_3]$

3.2.3.3.1. Carbonates of Cu^{2+}
3.2.3.3.1.1. Proper carbonates   3.2.3.3.1.1.1. Basic
Azurite (x= 1,5) $Cu_3(OH)_2[CO_3]_2$
Malachite family (x = 2)
Mcguinnessite $(Mg,Cu)_2(OH)_2[CO_3]$
Kolwezite $CuCo(OH)_2[CO_3]$
Glaukosphaerite $(Cu,Ni)_2(OH)_2[CO_3]$
Malachite $Cu_2(OH)_2[CO_3]$
*Huangodoyite $Na_2Cu[CO_3]_2$
Georgeite (x = 1,(6)) $Cu_2(OH)_2[CO_3]$
                                        3.2.3.3.1.1.2. Hydrates
                                        3.2.3.3.1.1.2.1. Basic

Callaghanite (x = 4) $Cu_2Mg_2(OH)_6[CO_3]\cdot 2H_2O$
                                                         3.2.3.3.1.1.2.2. Neutral
Chalconatronite (x = 1) $Na_2Cu[CO_3]_2\cdot 3H_2O$

3.2.3.3.1.2. Carbonato-sulfates (0,25:1) Cu^{2+}
                                                         3.2.3.3.1.2.1. Hydrates (basic)
Nakauriite (x = 1,6) $Cu_8(OH)_6[CO_3][SO_4]_4\cdot 48H_2O$

3.2.3.3.2. Carbonates of Hg^+ 3.2.3.3.2.1. Hydrates (basic)
Szymańskiite (x = 1,5) $(H_3O)_8Hg^+_{16}(Ni,Mg)_6(OH)_{12}[CO_3]_{12}\cdot 3H_2O$
*Clearcreekite mon. $Hg^+_3[CO_3](OH)\cdot 2H_2O$
*Peterbaylissite orth., pseudohex. $Hg^+_3[CO_3](OH)\cdot 2H_2O$

3.2.3.3.3. Carbonates of Zn and Cd
3.2.3.3.3.1. Proper carbonates   3.2.3.3.3.1.1. Neutral
   Smithsonite group (compare with calcite (group))
   Smithsonite (x = 1) $Zn[CO_3]$
   Otavite $Cd[CO_3]$
   Minrecordite (x = 1) $CaZn[CO_3]_2$
(compare with dolomite (group)
                                                         3.2.3.3.3.1.2. Basic
   Rosasite group (x = 2)
   Rosasite $(Cu,Zn)_2(OH)_2[CO_3]$
   Zincrosasite $(Zn,Cu)_2(OH)_2[CO_3]$
   Hydrozincite family (x = 2,5)
   Hydrozincite $Zn_5(OH)_6[CO_3]_2$
*Parádsasvárite $Zn_2(OH)_2[CO_3]$
   Aurichalcite $(Zn,Cu)_5(OH)_6[CO_3]_2$
   Loseyite group (x = 3,5)
   Sclarite $(Zn,Mg,Mn)_4Zn_3(OH)_{10}[CO_3]_2$
   Loseyite $(Mn,Zn)_7(OH)_{10}[CO_3]_2$
                                                         3.2.3.3.3.1.3. Hydrates (basic)
Claraite(x = 3) $(Cu,Zn)_3(OH)_4[CO_3]\cdot 4H_2O$
*Zaccagnaite $Zn_4Al_2(OH)_{12}[CO_3]\cdot 3H_2O$

3.2.3.3.3.2. Sulfato-carbonates (0,5:1)
                                                         3.2.3.3.3.2.1. Basic
*Brianyoungite $Zn_{12}[CO_3]_3[SO_4](OH)_{16}$
Hauckite (x = 8) $(Mg,Mn)_{24}Zn_{18}Fe^{3+}_3(OH)_{81}[CO_3]_2[SO_4]_4$

3.2.3.3.4. Carbonates of Pb^{2+}
3.2.3.3.4.1. Proper carbonates
                                                         3.2.3.3.4.1.1. Neutral
Cerussite (x = 1) $Pb[CO_3]$
(compare with aragonite (group))
*Sanromanite $Na_2CaPb_3[CO_3]_5$
                                                         3.2.3.3.4.1.2. Basic and carbonato-
                                                         chlorides
Hydrocerussite (x = 1,5) $Pb_5O(OH)_2[CO_3]_3$
Plumbonacrite (x = 1,(6)) $Pb_5(OH)_2O[CO_3]_3$

Phosgenite (x = 2) $Pb_2Cl_2[CO_3]$
*Barstowite (x = 4) $Pb_4Cl_6[CO_3] \cdot H_2O$
Schuilingite-(Nd) (x = 1,1(6)) $PbCuNd(OH)[CO_3]_3 \cdot H_2O$
Gysinite-(Nd) (x = 1,25) $PbNd(OH)[CO_3]_2 \cdot H_2O$
Dundasite (x = 2) $PbAl_2(OH)_4[CO_3]_2 \cdot H_2O$
*Petterdite (x = 2) $PbCr_2(OH)_4[CO_3]_2 \cdot H_2O$

*3.2.3.3.4.1.3.Oxido-carbonates
*Shannonite $Pb_2O[CO_3]$

3.2.3.3.4.2. Carbonato-sulfates
3.2.3.3.4.2.1. Carbonato-sulfates with $CO_3 : SO_4 = 4$
                             3.2.3.3.4.2.1.1. Oxido-hydrates
Nasledovite (x = 2) $PbMn_3Al_4O_5[CO_3]_4[SO_4] \cdot 5H_2O$

3.2.3.3.4.2.2. Carbonato-sulfates with $CO_3 : SO_4 = 2$
                             3.2.3.3.4.2.2.1. Basic
Leadhillite family(x = 1,(3))
Susannite $Pb_4(OH)_2[CO_3]_2[SO_4]$
Macphersonite $Pb_4(OH)_2[CO_3]_2[SO_4]$
Leadhillite $Pb_4(OH)_2[CO_3]_2[SO_4]$

3.2.3.3.4.2.3. Carbonato-sulfates with $CO_3 : SO_4 = 0,(3)$
                             3.2.3.3.4.2.3.1. Basic
Caledonite (x = 1,75) $Pb_5Cu_2(OH)_6[CO_3][SO_4]_3$

*3.2.3.3.4.2.4. Oxido-carbonato-sulfato-chlorides $CO_3 : SO_4: Cl = 1: 4: 4$
                           *3.2.3.3.4.2.4.1. Basic
*Philolithite $Pb_{12}O_6Mn(Mg,Mn)_2(Mn,Mg)_4[CO_3]_4[SO_4]Cl_4(OH)_{12}$

*3.2.3.3.4.2.5. Carbonato-tiosulfates
*Fassinaite $Pb_2[S^{2+}_2O_3][CO_3]$

3.2.3.3.5. Carbonates of As^{3+}, Sb^{3+} and Bi^{3+}
3.2.3.3.5.1. Oxido- and oxido-carbonato-fluorides
Beyerite (x = 2) $(Ca,Pb)Bi_2O_2[CO_3]_2$
Kettnerite (x = 2,5) $CaBiOF[CO_3]$
Bismutite (x = 3) $CaBiOF[CO_3]$

3.2.3.3.5.2. Oxido-hydroxido-carbonates
Armangite (x = 53) $Mn^{2+}_{26}As^{3+}_{18}(OH)_4O_{50}[CO_3]$
(rather arsenito-carbonate !!)

3.2.3.4. Subclass: Carbonates of light p-anionformers (only Si^{4+})
3.2.3.4.1. Carbonato-sulfates (1:1)
                             3.2.3.4.1.1. Hydrates (basic)
Thaumasite (x = 2,5) $Ca_3Si(OH)_6[CO_3][SO_4] \cdot 12H_2O$

3.2.4. *Class:* Phosphates
3.2.4.1.Quasiclass: Orthophosphates

3.2.4.1.1. Subclass: Orthophosphates of cations with low FC
3.2.4.1.1.1. Orthophosphates of s-, d_s- and p_s-cations
3.2.4.1.1.1.1. Orthophosphates of s-, d_s- and p_s-cations without Li and Be
3.2.4.1.1.1.1.1. Proper orthophosphates   3.2.4.1.1.1.1.1. Neutral

Alluaudite family (x = MO : PO_4 = 1,(3))
(compare with berzeliite (family); garnet (series))

Alluaudite	$Na\square Mn^{2+}Fe^{3+}{}_2[PO_4]_3$
Ferro-alluaudite	$Na\square(Fe^{2+},Mn^{2+})Fe^{3+}{}_2[PO_4]_3$
Maghagendorfite	$NaMgMn^{2+}(Fe^{2+},Fe^{3+})_2[PO_4]_3$
Hagendorfite	$NaCaMn^{2+}(Fe,Mn)^{2+}{}_2[PO_4]_3$
Varulite	$NaCaMnMn^{2+}{}_2[PO_4]_3$
Berlinite (x = 1,5)	$\{Al[PO_4]\}^{\infty3}$
*Pretulite	$Sc[PO_4]$
*Rodolicoite	$Fe[PO_4]$
*Grattarolaite	$Fe_3{}^{3+}O_3[PO_4]$

Purpurite series (x = 1,5)

Heterosite	$(Fe,Mn)^{3+}[PO_4]$
Purpurite	$(Mn,Fe)^{3+}[PO_4]$

Graftonite series (x = 1,5)

Farringtonite	$Mg_3[PO_4]_2$
Sarcopside	$Fe^{2+}{}_3[PO_4]_2$
*Zavaliaite	$(Mn^{2+},Fe^{2+}, Mg)_3[PO_4]_2$
*Chopinite	$Mg_3[PO_4]_2$
Graftonite	$(Fe^{2+},Mn^{2+},Ca)_3[PO_4]_2$
Beusite	$(Mn,Fe,Ca,Mg)_3[PO_4]_2$
*Tuite polymorph of high pressure	$\gamma\text{-}Ca_3[PO_4]_2$
*Merrillite	$Ca_9Na(Mg,Fe^{2+})[PO_4]_7$
*Ferromerrillite	$Ca_9NaFe^{2+}[PO_4]_7$
Stanfieldite (x = 1,5)	$Ca_4(Mg,Fe,Mn)_5[PO_4]_6$
Bobfergusonite (x = 1,5)	$Na_2Mn_5Fe^{3+}Al[PO_4]_6$
*Manitobaite	$Na_{16}Mn_{15}Al_8[PO_4]_{30}$
Johnsomervilleite (x = 1,5)	$Na_{10}Ca_6Mg_{18}(Fe,Mn)_{25}[PO_4]_{36}$

Brianite family(x = 1,5)

Panethite	$(Na,Ca,K)_2(Mg,Fe,Mn)_2[PO_4]_2$
Brianite	$Na_2CaMg[PO_4]_2$

Natrophilite family (x = 1,5)

Maricite	$NaFe[PO_4]$
*Karenwebberite	$Na(Fe,Mn)^{2+}[PO_4]$
Natrophilite	$NaMn[PO_4]$
Buchwaldite	$NaCa[PO_4]$

Fillowite family (x = 1,5)

Fillowite	$Na_2Ca(Mn,Fe)^{2+}{}_7[PO_4]_6$
*Chladniite	$Na_2CaMg_7[PO_4]_6$
*Galileiite	$(Na,K)_2(Fe,Mn,Cr)^{2+}{}_8[PO_4]_6$
*Unnamed	$(K,Na)_2(Fe,Mn,Cr)^{2+}{}_8[PO_4]_6$
Olgite	$(Ba,Sr)(Na,Sr,TR)_2Na[PO_4]_2$
*Bario-olgite	$Na(BaSr)Na[PO_4]_2$
Olympite (x = 1,5)	$LiNa_5[PO_4]_2$

Wyllieite group (x = 1,(6))

Rosemaryite	$NaMn^{2+}Fe^{3+}Al[PO_4]_3$

*Ferrorosemaryite	$\square NaFe^{2+}Fe^{3+}Al[PO_4]_3$
Wyllieite	$Na_2Mn^{2+}Fe^{2+}Al[PO_4]_3$
Ferrowyllieite	$Na_2(Fe^{2+},Mg)_2Al[PO_4]_3$
Qingheiite	$Na_2Mn^{2+}Mg(Al,Fe^{3+})[PO_4]_3$
*Qingheiite-(Fe^{2+})	$Na_2MgFe^{2+}Al[PO_4]_3$

*3.2.4.1.1.1.1.1.2. Oxido-and phosphato-halogenides
*3.2.4.1.1.1.1.1.2.1. Neutral

*Moraskoite	$Na_2MgF[PO_4]$
	3.2.4.1.1.1.1.1.2.2. Basic
Melonjosephite (x = 1,5)	$CaFe^{2+}Fe^{3+}(OH)[PO_4]_2$
Arrojadite group (x = 1,58)	
Dickinsonite-(K,Mn,Na)	$K(Na,Mn)Na_3CaAlMn_{13}[PO_4]_{12}(OH)_2$
Arrojadite-(BaFe)	$BaFe(CaNa_2)Fe^{2+}_{13}Al[PO_4]_{11}[PO_3(OH)](OH)_2$
*Arrojadite-(BaNa)	$BaNa_2(CaNa_2)Fe^{2+}_{13}Al[PO_4]_{11}[PO_3(OH)](OH)_2$
*Arrojadite-(KFe)	$(KNa)Fe^{2+}(CaNa_2)Fe^{2+}_{13}Al[PO_4]_{11}[PO_3(OH)](OH)_2$
*Arrojadite-(KNa)	$KNa_5CaFe^{2+}_{13}Al[PO_4]_{11}[PO_3(OH)](OH)_2$
*Arrojadite-(NaFe)	$Na_2Fe^{2+}(CaNa_2)Fe^{2+}_{13}Al[PO_4]_{11}[PO_3(OH)](OH)_2$
*Arrojadite-(PbFe)	$PbFe^{2+}(CaNa_2)Fe^{2+}_{13}Al[PO_4]_{11}[PO_3(OH)](OH)_2$
*Arrojadite-(SrFe)	$SrFe^{2+}(CaNa_2)Fe^{2+}_{13}Al[PO_4]_{11}[PO_3(OH)](OH)_2$
*Arrojadite-(SrNa)	$SrNa_2(CaNa_2)Fe^{2+}_{13}Al[PO_4]_{11}[PO_3(OH)](OH)_2$
*Fluorarrojadite-(BaFe)	$Na_2CaBaFe^{2+}Fe^{2+}_{13}Al[PO_4]_{11}[PO_3(OH)]F_2$
*Fluorarrojadite-(BaNa)	$BaNa_2(CaNa_2)Fe^{2+}_{13}Al[PO_4]_{11}[PO_3(OH)]F_2$
*Fluorarrojadite-(KNa)	$KNa_3CaNa_2Fe^{2+}_{13}Al[PO_4]_{11}[PO_3(OH)]F_2$
*Fluorarrojadite-(NaFe)	$Na_2Fe^{2+}(CaNa_2)Fe^{2+}_{13}Al[PO_4]_{11}[PO_3(OH)]F_2$
Samuelsonite (x = 1,6)	$(Ca,Ba)Fe^{2+}_2Mn^{2+}_2Ca_8Al_2(OH)_2[PO_4]_{10}$
Nefedovite (x = 1,625)	$Na_5Ca_4F[PO_4]_4$
Apatite family (x = 1,(6))	
Oxyapatite hipothet. minal of apatite series	$Ca_{10}O[PO_4]_6$
Fluorapatite	$Ca_5F[PO_4]_3$
Carbonate-fluorapatite	$Ca_5(F,OH)[(PO_4),(CO_3)]_3$
Hydroxylapatite	$Ca_5(OH)[PO_4]_3$
Carbonate-hydroxylapatite	$Ca_5(OH,F)[(PO_4),(CO_3)]_3$
Chlorapatite	$Ca_5Cl[PO_4]_3$
*Clinohydroxylapatite = apatite-(CaOH)-M	$Ca_5OH[PO_4]_3$
*Stronadelphite	$Sr_5F[PO_4]_3$
*Miyahisaite	$(Sr,Ca)_2Ba_3F[PO_4]_3$
*Alforsite	$Ba_5Cl[PO_4]_3$
Belovite series	
*Belovite-(Ce)	$NaCe\,Sr_3F[PO_4]_3$
*Belovite-(La)	$NaLaSr_3F[PO_4]_3$
*Fluorcaphite	$SrCaCa_3F[PO_4]_3$
Apatite-(SrOH)	$Sr_5OH[PO_4]_3$
Kuannersuite-(Ce)	$Ba_6Na_2Ce_2FCl[PO_4]_6$
Goedkenite (x = 1,75)	$(Sr,Ca)_2Al(OH)[PO_4]_2$
(comp. with brackebuschite gr.)	
Arctite	$Na_5Ca_7BaF_3[PO_4]_6$
*Bearthite	$Ca_2Al(OH)[PO_4]_2$
Trolleite (x = 2)	$Al_4(OH)_3[PO_4]_3$
Lazulite group (x = 2)	

Lazulite	$(Mg,Fe^{2+})Al_2(OH)_2[PO_4]_2$
Scorzalite	$(Fe^{2+},Mg)Al_2(OH)_2[PO_4]_2$
Barbosalite	$Fe^{2+}Fe^{3+}_2(OH)_2[PO_4]_2$
Lipscombite (x = 2)	$(Fe,Mn)^{2+}Fe^{3+}_2(OH)_2[PO_4]_2$
*Staněkite	$Fe^{3+}(Mn,Fe^{2+},Mg)O[PO_4]$
*Lulzacite	$Sr_2Fe^{2+}(Fe^{2+},Mg)_2Al_4(OH)_{10}[PO_4]_4$
Penikisite group (x = 2)	
Penikisite	$Ba(Mg,Fe^{2+})_2Al_2(OH)_3[PO_4]_3$
Kulanite	$BaFe^{2+}_2Al_2(OH)_3[PO_4]_3$
Bjarebyite group (x = 2)	
Bjarebyite	$BaMn_2Al_2(OH)_3[PO_4]_3$
Perloffite	$BaMn_2Fe^{3+}_2(OH)_3[PO_4]_3$
*Johntomaite	$BaFe^{2+}_2Fe^{3+}_2(OH)_3[PO_4]_3$
Cirrolite (x = 2)	$Ca_3Al_2(OH)_3[PO_4]_3$
Satterlyite group (x = 2)	
Holtedahlite	$Mg_{12}(OH,O)_6(PO_3,OH,CO_3)[PO_4]_5$
Satterlyite	$(Fe,Mg)_{12}(OH,O)_6(PO_3,OH)[PO_4]_5$
Althausite group (x = 2)	
Althausite	$Mg_4(OH,O)(F,\square)[PO_4]_2$
Thadeuite	$CaMg_3(OH,F)_2[PO_4]_2$
Triplite group (x = 2)	
Zwieselite	$Fe^{2+}_2F[PO_4]$
Triplite	$Mn^{2+}_2F[PO_4]$
Wagnerite group (x = 2)	
Magniotriplite	$(Mg,Fe,Mn)_2F[PO_4]$
Wagnerite	$(Mg,Fe^{2+})_2F[PO_4]$
*Wagnerite Ma5bc polytype	$Mg_2(F,OH)[PO_4]$
*Hydroxylwagnerite	$Mg_2(OH)[PO_4]$
Wolfeite	$(Fe,Mn)^{2+}_2(OH)[PO_4]$
Triploidite	$(Mn,Fe)^{2+}_2(OH)[PO_4]$
Panasqueiraite family (x = 2)	
*Isokite	$CaMgF[PO_4]$
Panasqueiraite	$CaMg(OH,F)[PO_4]$
Lacroixite	$NaAl(F,OH)[PO_4]$
Nacaphite (x = 2)	$Na_2CaF[PO_4]$
Richellite (x = 2)	$CaFe^{3+}_2(OH,F)_2[PO_4]_2$
*Jagowerite (x = 2)	$BaAl_2(OH)_2[PO_4]_2$
Rockbridgeite series (x = 2,(3))	
Rockbridgeite	$(Fe,Mn)^{2+}Fe^{3+}_4(OH)_5[PO_4]_3$
Frondelite	$(Mn,Fe)^{2+}Fe^{3+}_4(OH)_5[PO_4]_3$
Brazilianite (x = 2,5)	$NaAl_3(OH)_4[PO_4]_2$
*Getehouseite (x = 2,5)	$Mn_5(OH)_4[PO_4]_2$
Augelite (x = 3)	$Al_2(OH)_3[PO_4]$
Laubmannite (x = 3)	$Fe^{2+}_3Fe^{3+}_6(OH)_{12}[PO_4]_4$
Viitaniemiite (x = 3)	$Na(Ca,Mn)Al(F,OH)_3[PO_4]$
*Raadeite (x = 3,5)	$Mg_7(OH)_8[PO_4]_2$
*Waterhauseite	$Mn_7(OH)_8[PO_4]_2$
*Unnamed	$(Fe,Mn)_3Al(OH)_6[PO_4]$

3.2.4.1.1.1.1.1.3.1. Basic, oxido-phosphato-fluorides

3.2.4.1.1.1.1.1.2.1.1. Hydrates

*Bederite (x = 1,(3)) $\square Ca_2Mn^{2+}_2Fe^{3+}_2Mn^{2+}_2[PO_4]_6\cdot(H_2O)_2$
Senegalite $Al_2(OH)_3[PO_4]\cdot H_2O$
Englishite (x = 1,(6)) $K_3Na_2Ca_{10}Al_{15}(OH)_7[PO_4]_{21}\cdot26H_2O$
Landesite (x = 1,69) $Mn_9Fe^{3+}_3(OH)_3[PO_4]_8\cdot9H_2O$
Giniite (x = 1,75) $Fe^{2+}Fe^{3+}_4(OH)_2[PO_4]_4\cdot2H_2O$
Xanthoxenite (x = 1,75) $Ca_4Fe^{3+}_2(OH)_2[PO_4]_4\cdot3H_2O$
  Overite group (x = 1,75)
  Overite $CaMgAl(OH)[PO_4]_2\cdot4H_2O$
  *Juonniite $CaMgSc(OH)[PO_4]_2\cdot4H_2O$
  Segelerite $CaMgFe^{3+}(OH)[PO_4]_2\cdot4H_2O$
  *Manganosegelerite $(Mn^{2+},Ca)(Mn^{2+},Fe^{2+},Mg)Fe^{3+}(OH)[PO_4]_2\cdot4H_2O$
  *Falsterite $Ca_2MgMn^{2+}_2Fe^{2+}_2Fe^{3+}_2Zn_4(OH)_4[PO_4]_8\cdot4H_2O$
*Ferraioloite (x = 1,75) $MgMn^{2+}_4(Fe^{2+}_{0.5}Al_{0.5})_4Zn_4[PO_4]_8(OH)_4(H_2O)_{20}$
  Lun'okite group (x = 1,75)
  Lun'okite $MgMnAl(OH)[PO_4]_2\cdot4H_2O$
  Wilhelmvierlingite $(Ca,Zn)Mn^{2+}Fe^{3+}(OH)[PO_4]_2\cdot4H_2O$
  Whiteite series (x = 1,75)
  Whiteite-(CaFeMg) $Ca(Fe,Mn)^{2+}Mg_2Al_2(OH)_2[PO_4]_4\cdot8H_2O$
  Whiteite-(CaMnMg) $CaMn^{2+}Mg_2Al_2(OH)_2[PO_4]_4\cdot8H_2O$
  Whiteite-(CaMnMn) $CaMn^{2+}Mn_2Al_2(OH)_2[PO_4]_4\cdot8H_2O$
  Whiteite-(MnFeMg) $(Mn^{2+},Ca)(Fe,Mn)^{2+}Mg_2Al_2(OH)_2[PO_4]_4\cdot8H_2O$
  Jahnsite series (x = 1,75)
  *Jahnsite-(CaMgMg) $\{Ca\}\{Mg\}\{Mg_2\}\{Fe_2^{3+}\}(OH)_2[PO_4]_4\cdot8H_2O$
  *Jahnsite-(CaFeFe) $\{Ca\}\{Fe^{2+}\}\{Fe_2^{2+}\}\{Fe_2^{3+}\}(OH)_2[PO_4]_4\cdot8H_2O$
  Jahnsite-(MnMnFe) (Rittmannite) $MnMnFe^{2+}_2Fe^{3+}_2(OH)_2[PO_4]_4\cdot8H_2O$
  Jahnsite-(MnMnMn) $Mn^{2+}Mn^{2+}Mn_2^{2+}Fe_2^{3+}(OH)_2[PO_4]_4\cdot8H_2O$
  Jahnsite-(CaMnMg) $CaMn^{2+}Mg_2Fe_2^{3+}(OH)_2[PO_4]_4\cdot8H_2O$
  Jahnsite-(CaMnFe) $CaMn^{2+}Fe_2^{2+}Fe_2^{3+}(OH)_2[PO_4]_4\cdot8H_2O$
  Jahnsite-(CaMnMn) $CaMn^{2+}Mn_2^{2+}Fe_2^{3+}(OH)_2[PO_4]_4\cdot8H_2O$
  *Jahnsite-(NaFeMg) $NaFe^{3+}Mg_2Fe_2^{3+}(OH)_2[PO_4]_4\cdot8H_2O$
*Gladiusite $(Fe^{2+},Mg)_4Fe^{3+}_2(OH)_{11}[PO_4]\cdot H_2O$
  Leucophosphite series (x = 1,75)
  Tinsleyite $KAl_2(OH)[PO_4]_2\cdot2H_2O$
  *Potassium-rich tinsleyite $\left| K_{1.5}(H_2O)_{0.5} \right| [Al_2(OH)\{(OH)_{0.5}(H_2O)_{0.5}\}[PO_4]_2]$
  Leucophosphite $KFe^{3+}_2(OH)[PO_4]_2\cdot2H_2O$
  Spheniscidite $(NH_4,K)(Fe^{3+},Al)_2(OH)[PO_4]_2\cdot2H_2O$
Minyulite (x = 1,75) $KAl_2(OH,F)[PO_4]_2\cdot4H_2O$
Natrophosphate (x = 1,75) $Na_7F[PO_4]_2\cdot19H_2O$
Vashegyite (x = 1,8) $Al_{11}(OH)_6[PO_4]_9\cdot38H_2O$
Calcioferrite family (x = 1,8(3))
Montgomeryite $Ca_4MgAl_4(OH)_4[PO_4]_6\cdot12H_2O$
Kingsmountite $(Ca,Mn)_4(Fe,Mn)^{2+}Al_4(OH)_4[PO_4]_6\cdot12H_2O$
Zodacite $Ca_4MnFe^{3+}_4(OH)_4[PO_4]_6\cdot12H_2O$
Keckite (x = 1,6) $CaMn^{2+}(Fe^{3+}Mn^{2+})Fe_2^{3+}(OH)_3[PO_4]_4\cdot7H_2O$
*Kapundaite $NaCaFe^{3+}_4(OH)_3[PO_4]_4\cdot5H_2O$
Vantasselite (x = 2) $Al_4(OH)_3[PO_4]_3\cdot9H_2O$
Cacoxenite (x = 2) $AlFe^{3+}_{24}(OH)_{12}O_6[PO_4]_{17}\cdot\sim75H_2O$
  Kryzhanovskite group (x = 2)
  Kryzhanovskite $MnFe^{3+}_2(OH)_2[PO_4]_2\cdot H_2O$
  Garyansellite $(Mg,Fe^{3+})_3(OH)[PO_4]_2\cdot2H_2O$

*Angarfite (x = 2) $NaFe^{3+}_5(OH)_4[PO_4]_4 \cdot 4H_2O$

Gatumbaite (x = 2) $CaAl_2(OH)_2[PO_4]_2 \cdot H_2O$

Isoclasite (x = 2) $Ca_2(OH)[PO_4] \cdot 2H_2O$

    Whitmoreite group (x = 2) (compare with arhurite (group))

    Whitmoreite $Fe^{2+}Fe^{3+}_2(OH)_2[PO_4]_2 \cdot 4H_2O$

    Earlshannonite $Mn^{2+}Fe^{3+}_2(OH)_2[PO_4]_2 \cdot 4H_2O$

    Bermanite (x = 2) $Mn^{2+}Mn^{3+}_2(OH)_2[PO_4]_2 \cdot 4H_2O$

*Ercitite(x = 2) $NaMn^{3+}(OH)[PO_4] \cdot 2H_2O$

Sigloite (x = 2) $Fe^{3+}Al_2(OH)_3[PO_4]_2 \cdot 7H_2O$

    Vauxite group (x = 2)

    Vauxite $Fe^{2+}Al_2(OH)_2[PO_4]_2 \cdot 6H_2O$

    *Ferrivauxite $Fe^{3+}Al_2(OH)_3[PO_4]_2 \cdot 5H_2O$

    *Ferristrunzite $Fe^{3+}Fe^{3+}_2(OH)_3[PO_4]_2 \cdot 5H_2O$

    *Ferrostrunzite $Fe^{2+}Fe^{3+}_2(OH)_2[PO_4]_2 \cdot 6H_2O$

    Strunzite $Mn^{2+}Fe^{3+}_2(OH)_2[PO_4]_2 \cdot 6H_2O$

    Metavauxite group (x = 2)

    Metavauxite $Fe^{2+}Al_2(OH)_2[PO_4]_2 \cdot 8H_2O$

    Pseudolaueite $Mn^{2+}Fe^{3+}_2(OH)_2[PO_4]_2 \cdot 8H_2O$

    Paravauxite group (x = 2)

    Gordonite $MgAl_2(OH)_2[PO_4]_2 \cdot 8H_2O$

    *Mangangordonite $Mn^{2+}Al_2(OH)_2[PO_4]_2 \cdot 8H_2O$

    *Kastningite $(Mn,Fe,Mg)Al_2(OH)_2[PO_4]_2 \cdot 8H_2O$

    *Nordgauite $MnAl_2(F,OH)_2[PO_4]_2 \cdot 5H_2O$

    *Kayrobertsonite $MnAl_2(OH)_2[PO_4]_2 \cdot 6H_2O$

    Paravauxite $Fe^{2+}Al_2(OH)_2[PO_4]_2 \cdot 8H_2O$

    Ushkovite $MgFe^{3+}_2(OH)_2[PO_4]_2 \cdot 8H_2O$

    Laueite $Mn^{2+}Fe^{3+}_2(OH)_2[PO_4]_2 \cdot 8H_2O$

    *Kummerite $Mn^{2+}Fe^{3+}Al(OH)_2[PO_4]_2 \cdot 8H_2O$

    *Ferrolaueite $Fe^{2+}Fe^{3+}_2(OH)_2[PO_4]_2 \cdot 8H_2O$

Stewartite (x = 2) $Mn^{2+}Fe^{3+}_2(OH)_2[PO_4]_2 \cdot 8H_2O$

Kovdorskite (x = 2) $Mg_2(OH)[PO_4] \cdot 3H_2O$

Beraunite (x = 2, 2) $Fe^{2+}Fe^{3+}_5(OH)_5[PO_4]_4 \cdot 4H_2O$

Pararobertsite (x = 2,1(6)) $Ca_2Mn^{3+}_3O_2[PO_4]_3 \cdot 3H_2O$

    Wavellite family (x = 2,25)

 *Kobokoboite $Al_6(OH)_6[PO_4]_4 \cdot 11H_2O$

 Wavellite $Al_3(OH,F)_3[PO_4]_2 \cdot 5H_2O$

 Kingite $Al_3(OH,F)_3[PO_4]_2 \cdot 8H_2O$

 Tinticite $Fe^{3+}_3(OH)_3[PO_4]_2 \cdot 3H_2O$

*Allanpringite $Fe^{3+}_3(OH)_3[PO_4]_2 \cdot 5H_2O$

    Gormanite group (x = 2,25)

    Gormanite $Fe^{2+}_3Al_4(OH)_6[PO_4]_4 \cdot 2H_2O$

    *Eleonorite $Fe^{3+}_6O(OH)_4[PO_4]_4 \cdot 6H_2O$

    Souzalite $Mg_3Al_4(OH)_6[PO_4]_4 \cdot 2H_2O$

    Mitridatite group (x = 2,17)

    Mitridatite $Ca_2Fe^{3+}_3O_2[PO_4]_3 \cdot 3H_2O$

    Robertsite $Ca_2Mn^{3+}_3O_2[PO_4]_3 \cdot 3H_2O$

Natrodufrenite (x = 2,25) $(Na,\square)(Fe^{3+},Fe^{2+})(Fe^{3+},Al)_5(OH)_6[PO_4]_4 \cdot 2H_2O$

Dufrenite $Ca_{0.5}Fe^{2+}Fe^{3+}_5(OH)_6[PO_4]_4 \cdot 2H_2O$

Burangaite $(Na,Ca)(Fe^{2+},Mg)Al_5(OH,O)_6[PO_4]_4 \cdot 2H_2O$

*Matioliite $Na_2Mg_2Al_{10}(OH)_{12}[PO_4]_8 \cdot 4H_2O$

Kidwellite (x = 2,3)	$NaFe^{3+}_9(OH)_{11}[PO_4]_6 \cdot 3H_2O$
*Meurigite	$KFe^{3+}_7(OH)_7[PO_4]_5 \cdot 8H_2O$
*Meurigite-(Na)	$[Na(H_2O)_{2.5}][Fe^{3+}_8(OH)_7[PO_4]_6 \cdot (H_2O)_4$
Phosphofibrite	$(K_{0.5}(H_2O)_3)Fe_8^{3+}[PO_4]_6(OH)_{6.5} \cdot 6.5H_2O$

Eosphorite series (x = 2,5)

Childrenite	$(Fe,Mn)^{2+}Al(OH)_2[PO_4] \cdot H_2O$
Ernstite	$(Mn^{2+},Fe^{3+})Al(OH)_2[PO_4] \cdot H_2O$
Eosphorite	$(Mn,Fe)^{2+}Al(OH)_2[PO_4] \cdot H_2O$
Foggite (x = 2,5)	$CaAl(OH)_2[PO_4] \cdot H_2O$

Wardite family (x = 2,5) (compare with turquoise (series); faustite (series))

Aheylite	$(Fe,Zn)Al_6(OH)_8[PO_4]_4 \cdot 4H_2O$
Wardite	$NaAl_3(OH)_4[PO_4]_2 \cdot 2H_2O$
Cyrilovite	$NaFe^{3+}_3(OH)_4[PO_4]_2 \cdot 2H_2O$
*Angastonite	$CaMgAl_2(OH)_4[PO_4]_2 \cdot 7H_2O$
Millisite (x = 2,625)	$(Na,K)CaAl_6(OH)_9[PO_4]_4 \cdot 3H_2O$
Attakolite (x = 2,6(6))	$(Ca,Sr)Mn(Al,Fe^{3+})_4[PO_4]_3[HSiO_4](OH)_4$
Goyazite (x = 2,75)	$SrAl_3(OH)_5[PO_4]_2 \cdot H_2O$
*Springcreekite	$BaV^{3+}_3[(OH)_5(H_2O)]_{\Sigma6}[PO_4]_2$
Delvauxite (x = 2,75)	$(Ca,Mg)(Fe^{3+},Al)_4(OH)_8[(PO_4),(SO_4),(CO_3)]_2 \cdot (4-6)H_2O$
Morinite (x = 2,75)	$NaCa_2Al_2(F,OH)_5[PO_4]_2 \cdot 2H_2O$
Aldermanite (x = 2,875)	$Mg_5Al_{12}(OH)_{22}[PO_4]_8 \cdot 32H_2O$

Fluellite family (x = 3)

Fluellite	$Al_2(OH)F_2[PO_4] \cdot 7H_2O$
Bolivarite	$Al_2(OH)_3[PO_4] \cdot (4-5)H_2O$
Evansite (x = 4,5)	$Al_3(OH)_6[PO_4] \cdot 8H_2O$

3.2.4.1.1.1.1.1.2.2. Neutral

Variscite family (x = 1,5)

Variscite	$Al[PO_4] \cdot 2H_2O$
*Serrabrancaite	$Mn[PO_4] \cdot H_2O$
Strengite	$Fe^{3+}[PO_4] \cdot 2H_2O$
Metavariscite	$Al[PO_4] \cdot 2H_2O$
Phosphosiderite	$Fe^{3+}[PO_4] \cdot 2H_2O$
Kolbeckite	$Sc[PO_4] \cdot 2H_2O$
Koninckite (x = 1,5)	$Fe[PO_4] \cdot 3H_2O$
*Santabarbaraite	$Fe^{3+}_3(OH)_3[PO_4]_2 \cdot 5H_2O$
*Pakhomovskyite	$Co^{2+}_3[PO_4]_2 \cdot 8H_2O$

Fairfieldite group (x = 1,5)

Cassidyite	$Ca_2Ni^{2+}[PO_4]_2 \cdot 2H_2O$
Collinsite	$Ca_2Mg[PO_4]_2 \cdot 2H_2O$
Messelite	$Ca_2(Fe,Mn)^{2+}[PO_4]_2 \cdot 2H_2O$
Fairfieldite	$Ca_2(Mn,Fe)^{2+}[PO_4]_2 \cdot 2H_2O$
*Hillite	$Ca_2Zn[PO_4]_2 \cdot 2H_2O$

Phosphoferrite group (x = 1,5)

Phosphoferrite	$Fe^{2+}Fe^{2+}_2[PO_4]_2 \cdot 3H_2O$
*Correianevesite	$Fe^{2+}Mn^{2+}_2[PO_4]_2 \cdot 3H_2O$
Reddingite	$Mn^{2+}Mn^{2+}_2[PO_4]_2 \cdot 3H_2O$

Ludlamite group (x = 1,5)

Ludlamite	$Fe^{2+}_3[PO_4]_2 \cdot 4H_2O$
Metaswitzerite	$(Mn^{2+},Fe^{2+})_3[PO_4]_2 \cdot 4H_2O$
Switzerite (x = 1,5)	$(Mn,Fe)^{2+}_3[PO_4]_2 \cdot 7H_2O$

Anapaite (x = 1,5) $Ca_2Fe^{2+}[PO_4]_2 \cdot 4H_2O$
Vivianite family (x = 1,5)
Bobierrite $Mg_3[PO_4]_2 \cdot 8H_2O$
*Cattiite $Mg_3[PO_4]_2 \cdot 22H_2O$
*Rimkorolgite $BaMg_5[PO_4]_4 \cdot 8H_2O$
  Vivianite group
  Barićite $(Mg,Fe^{2+})_3[PO_4]_2 \cdot 8H_2O$
  Vivianite $Fe^{2+}_3[PO_4]_2 \cdot 8H_2O$
  *Arupite $Ni_3[PO_4]_2 \cdot 8H_2O$
Metavivianite $Fe^{2+}Fe_2^{3+}(OH)_2[PO_4]_2 \cdot 6H_2O$
Wicksite (x = 1,5) $NaCa_2(Fe,Mn)^{2+}_4MgFe^{3+}[PO_4]_6 \cdot 2H_2O$
*Phase like to wicksite $(Ca,Na)_2(Mn,Fe,Mg)_4(Fe,Al)_2[PO_4]_6 \cdot 4H_2O$
*Tassieite $(Na,\square)Ca_2(Mg,Fe^{2+},Fe^{3+})_2(Fe^{3+},Mg)_2(,Fe^{2+},Mg)_2[PO_4]_6 \cdot 2H_2O$ or
                                                       $NaCa_2Mg_3Fe^{2+}_2Fe^{3+}[PO_4]_6 \cdot 2H_2O$

  Nastrophite group (x = 1,5)
  Nastrophite $Na(Sr,Ba)[PO_4] \cdot 9H_2O$
  Nabaphite $NaBa[PO_4] \cdot 9H_2O$
Dittmarite family (x = 1,5)
Dittmarite $(NH_4)Mg[PO_4] \cdot H_2O$
Niahite $(NH_4)Mn[PO_4] H_2O$
  *Struvite group (x = 1,5)
  Struvite $(NH_4)Mg[PO_4] \cdot 6H_2O$
  *Struvite-(K) $KMg[PO_4] \cdot 6H_2O$
  *Hazenite $KNaMg_2[PO_4]_2 \cdot 14H_2O$
  *Bakhchisaraitsevite $Na_2Mg_5[PO_4]_4 \cdot 7H_2O$
  *Apexite $NaMg[PO_4] \cdot 9H_2O$

*3.2.4.1.1.1.1.1.3. Orthophosphato-carbonates   *3.2.4.1.1.1.1.1.3.1. Hydrates
*Krasnovite $Ba(Al,Mg)[(PO_4),(CO_3)](OH)_2 \cdot H_2O$
*Parwanite $(Na,K)(Mg,Ca)_4Al_8(OH)_7[PO_4]_8[CO_3] \cdot 30H_2O$

3.2.4.1.1.1.1.1.4. Orthophosphato-sulfates
3.2.4.1.1.1.1.1.4.1. Orthophosphato-sulfates with $PO_4 : SO_4 = 11$
                                  3.2.4.1.1.1.1.1.4.1.1. Hydrates (basic)
Sasaite $Al_6(OH)_3[PO_4]_5 \cdot 36H_2O$

3.2.4.1.1.1.1.1.4.2. Orthophosphato-sulfates with $PO_4 : SO_4 = 5$
                                  3.2.4.1.1.1.1.1.4.2.1. Hydrates (basic)
Peisleyite $Na_2Al_9(OH)_6[(P,S)O_4]_8 \cdot 28H_2O$

3.2.4.1.1.1.1.1.4.3. Orthophosphato-sulfates with $PO_4 : SO_4 = 3$
                                  3.2.4.1.1.1.1.1.4.3.1. Hydrates (basic)
Kribergite $Al_5(OH)_4[PO_4]_3[SO_4] \cdot 4H_2O$

3.2.4.1.1.1.1.1.4.4. Orthophosphato-sulfates with $PO_4 : SO_4 = 1$ (basic)
Svanbergite $SrAl_3(OH)_6[PO_4][SO_4]$
*Woodhouseite $CaAl_3(OH)_6[PO_4][SO_4]$
*Unnamed $(Ba,Ca,K,Na,REE,Sr)(Al,Fe)_3(OH,F)_6[PO_4][SO_4]$
                                  3.2.4.1.1.1.1.1.4.4.1. Hydrates (basic)
  Diadochite family

Sanjuanite	$Al_2(OH)[PO_4][SO_4] \cdot 9H_2O$
Diadochite	$Fe^{3+}_2(OH)[PO_4][SO_4] \cdot 6H_2O$
*Destinezite	$Fe^{3+}_2(OH)[PO_4][SO_4] \cdot 5H_2O$

3.2.4.1.1.1.1.1.4.5. Orthophosphato-sulfates with $PO_4 : SO_4 = 0,(6)$
3.2.4.1.1.1.1.1.4.5.1. Hydrates (basic)
Hotsonite $Al_5(OH)_{10}[PO_4][SO_4] \cdot 8H_2O$

*3.2.4.1.1.1.1.1.4.6. Orthophosphato-sulfates with $PO_4 : SO_4 = 0,5$
*Rossiantonite $Al_3(OH)_2[PO_4][SO_4]_2(H_2O)_{10} \cdot 4H_2O$

3.2.4.1.1.1.1.1.5. Orthophosphato-fluoraluminates
3.2.3.1.1.1.1.1.5.1. Neutral
Boggildite $Na_2Sr_2[PO_4][Al_2F_9]$

*3.2.4.1.1.1.1.1.5.2. Orthophosphato-sulfato- fluoraluminates
*3.2.4.1.1.1.1.1.5.2.1. Hydrates
Mitryaevaite $Al_4[PO_4]_2[(P,S)O_3(OH,O)]_2AlF_2(OH)_2 \cdot 14.5H_2O$

*3.2.4.1.1.1.1.1.6. Hydro-orthophosphates
*3.2.4.1.1.1.1.1.6.1. Hydro-orthophosphates with $[HPO_4] : [PO_4] = 0,1(6)$
*3.2.4.1.1.1.1.1.6.1.1. Neutral

Whitlockite	$Ca_9(Mg,Fe^{2+})[PO_3OH][PO_4]_6$
*Strontiowhitlockite	$Sr_9Mg[PO_3OH][PO_4]_6$
*Hedegaardiite	$(Ca,Na)_9(Ca,Na)Mg[PO_3OH][PO_4]_6$
*Bobdownsite	$Ca_9Mg[PO_3F][PO_4]_6$

3.2.4.1.1.1.1.1.6.2. Hydro-orthophosphat- halogenides with $[HPO_4] : [PO_4] = 0,(3)$
3.2.4.1.1.1.1.1.6.2.1. Hydrates
3.2.4.1.1.1.1.1.6.2.1.1. Neutral (fluorides)
Mcauslanite $Fe^{2+}_3Al_2F[HPO_4][PO_4]_3 \cdot 18H_2O$

*3.2.4.1.1.1.1.1.6.3. Hydro-orthophosphates with $[HPO_4] : [PO_4] = 0.5$
*3.2.4.1.1.1.1.1.6.3.1. Hydrates (basic)
*Matulaite (x = 2) $Fe^{3+}Al_7(OH)_8[PO_3OH]_2[PO_4]_4 (H_2O)_8 \cdot 8H_2O$

3.2.4.1.1.1.1.1.6.4. Hydro-orthophosphates with $[PO_3(OH)] : [PO_4] = 1$
3.2.4.1.1.1.1.1.6.4.1. Basic

Crandallite group(x = 2,25)

Crandallite	$CaAl_3(OH)_6[PO_3(OH)][PO_4]$
*Benauite	$SrFe^{3+}_3(OH)_6[PO_3(OH)][PO_4]$
Lusungite	$SrAl_3(OH)_6[PO_3(OH)][PO_4]$
Gorceixite	$BaAl_3(OH)_6[PO_3(OH)][PO_4]$

3.2.4.1.1.1.1.1.6.4.2. Hydrates

*Mejillonesite	$NaMg_2(OH)_4[PO_3(OH)][PO_4] \cdot H_2O$
*Afmite	$Al_3(OH)_4[PO_3(OH)][PO_4](H_2O)_3 \cdot 8H_2O$
*Planerite	$Al_6(OH)_8[PO_3(OH)]_2[PO_4]_2 \cdot 4H_2O$
*Iangreyite	$Ca_2Al_7(OH,F)_{15}[PO_3(OH)]_2[PO_4]_2 \cdot 8H_2O$
Hureaulite	$Mn_5[PO_3(OH)]_2[PO_4]_2 \cdot 4H_2O$

*3.2.4.1.1.1.1.1.6.5. *Hydro-orthophosphates with [HPO$_4$] : [PO$_4$] = 1,(3)
*Wopmayite   Ca$_6$Na$_3$□Mn[HPO$_4$]$_4$[PO$_4$]$_3$

*3.2.4.1.1.1.1.1.6.6. Hydro-orthophosphates with [HPO$_4$] : [PO$_4$] = 2
*Groatite   NaCaMn^{2+}$_2$[PO$_3$(OH)]$_2$[PO$_4$]

3.2.4.1.1.1.1.1.6.7. Hydro-orthophosphates with [HPO$_4$]:[PO$_4$] = 3
                                                          3.2.4.1.1.1.1.1.6.7.1. Hydrates
Francoanellite   K$_3$Al$_5$[HPO$_4$]$_6$[PO$_4$]$_2$·12H$_2$O
Taranakite   K$_3$Al$_5$[HPO$_4$]$_6$[PO$_4$]$_2$·18H$_2$O

*3.2.4.1.1.1.1.1.6.8. Hydro-orthophosphato-carbonates with [HPO$_4$] : [PO$_4$] = 0,(3)
*Phosphoellenbergerite   Mg$_{14}$[PO$_4$]$_6$[(HPO$_4$),(CO$_3$)]$_2$(OH)$_6$

3.2.4.1.1.1.1.1.7. Hydrophosphates
                                                          3.2.4.1.1.1.1.7.1. Neutral (all with MO : HPO$_4$ = 1)
Monetite family
Monetite   Ca[HPO$_4$]
Nahpoite   Na$_2$[HPO$_4$]
Phosphammite   (NH$_4$)$_2$[HPO$_4$]
                                                          3.2.4.1.1.1.1.7.2. Hydrates
                                                          3.2.4.1.1.1.1.7.2.1. Basic
Sinkankasite   MnAl(OH)[HPO$_4$]$_2$·6H$_2$O
                                                          3.2.4.1.1.1.1.7.2.2. Neutral
Newberyite family
Newberyite   Mg[HPO$_4$]·3H$_2$O
Phosphorrösslerite   Mg[HPO$_4$]·7H$_2$O
Brushite   Ca[HPO$_4$]·2H$_2$O
Dorfmanite   Na$_2$[HPO$_4$]·2H$_2$O
*Catalanoite   Na$_2$[HPO$_4$]·8H$_2$O
Hannayite family
Hannayite   (NH$_4$)$_2$Mg$_3$[HPO$_4$]$_4$·8H$_2$O
Schertelite   (NH$_4$)$_2$Mg[HPO$_4$]$_2$·4H$_2$O
Mundrabillaite   (NH$_4$)$_2$Ca[HPO$_4$]$_2$·H$_2$O
*Swaknoite   (NH$_4$)$_2$Ca[HPO$_4$]$_2$·H$_2$O
Stercorite   (NH$_4$)Na[HPO$_4$]·4H$_2$O

3.2.4.1.1.1.1.1.8. Hydrophosphsto-sulfates
                                                          3.2.4.1.1.1.1.1.8.1. Hydrates
Ardealite   Ca$_2$[HPO$_4$][SO$_4$]·4H$_2$O
*Camaronesite   [Fe^{3+}(H$_2$O)$_2$(PO$_3$OH)]$_2$[SO$_4$]·1-2H$_2$O

*3.2.4.1.1.1.1.1.9. Hydro-dihydrophosphates   *3.2.4.1.1.1.1.1.9.1. Hydrates

*Haigerachite   KFe^{3+}$_3$[H$_2$PO$_4$]$_6$[HPO$_4$]$_2$·4H$_2$O
*Unnamed   KFe^{3+}$_3$[H$_2$PO$_4$]$_2$[HPO$_4$]$_4$·6H$_2$O

3.2.4.1.1.1.1.1.10. Dihydrophosphstes   3.2.4.1.1.1.1.1.10.1. Neutral
    Biphosphammite group
    Archerite   (K,NH$_4$)[H$_2$PO$_4$]

Biphosphammite $(NH_4,K)[H_2PO_4]$

3.2.4.1.1.1.2. Orthophosphates of Li
3.2.4.1.1.1.2.1. Proper orthophosphates   3.2.4.1.1.1.2.1.1. Neutral (x = 1,5)
  Sicklerite series
  Ferrisicklerite $Li_{1-x}(Fe,Mn)[PO_4]$
  Sicklerite $Li_{1-x}(Mn,Fe)[PO_4]$
  Triphylite series
  Triphylite $LiFe^{2+}[PO_4]$
  Lithiophilite $LiMn^{2+}[PO_4]$
  *Simferite $Li(MgMn^{3+}_{0,4}Fe^{3+}_{0,6})_{\Sigma 2}[PO_4]_2$
  Lithiophosphate $Li_3[PO_4]$
  *Nalipoite $Li_2Na[PO_4]$
                                        3.2.4.1.1.1.2.1.2. Basic
Griphite (x = 1,(6)) $Ca(Mn^{2+},Na,Li)_6Fe^{2+}Al_2(F,OH)_2[PO_4]_6$
  Palermoite series (x = 2)
  Bertossaite $(Li,Na)_2CaAl_4(OH,F)_4[PO_4]_4$
  Palermoite $(Li,Na)_2(Sr,Ca)Al_4(OH)_4[PO_4]_4$
 Amblygonite family (x = 2)
  Amblygonite series
  Amblygonite $(Li,Na)Al(F,OH)[PO_4]$
  Montebrasite $LiAl(OH)[PO_4]$
  Tavorite $LiFe^{3+}(OH)[PO_4]$

3.2.4.1.1.1.2.2. Hydrophosphato-phosphates Li
                                        3.2.4.1.1.1.2.2.1. Basic
Tancoite $Na_2LiAl(OH)[HPO_4][PO_4]$

*3.2.4.1.1.1.2.3. Phosphato-carbonates
*3.2.4.1.1.1.2.3.1. Basic
*Peatite-(Y) $LiNa_3(Y,Na,Ca,HREE)_3[PO_4]_3[CO_3](F,OH)_2$
*Ramikite-(Y) $Li_2Na_6(Y,Ca,REE)_3Zr_3[PO_4]_6[CO_3]_2O_2(OH,F)_2$

3.2.4.1.1.1.3. Phosphates of Be → beryllophosphates
3.2.4.1.1.1.3.1. Proper phosphates → beryllophosphates
                                        3.2.4.1.1.1.3.1.1. Neutral (x = 1,5)
Beryllonite $Na[Be(PO_4)]^{\infty}$
Hurlbutite $CaBe_2[PO_4]_2 \rightarrow Ca[Be_2O(P_2O_7)]^{\infty 3}$
*Stronriohurlbutite $SrBe_2[PO_4]_2$
Pahasapaite $(Ca_{5,5}Li_{3,6}K_{1,2}Na_{0,2}\square_{13,5})\{Li_8[Be_{24}(PO_4)_{24}]^{\infty 3}\}^{\infty 3}\cdot(H_2O)_{38}$
                                        3.2.4.1.1.1.3.1.2. Basic (x = 2)
Babefphite $Ba(O,F)[Be(PO_4)]^{\infty 3}$
                                        3.2.4.1.1.1.3.1.3. Acid (x = 2)

  Herderite series
  Väyrynenite $Mn[Be(OH)(PO_4)]^{\infty 2}$
  Hydroxyl-herderite $Ca[Be(OH)(PO_4)]^{\infty 2}$
  Fluorherderite $Ca[Be(F,OH)(PO_4)]^{\infty 2}$
  *Herderite $CaBeF[PO_4]$
                                        3.2.4.1.1.1.3.1.4. Hydrates

*Faheyite $MnFe^{3+}_2Be_2[PO_4]_4 \cdot 6H_2O$

3.2.4.1.1.1.3.1.4.1. Hydrates (acid)

Moraesite family

Moraesite $Be_2(OH)[PO_4] \cdot 4H_2O \rightarrow$

$\rightarrow Be(H_2O)_4[Be(OH)(PO_4)]$

*Weinebeneite $CaBe_3(OH)_2[PO_4]_2 \cdot 4H_2O \rightarrow$

$\rightarrow CaBe(H_2O)_4[Be(OH)(PO_4)]\}^{\infty}_2$

Uralolite $Ca_2Be_4(OH)_3[PO_4]_3 \cdot 5H_2O$

Roscherite family

Zanazziite (x = 1,8(3)) $Ca_2(Mg,Fe^{2+})(Mg,Fe^{2+},Al)_4(OH)_4Be_4[PO_4]_6 \cdot 6H_2O$

*Atencioite $Ca_2Fe^{2+}\square Mg_2Fe^{2+}_2(OH)_4Be_4[PO_4]_6 \cdot 6H_2O$

*Footemineite $Ca_2Mn^{2+}\square Mn^{2+}_2Mn^{2+}_2(OH)_4Be_4[PO_4]_6 \cdot 6H_2O$

*Greifensteinite $Ca_2(Fe^{2+},Mn^{2+})_5(OH)_4Be_4[PO_4]_6 \cdot 6H_2O$

Roscherite (x = 2) $Ca_2Mn_5^{2+}Be_4(OH)_4[PO_4]_6 \cdot 6H_2O$

*Ruifrancoite $Ca_2\square_2(Fe^{3+},Mn^{2+},Mg)_4(OH)_4(OH,H_2O)_2Be_4[PO_4]_6 \cdot 4H_2O$

*Gimaräesite $Ca_2(Zn,Mg,Fe)_5(OH)_4Be_4[PO_4]_6 \cdot 6H_2O$

3.2.4.1.1.1.3.2. Hydrophosphato-orthophosphates→beryllophosphato-hydrophosphates
(?)   3.2.4.1.1.1.3.2.1. Hydrates

Fransoletite family

Fransoletite $Ca_3Be_2[HPO_4]_2[PO_4]_2 \cdot 4H_2O \rightarrow Ca_3[HPO_4]_2[BePO_4]_2 \cdot 4H_2O$

*Parafransoletite $Ca_3Be_2[PO_3(OH)]_2[PO_4]_2 \cdot 4H_2O$

Ehrleite $Ca_2ZnBe[PO_3OH][PO_4]_2 \cdot 4H_2O$

3.2.4.1.1.1.3.3. Beryllophosphates with unknown structure

3.2.4.1.1.1.3.3.1. Hydrates

Tiptopite (x = 2) $*K_2Na_3Li_3(OH)_2Be_6[PO_4]_6 \cdot H_2O$

Glucine (x = 2,5) $Ca(OH)_4Be_4[PO_4]_2 \cdot 0,5H_2O$

3.2.4.1.1.2.Phosphates of f-elements

3.2.4.1.1.2.1. Proper phosphates   3.2.4.1.1.2.1.1. Neutral (x = 1,5)

Monazite family

Monazite series

Monazite-(Ce) $Ce[PO_4]$

Monazite-(La) $La[PO_4]$

Monazite-(Nd) $Nd[PO_4]$

*Монацит-(Sm) $Sm[PO_4]$

Cheralite $CaTh[PO_4]_2$

Xenotime-(Y) $Y[PO_4]$

*Xenotime-(Yb) $Yb[PO_4]$

Vitusite-(Ce) $Na_3Ce[PO_4]_2$

*Stornesite-(Y) $Y\square_2Na_6(Ca_5Na_3)(Mg,Fe)_{43}[PO_4]_{36}$

*Deloneite-(Ce) $(Na_{0.5}REE_{0.25}Ca_{0.25})(Ca_{0.75}REE_{0.25})Sr_{1.5}(CaNa_{0.25}REE_{0.25})$

$[PO_4]_{30.5}(F, OH)$

*Karlgizekite-(Nd) $NaNdCa_3[PO_4]_3F$

Florencite group   3.2.4.1.1.2.1.2. Basic (x = 1,5)

Florencite-(Ce) $CeAl_3(OH)_6[PO_4]_2$

Florencite-(La) $LaAl_3(OH)_6[PO_4]_2$

Florencite-(Nd) $NdAl_3(OH)_6[PO_4]_2$

*Florencite-(Sm) $(Sm,Nd)Al_3(OH)_6[PO_4]_2$

Eylettersite $Th_{0.75}Al_3(OH)_6[PO_4]_2$
                                                  3.2.4.1.1.2.1.3. Hydrates(basic)
Vyacheslavite family (x = 2)
Vyacheslavite $U^{4+}(OH)[PO_4]\cdot nH_2O$
Lermontovite $U^{4+}(OH)[PO_4]\cdot H_2O$
                                                  3.2.4.1.1.2.1.3.2. Neutral

  Rhabdophane group (x = 1,5)
Rhabdophane -(Ce)` $Ce[PO_4]\cdot H_2O$
Rhabdophane -(La) $La[PO_4]\cdot H_2O$
 Rhabdophane -(Nd) $Nd[PO_4]\cdot H_2O$
 Rhabdophane-(Y) $Y[PO_4]\cdot H_2O$
 Grayite $(Th,Pb,Ca)[PO_4]\cdot H_2O$
Brockite $(Ca,Th,Ce^{3+})[PO_4]\cdot H_2O$
Tristramite $(Ca,U^{4+},Fe^{3+})[(PO_4),(SO_4)]\cdot 2H_2O$
Ningyoite $(U^{4+},Ca_2[PO_4]_2\cdot 1-2H_2O$
Churchite-(Y) $Y[PO_4]\cdot 2H_2O$
Churchite-(Nd) $Nd[PO_4]\cdot 2H_2O$

3.2.4.1.2. Subclass: Phosphates of cations with middle FC
3.2.4.1.2.1.Phosphates of Zr
3.2.4.1.2.1.1. Proper phosphates   3.2.4.1.2.1.1.1. Neutral (x = 1,5)
*Kosnarite $KZr_2[PO_4]_3$
                                                  *3.2.4.1.2.1.1.1.1. Hydrates
Gainesite $Na_2Zr_2Be[PO_4]_4\cdot 1,5H_2O$
*Zigrasite $MgZr[PO_4]_2(H_2O)_4$
*Mahlmoodite $FeZr[PO_4]_2\cdot 4H_2O$
*Mccrillisite $NaCsZr_2(Be,Li)[PO_4]_4\cdot 1-2H_2O$
*Selwynite $NaKZr_2(Be,Al)[PO_4]_4\cdot 2H_2O$
                                                  *3.2.4.1.2.1.1.1.2. Hydrates (basic)
*Wycheproofite $NaAlZr(OH)_2[PO_4]_2\cdot H_2O$

3.2.4.1.2.1.2. Phosphato-carbonates   3.2.4.1.2.1.2.1. Hydrates (basic)
Voggite $Na_2Zr (OH)[PO_4][CO_3]\cdot 2H_2O$

3.2.4.1.2.3. Phosphates of Ti   3.2.4.1.2.3.1. Basic
Curetonite $Ba_4Al_3Ti(O,OH)_4F[PO_4]_4$
                                                  3.2.4.1.2.3.1.1. Hydrates (basic)
 Mantienneite group
Mantienneite $KMg_2Al_2Ti^{4+}(OH)_3[PO_4]_4\cdot 15H_2O$
Paulkerrite $K(Mg,Mn)_2(Fe^{3+},Al)_2Ti^{4+}(OH)_3[PO_4]_4\cdot 15H_2O$
*Benyacarite $(H_2O,K)_2Ti(Mn,Fe)^{2+}_2(Fe^{3+},Ti)_2[PO_4]_4(O,F)_2\cdot 14H_2O$

*3.2.4.1.2.3.2. Titano-(niobo)-oxido-dihydrophosphates
*Tazzoliite $Ba_{4-x}Na_xTi_2Nb_3SiO_{17}[PO_2(OH)_2]_x(OH)_{1-2x}$; $(0\leq x\leq 0.5)$

3.2.4.1.2.4. Phosphates of Nb and Ta   3.2.4.1.2.4.1. Oxido-hydrates
 Olmsteadite group
Olmsteadite $KFe^{2+}_2(Nb,Ta)O_2[PO_4]_2\cdot 2H_2O$
Johnwalkite $K(Mn,Fe^{3+})_2(Nb,Ta)O_2[PO_4]_2\cdot 2(H_2O,OH)$

3.2.4.1.2.5. Phosphates of V^{4+} 3.2.4.1.2.5.1 Oxido-hydrates
Sincosite $Ca(V^{4+}O)_2[PO_4]_2 \cdot 5H_2O$
*Bariosincosite $Ba(V^{4+}O)_2[PO_4]_2 \cdot 4H_2O$
*Cloncurryite $Cu^{2+}V^{4+}Al_4OF_4[PO_4]_4 \cdot 10H_2O$

3.2.4.1.3. Subclass: Orthophosphates of chalcophylic elements
3.2.4.1.3.1.Orthophosphates of Cu
3.2.4.1.3.1.1. M^{2+}
3.2.4.1.3.1.1.1.Proper phosphates   3.2.4.1.3.1.1.1.1. Hydrates
                                              (basic and phosphato-halogenides)
Nissonite (x=2) $Cu_2Mg_2(OH)_2[PO_4]_2 \cdot 5H_2O$

3.2.4.1.3.1.2. M^{2+} and M^{3+}
3.2.4.1.3.1.2.1. Proper phosphates   3.2.4.1.3.1.2.1.1. Основные
Hentschelite (x=1,5) (comp. with lazulite (gr.) $Cu^{2+}Fe^{3+}_2(OH)_2[PO_4]_2$
*Zincolibethenite $CuZn(OH)[PO_4]$
Libethenite $Cu_2(OH)[PO_4]$
Pseudomalachite $Cu_5(OH)_4[PO_4]_2$
Reichenbachite $Cu_5(OH)_4[PO_4]_2$
Ludjibaite $Cu_5(OH)_4[PO_4]_2$
Cornetite $Cu_3(OH)_3[PO_4]$
                                              3.2.4.1.3.1.2.1.2. Hydrates (basic)
*Kunatite $Cu^{2+}Fe^{3+}_2(OH)_2[PO_4]_2 \cdot 4H_2O$
Petersite-(Y) (x=2,5) (comp. with mixite (gr.)) $Cu^{2+}_6(Y,Ce,Nd,Ca)(OH)_6[PO_4]_3 \cdot 3H_2O$
   Turquoise series (x=2,5) (compare with wardire (family); faustite (series)
      Coeruleolactite $(Ca,Cu)Al_6(OH)_8[PO_4]_4 \cdot (4-5)H_2O$
      Turquoise $CuAl_6(OH)_8[PO_4]_4 \cdot 4H_2O$
      Chalcosiderite $CuFe^{3+}_6(OH)_8[PO_4]_4 \cdot 4H_2O$
*Bleasdaleite $(Ca,Fe^{3+})_2Cu_5(Bi,Cu)[PO_4]_4(H_2O,OH,Cl)_{13}$
Zapatalite (x=3) $Cu_3Al_4(OH)_9[PO_4]_3 \cdot 4H_2O$
Sieleckiite (x=4,5) $Cu_3Al_4(OH)_{12}[PO_4]_2 \cdot 2H_2O$

*3.2.4.1.3.1.2. Hydro-orthophosphates
*Calciopetersite $CaCu^{2+}_6(OH)_6[PO_3OH][(PO_4]_2 \cdot 3H_2O$

*3.2.4.1.3.1.2.3. Phosphato-oxides   *3.2.4.1.3.1.2.3.1. Hydrates
*Mrázekite $Cu^{2+}_3Bi^{3+}_2(OH)_2O_2[PO_4]_2 \cdot 2H_2O$

3.2.4.1.3.1.2.4. Phosphato-sulfates
*Birchite $Cd_2Cu_2[PO_4]_2[SO_4] \cdot 2H_2O$

*3.2.4.1.3.1.2.5. Phosphato -halogenides   *3.2.4.1.3.1.2.5.1. Hydrates
*Goldquarryite $(Cu^{2+},\square)(Cd,Ca)_2Al_3[PO_4]_4F_2(H_2O)_{10}\{(H_2O),F\}_2$
*Nevadaite $(Cu^{2+}\square AlV^{3+})_{\Sigma 6}\{Al_8[PO_4]_8F_8\}(OH)_2(H_2O)_{22}$

3.2.4.1.3.1.3. M^+ and M^{2+}
3.2.4.1.3.1.3.1. Phosphato-halogenides   3.2.4.1.3.1.3.1.1. Hydrates (basic)

Sampleite (x = 1,625) $NaCaCu_5Cl[PO_4]_4 \cdot 5H_2O$

3.2.4.1.3.2. Orthophosphates of Zn^{2+}

3.2.4.1.3.2.1. Proper phosphates	3.2.4.1.3.2.1.1. Basic
Tarbuttite (x = 2)	$Zn_2(OH)[PO_4]$
3.2.4.1.3.2.1.2. Hydrates	3.2.4.1.3.2.1.2.1. Basic
Spencerite (x = 2)	$Zn_4(OH)_2[PO_4]_2 \cdot 3H_2O$
Kipushite family (x = 3)	
Kipushite	$(Zn,Cu)_6(OH)_6[PO_4]_2 \cdot H_2O$
Veszelyite	$(Zn,Cu)_3(OH)_3[PO_4] \cdot 2H_2O$
3.2.4.1.3.2.1.2. Neutral	
	3.2.4.1.3.2.1.2.1. Hydrates
Hopeite family (x = 1,5)	
Phosphophyllite	$Zn_2(Fe,Mn)^{2+}[PO_4]_2 \cdot 4H_2O$
Hopeite	$Zn_3[PO_4]_2 \cdot 4H_2O$
Parahopeite	$Zn_3[PO_4]_2 \cdot 4H_2O$
*Nizamoffite	$Mn^{2+}Zn_2[PO_4]_2(H_2O)_4$
Scholzite	$CaZn_2[PO_4]_2 \cdot 2H_2O$
Parascholzite	$CaZn_2[PO_4]_2 \cdot 2H_2O$

3.2.4.1.3.2a. M^{2+} and M^{3+}
3.2.4.1.3.2a.1. Proper phosphates
3.2.4.1.3.2a.1.1. Basic

*Plimerite (x = 2.3)	$ZnFe^{3+}_4(OH)_5[PO_4]_3$
*Zinclipscombite (x = 2)	$ZnFe^{3+}_2(OH)_2[PO_4]_2$
	3.2.4.1.3.2a.1.1.1. Hydrates
*Zincoberaunite(x = 2.125)	$ZnFe^{3+}_5(OH)_5[PO_4]_4 \cdot 6H_2O$
Kleemanite family (x = 2)	
Kehoeite	$(Zn, Ca)Al_2(OH)_2[PO_4]_2 \cdot 5H_2O$ (?)
Kleemanite	$ZnAl_2(OH)_2[PO_4]_2 \cdot 3H_2O$
Jungite (x=2)	$Ca_2Zn_4Fe^{3+}_8(OH)_9[PO_4]_9 \cdot 16H_2O$
Schoonerite (x = 1,8(3))	$ZnMn^{2+}Fe^{2+}_2Fe^{3+}(OH)_2[PO_4]_3 \cdot 9H_2O$
*Flurlite (x = 1,8(3))	$Zn_3Mn^{2+}Fe^{3+}(OH)_2[PO_4]_3 \cdot 9H_2O$
*Ferraioloite (x = 1,75)	$MgMn^{2+}_4(Fe^{2+}_{0.5}Al_{0.5})_4Zn_4(OH)_4[PO_4]_8 \cdot 20H_2O$

Faustite series (x = 2,5) (compare with turquoise (series))

Aheylite	$(Fe,Zn)Al_6(OH)_8[PO_4]_4 \cdot 4H_2O$
Faustite	$(Zn,Cu)Al_6(OH)_8[PO_4]_4 \cdot 4H_2O$

*3.2.4.1.3.2a.2. Phosphato-carbonates	*3.2.4.1.3.2a.2.1. Hydrates(basic)
*Scorpionite	$Ca_3Zn_2(OH)_2[PO_4]_2[CO_3] \cdot H_2O$

3.2.4.1.3.3. Phosphtes of Pb
3.2.4.1.3.3.1. Proper phosphates and phosphato-halogenides
3.2.4.1.3.3.1.1. Basic and phosphato-halogenides
Pyromorphite series (x = 1,(6)) (compare with apatite (family); mimetite (group))

Pyromorphite	$Pb_5Cl[PO_4]_3$
Hydroxylpyromorphite	$Pb_5(OH)[PO_4]_3$
*Phosphohedyphane	$Ca_2Pb_3[PO_4]_3Cl$
*Fluorphosphohedyphane	$Ca_2Pb_3[PO_4]_3F$
	*3.2.4.1.3.3.1.2. Hydrates (basic)
*Phosphogartrellite(x = 1,5)	$PbCuFe[PO_4]_2(OH) \cdot H_2O$
*Kintoreite (x = 2)	$PbFe_3[PO_4][PO_3 OH] (OH)_6$

*Pattersonite $PbFe_3[PO_4]_2(OH)_5 \cdot H_2O$

3.2.4.1.3.3.2. Hydrophoshato-phosphates
                                      3.2.4.1.3.3.2.1. Basic
Drugmanite (x=1,75) $Pb_2(Fe^{3+},Al)(OH)_2[PO_3(OH)][PO_4]$
Plumbogummite (x=2,75) $PbAl_3(OH)_6[PO_3(OH)][PO_4]$

3.2.4.1.3.3.3. Hydrophoshato-phosphato-sulfates
                                      3.2.4.1.3.3.3.1. Hydrates (basic)
Orpheite (x=2,35)   discredited

3.2.4.1.3.3.4. Phosphato-sulfates   3.2.4.1.3.3.4.1. Basic
Tsumebite (x=1,5) $Pb_2Cu(OH)[PO_4][SO_4]$
  Hinsdalite group (x=2,75) (compare with beudantite (group)
  Hinsdalite $(Pb,Sr)Al_3(OH)_6[PO_4][SO_4]$
  Corkite $PbFe^{3+}_3(OH)_6[PO_4][SO_4]$

3.2.4.1.3.3.5. Phosphato-chromates   3.2.4.1.3.3.5.1. Basic
Vauquelinite (x=1,25) $Pb_2Cu(OH)[PO_4][CrO_4]$
Embreyite (x=1,25) $Pb_5[PO_4]_2[CrO_4]_2 \cdot H_2O$

*3.2.4.1.3.3.6. Phosphato-vanadates   *3.2.4.1.3.3.6.1. Basic
*Bushmakinite $Pb_2Al(OH)[PO_4][VO_4]$
*Ferribushmakinite $Pb_2Fe^{3+}(OH)[PO_4][VO_4]$

*3.2.4.1.3.4. Phosphates of Hg
*3.2.4.1.3.4.1. Proper phosphates
                                      3.2.4.1.3.4.1.1.Basic
*Artsmithite $Hg_4Al(OH)_{1+3x}[PO_4]_{2-x}$ $x = 0,26$

3.2.4.1.3.5. Phosphates of Va-cations (Bi^{3+})
                                      3.2.4.1.3.5.1. Neutral
Ximengite (x=1,5) $BiPO_4$

3.2.4.1.3.4.2. Oxido-hydroxido-phosphates
                                      3.2.4.1.3.5.2.1. Basic
Paulkellerite (x=4,5) $Bi_2Fe^{3+}(OH)_2O_2[PO_4]$
*Brendelite $(Bi,Pb)_2Fe^{3+}(OH)O_2[PO_4]$
*Petitjeanite $Bi^{3+}_3O(OH)[PO_4]_2$
*Smrkovecite $Bi_2O(OH)[PO_4]$
 Waylandite family (x=3)
 Zaïrite $BiFe^{3+}_3(OH)_6[PO_4]_2$
 Waylandite $BiAl_3(OH)_6[PO_4]_2$
 Hydroxylphosphabismite $Bi_2(OH)_3[PO_4]$

                                      *3.2.4.1.3.4.2.2. Hydrates
*Mrázekite $Bi_2Cu^{2+}_3(OH)_2O_2[PO_4]_2 \cdot 2H_2O$

*3.2.4.2. Quasiclass: Pyrophosphates
*3.2.4.2.1. Subclass: Pyrophosphates of *s-*, *d$_s$*- and *p$_s$* - cations

*3.2.4.2.1.1. Pyrophosphates of s-, d_s- and p_s - cations without Li and Be
*3.2.4.2.1.1.1. Proper pyrophosphates
*Pyrocoproite $Mg(K,Na)_2[P_2O_7]$
                                    *3.2.4.2.1.1.1.1. Hydrates
Canaphite $Na_2Ca[P_2O_7]·4H_2O$
*Arnhemite $(K,Na)_4Mg_2[P_2O_7]_2·5H_2O$

*3.2.4.2.2. Subclass: Pyrophosphates of chalcophylic elements
*3.2.4.2.2.1. Pyrophosphates of Cu
*3.2.4.2.2.1.1. M^+ и M^{2+}
*3.2.4.2.2.1.1.1. Proper pyrophosphates
*3.2.4.2.2.1.1.1.1. Hydrates
*Wooldridgeite $Na_2CaCu^{2+}_2[P_2O_7]_2(H_2O)_{10}$

*3.2.4.3. Quasiclass: Triphosphates
*3.2.4.3.1. Subclass: Triphosphates of cations with low FC
*3.2.4.3.1.1. Triphosphates s-, d_s- and p_s - cations
*3.2.4.3.1.1. Triphosphates s-, d_s- and p_s - cations without Li and Be
*3.2.4.3.1.1.1. Proper triphosphates   *3.2.4.3.1.1.1.1. Hydrates
*Kanonerovite $MnNa_3P_3O_{10}·12H_2O$

3.2.4a. Class: Arsenates
3.2.4a.1. Quasiclass: (6)-Arsenates
3.2.4a.1.1. (6)-Arsenates of d- cations
3.2.4a.1.1.1. (6)-Arsenates of Cu^{2+}
3.2.4a.1.1.1.1. Oxido-(6)-arsenates $x = MO/[AsO_4]$
3.2.4a.1.1.1.1.1. Hydrates
Geminite $Cu[AsO_3OH]·H_2O$

3.2.4a.1.2. (6)-Arsenates of p-cations
3.2.4a.1.2.1. (6)-Arsenates of Pb^{2+} 3.2.4a.1.2.1.1. Neutral
Ludlockite $PbFe_4^{3+}As_{10}^{3+}O_{22}$

3.2.4a.2. Quasiclass: (4)-Arsenates (orthoarsenates)
3.2.4a.2.1. Orthoarsenates of cations with low FC
3.2.4a.2.1.1. Orthoarsenates of s-, d_s- and p_s- cations
3.2.4a.2.1.1.1. Orthoarsenates of s-, d_s- and p_s- cations without Li and Be
3.2.4a.2.1.1.1.1. Proper orthoarsenates   3.2.4a.2.1.1.1.1.1. Neutral
*Alarsite $Al[AsO_4]$
Xanthiosite $Ni_3[AsO_4]_2$
 Berzeliite family
  Berzeliite series (compare with garnet (series)) ($x = 1.5$)
  Manganberzeliite $Ca_2NaMn_2[AsO_4]_3$
  Berzeliite $NaCa_2Mg_2[AsO_4]_3$
 Caryinite ($x = 1.5$) $NaCaCa(Mn,Mg)_2[AsO_4]_3$
(compare with alluaudite (family))
 *Arseniopleite ($x = 1.5$) $CaNaMn^{2+}Mn^{2+}_2[AsO_4]_3$
 *Ozerovaite ($x = 1.5$) $Na_2KAl_3[AsO_4]_4$
 *Calciojohillerite ($x = 1.5$) $NaCaMg_3[AsO_4]_3$

*Badalovite (x = 1.5) $Na_2Mg_2Fe^{3+}[AsO_4]_3$
*Yurmarinite (x = 1.6) $Na_7(Fe^{3+},Mg,Cu)_4[AsO_4]_6$
*Anatolyite (x = 2) $Na_6(Ca,Na)(Mg,Fe^{3+})_3Al[AsO_4]_6$
*Magnesiohatertite (x = 2.2) $(Na,Ca)_2Ca(Mg,Fe^{3+})_2[AsO_4]_3$
                                                    3.2.4a.2.1.1.1.1.1.1. Hydrates

Scorodite series
  Mansfieldite $Al[AsO_4] \cdot 2H_2O$
  Scorodite $Fe^{3+}[AsO_4] \cdot 2H_2O$
Kankite $Fe^{3+}[AsO_4] \cdot 3,5H_2O$
Grischunite $NaCa_2Mn^{2+}_5Fe^{3+}[AsO_4]_6 \cdot 2H_2O$
Sterlinghillite $Mn_3[AsO_4]_2 \cdot 3H_2O$
*Parascorodite $Fe^{3+}[AsO_4] \cdot 2H_2O$
*Yazganite $NaFe^{3+}_2Mg[AsO_4]_3 \cdot H_2O$
Erythrite family
  Erythrite series (compare with vivianite (group))
    Hörnesite $Mg_3[AsO_4]_2 \cdot 8H_2O$
    Annabergite $Ni_3[AsO_4]_2 \cdot 8H_2O$
    Erythrite $Co_3[AsO_4]_2 \cdot 8H_2O$
Manganese-hörnesite $(Mn,Mg)_3[AsO_4]_2 \cdot 8H_2O$
  *Castellaroite $Mn^{2+}_3[AsO_4]_2 \cdot 4.5H_2O$
Parasymplesite $Fe^{2+}_3[AsO_4]_2 \cdot 8H_2O$
Symplesite $Fe^{2+}_3[AsO_4]_2 \cdot 8H_2O$
Rauenthalite $Ca_3[AsO_4]_2 \cdot 10H_2O$
Phaunouxite $Ca_3[AsO_4]_2 \cdot 11H_2O$
Roselite family
  Roselite series
    Wendwilsonite $Ca_2Mg[AsO_4]_2 \cdot 2H_2O$
    Roselite $Ca_2Co[AsO_4]_2 \cdot 2H_2O$
    Zincroselite $Ca_2Zn[AsO_4]_2 \cdot 2H_2O$
    Gaitite $Ca_2Zn[AsO_4]_2 \cdot 2H_2O$
    Brandtite $Ca_2Mn[AsO_4]_2 \cdot 2H_2O$
Parabrandtite $Ca_2Mn[AsO_4]_2 \cdot 2H_2O$
  Talmessite series
    Talmessite $Ca_2Mg[AsO_4]_2 \cdot 2H_2O$
    *Nickeltalmessite $Ca_2Ni[AsO_4]_2 \cdot 2H_2O$
  Roselite-бета = betaroselite $Ca_2Co[AsO_4]_2 \cdot 2H_2O$
Smolianinovite (x = 1,5) $(Co,Ni,Mg,Ca)_3(Fe^{3+},Al)_2[AsO_4]_4 \cdot 11H_2O$
                                                    3.2.4a.2.1.1.1.1. Basic
Johnbaumite $Ca_5(OH)[AsO_4]_3$
Fermorite $Ca_5(OH)[AsO_4]_3$
*Grandaite (x = 1.75) $Sr_2Al(OH)[AsO_4]_2$
Eveite (x = 2) (comp. with adamite (gr.)) $Mn_2(OH)[AsO_4]$
Adelite family (x = 2) (compare with conichalcite (group); austinite (group)
  Adelite (x = 2) $CaMg(OH)[AsO_4]$
  Nickelaustinite $Ca(Ni,Zn)(OH)[AsO_4]$
  *Sewardite $CaFe^{3+}_2(OH)_2[AsO_4]_2$
  Sarkinite $Mn^{2+}_2(OH)[AsO_4]$
  Arsenoclasite (x = 2,5) $Mn^{2+}_5(OH)_4[AsO_4]_2$
  Flinkite (x = 3,5) $Mn^{2+}_2Mn^{3+}(OH)_4[AsO_4]$
  Allactite (x = 3,5) $Mn^{2+}_7(OH)_8[AsO_4]_2$

Jarosewichite (x = 4,5)	$Mn^{2+}_3Mn^{3+}(OH)_6[AsO_4]$
*Canosioite	$Ba_2Fe^{3+}(OH)[AsO_4]_2$

3.2.4a.2.1.1.1.1.2.1. Hydrates

*Cabalzarite	$Ca(Mg,Al,Fe)_2[AsO_4]_2(H_2O,OH)_2$
*Cobaltlotharmeyerite (x = 1,5)	$Ca(Co,Fe^{3+},Ni)_2[AsO_4]_2(OH,H_2O)_2$
*Nickellotharmeyerite (x = 1,5)	$Ca(Ni,Fe^{3+})_2[AsO_4]_2(H_2O,OH)_2$
*Barahonite-(Al) (x = 1,88)	$Ca_{12}Al_2(OH)_6[AsO_4]_8 \cdot 6H_2O$
*Barahonite-(Fe)	$Ca_{12}Fe^{3+}_2(OH)_6[AsO_4]_8 \cdot 6H_2O$
*Cobaltarthurite (x = 2)	$Co^{2+}Fe^{3+}_2(OH)_2[AsO_4]_2 \cdot 4H_2O$
*Maghrebite	$MgAl_2(OH)_2[AsO_4]_2 \cdot 8H_2O$
*Bendadaite	$Fe^{2+}Fe^{3+}_2(OH)_2[AsO_4]_2 \cdot 4H_2O$
*Césarferreiraite	$Fe^{2+}Fe^{3+}_2(OH)_2[AsO_4]_2 \cdot 8H_2O$
Camgasite (x = 2)	$CaMg(OH)[AsO_4] \cdot 5H_2O$
*Coralloite (x = 2)	$Mn^{2+}Mn^{3+}_2(OH)_2[AsO_4]_2 \cdot 4H_2O$
*Tapiaite	$Ca_5Al_2(OH)_4[AsO_4]_4 \cdot 12H_2O$

Alumopharmacosiderite family (x = 2,1(6))

Alumopharmacosiderite	$KAl_4(OH)_4[AsO_4]_3 \cdot 6,5H_2O$
Pharmacosiderite	$KFe^{3+}_4(OH)_4[AsO_4]_3 \cdot (6-7)H_2O$
*Bariuopharmacosiderite	$BaFe^{3+}_8(OH)_8[AsO_4]_6 \cdot 10H_2O$
*Hydroniumpharmacosiderite	$(H_3O)\,Fe^{3+}_4(OH)_4[AsO_4]_3 \cdot 4H_2O$
*Hydroniumpharmacoalumite	$(H_3O)Al_4(OH)_4[AsO_4]_3 \cdot 4H_2O$
*Cesiumpharmacosiderite	$CsFe^{3+}_4(OH)_4[AsO_4]_3 \cdot 4H_2O$
Talliumpharmacosiderite	$TlFe_4[AsO_4]_3 \cdot 4H_2O$
*Bariopharmacoalumite	$Ba_{0.5}Al_4(OH)_4[AsO_4]_3 \cdot 4H_2O$
*Natropharmacoalumite	$NaAl_4(OH)_4[AsO_4]_3 \cdot 4H_2O$
Ferrisymplesite (x = 2,25)	$Fe^{3+}_3(OH)_3[AsO_4]_2 \cdot 5H_2O$
*Kamaricaite(x = 2,25	$Fe^{3+}_3(OH)_3[AsO_4]_2 \cdot 3H_2O$
*Natropharmacosiderite (x = 2,(3))	$Na_2Fe^{3+}_4(OH)_5[AsO_4]_3 \cdot 7H_2O$
Calcium-pharmacosiderite (x = 2,(3))	$CaFe^{3+}_4(OH)_5[AsO_4]_3 \cdot 5H_2O$
*Ba-Zn-alumofarmacosiderite (x = 2,(3))	
	$(Ba,K)_{0,5}(Zn,Cu)_{0,5}(Al,Fe)_4(OH)_5[AsO_4]_3 \cdot 5H_2O$
Yukonite (x = 2,375)	$Ca_2Fe^{3+}_5(OH)_7[AsO_4]_4 \cdot 7H_2O$

Akrochordite family (x = 2,5)

Akrochordite	$MgMn^{2+}_4(OH)_4[AsO_4]_2 \cdot 4H_2O$
Wallkilldellite-Mn	$Ca_2Mn^{2+}_3(OH)_4[AsO_4]_2 \cdot 9H_2O$
*Wallkilldellite-Fe	$(Ca,Cu)_4Fe_6(OH)_8[(As,Si)O_4]_4 \cdot 18H_2O$
Ogdensburgite (x = 2,6)	$(Ca,Zn,Mn)_4Fe^{3+}_6(OH)_{11}[AsO_4]_5 \cdot 5H_2O$
*Esperansaite (x = 2,75)	$NaCa_2Al_2(OH)F_4[AsO_4]_2 \cdot 2H_2O$
Bulachite (x = 3)	$Al_2(OH)_3[AsO_4] \cdot 3H_2O$
*Bettertonite	$Al_6(OH)_9[AsO_4]_3 \cdot 16H_2O$
*Penberthycroftite	$Al_6(OH)_9(H_2O)_5[AsO_4]_3 \cdot 8H_2O$
Liskeardite (x = 4,5)	$(Al,Fe)_3(OH)_6[AsO_4] \cdot 5H_2O$

3.2.4a.2.1.1.1.2. Oxido-orthoarsenates (neutral)

*Wrightite (x = 2)	$K_2Al_2O[AsO_4]_2$
*Katiarsite (x = 2.5)	$KTiO[AsO_4]$
*Arsenatrotitanite(x = 2.5)	$NaTiO[AsO_4]$
Angelellite (x = 3)	$Fe^{3+}_4O_3[AsO_4]_2$

3.2.4a.2.1.1.1.2.1. Hydrates

Arseniosiderite family (x = 2,1(6))

Kolfanite $Ca_2Fe^{3+}_3O_2[AsO_4]_3 \cdot 2H_2O$
Arseniosiderite $Ca_2Fe^{3+}_3O_2[AsO_4]_3 \cdot 3H_2O$

3.2.4a.2.1.1.1.2.1.1 Oxido-orthoarsenato-arsenito-silicates (basic)
*Turtmannite $(Mn,Mg)_{22,5}Mg_{3-3x}O_{5-5x}(OH)_{20+x}[(V,As)O_4]_3[AsO_3]_x[SiO_4]_3$

*3.2.4a.2.1.1.1.2.1.2. Oxido-orthoarsenato-carbonates
*Sailaufite $(Ca,Na,\square)_2Mn^{3+}_3O_2[AsO_4]_2[CO_3] \cdot 3H_2O$

3.2.4a.2.1.1.1.3. Orthoarsenato-halogenides 3.2.4a.2.1.1.1.3.1. Neutral
  Svabite series (x = 1,(6)) (compare with apatite (series))
  Turneaureite $Ca_5Cl[AsO_4]_3$
  Svabite $Ca_5F[AsO_4]_3$
  Morelandite $Ca_2Ba_3[AsO_4]_3Cl$
Durangite family (x = 2)
Tilasite $CaMgF[AsO_4]$
 *Arsenowagnerite $Mg_2F[AsO_4]$
 *Maxwellite $NaFe^{3+}F[AsO_4]$
Durangite $NaAlF[AsO_4]$

3.2.4a.2.1.1.1.4. Arsenato-sulfates   3.2.4a.2.1.1.1.4.1. Basic
Weilerite family (x = 2,75)
Kemmlitzite $SrAl_3(OH)_6[AsO_4][SO_4]$
Weilerite $BaAl_3(OH)_6[AsO_4][SO_4]$
                              3.2.4a.2.1.1.1.4.1.1. Hydrates

3.2.4a.2.1.1.1.4.1.1.1. As : S = 3
Zykaite (x=1,5)(x = 1,5) $Fe^{3+}_4(OH)[AsO_4]_3[SO_4] \cdot 15H_2O$
3.2.4a.2.1.1.1.4.1.1.2. As : S = 1
Pitticite family (x=1,5)
Sarmientite $Fe^{3+}_2(OH)[AsO_4][SO_4] \cdot 5H_2O$
*Hilarionite $Fe^{3+}_2(OH)[AsO_4][SO_4] \cdot 6H_2O$
Bukovskyite $Fe^{3+}_2(OH)[AsO_4][SO_4] \cdot 7H_2O$
Pitticite $Fe^{3+}_2(OH)[AsO_4][SO_4] \cdot nH_2O$ (?)

3.2.4a.2.1.1.1.5. Гидроарсенато-арсенаты
*3.2.4a.2.1.1.1.5.1. Гидроарсенато-арсенаты с $AsO_3OH : AsO_4 = 1 : 2$
*Magnesiocanutite $NaMnMg_2[AsO_2(OH)_2][AsO_4]_2$
                              *3.2.4a.2.1.1.1.5.1.1. Кристаллогидраты
*Vladimirite (x = 2) $Ca_4[AsO_3(OH)][AsO_4]_2 \cdot 4H_2O$
*Chongite (x = 2.5) $Mg_2Ca_3[AsO_3(OH)][AsO_4]_2 \cdot 4H_2O$ *

3.2.4a.2.1.1.1.5.2. Hydroarsenato-arsenates with $AsO_3OH : AsO_4 = 1 : 1$
                              3.2.4a.2.1.1.1.5.2.1. Basic
  Dussertite group (x = 2,75)
  Arsenocrandallite $(Ca,Sr)Al_3(OH)_6[AsO_3(OH)][AsO_4]$
  Arsenogoyazite $(Sr,Ca,Ba)Al_3(OH)_6[AsO_3(OH)][AsO_4]$
  *Arsenogorceixite $BaAl_3(OH)_6[AsO_3(OH)][AsO_4]$
  Dussertite $BaFe^{3+}_3(OH)_6[AsO_3(OH)][AsO_4]$
                              3.2.4a.2.1.1.1.5.2.1.1. Hydrates
  Sainfeldite group (x = 1,25)

Villyaellenite	$(Mn,Ca)Mn_2Ca_2[AsO_3(OH)]_2[AsO_4]_2 \cdot 4H_2O$
Irhtemite	$Ca_4Mg[AsO_3(OH)]_2[AsO_4]_2 \cdot 4H_2O$
Sainfeldite	$Ca_5[AsO_3(OH)]_2[AsO_4]_2 \cdot 4H_2O$

Picropharmacolite family

Guérinite	$Ca_5[AsO_3(OH)]_2[AsO_4]_2 \cdot 9H_2O$
Picropharmacolite	$Ca_4Mg[AsO_3(OH)]_2[AsO_4]_2 \cdot 11H_2O$

  Chudobaite group

Chudobaite	$(Mg,Zn)_5[AsO_3(OH)]_2[AsO_4]_2 \cdot 10H_2O$
Geigerite	$Mn_5[AsO_3(OH)]_2[AsO_4]_2 \cdot 10H_2O$
*Miguelromeroite	$MnMn_2Mn_2[AsO_3(OH)]_2[AsO_4]_2 \cdot 4H_2O$
Ferrarisite	$Ca_5(H_2O)_8[AsO_3(OH)]_2[AsO_4]_2 \cdot H_2O$

*3.2.4a.2.1.1.1.5.3. Hydroarsenato-arsenates with $AsO_3OH : AsO_4 = 2 : 1$

*Canutite	$NaMn_3[AsO_3(OH)]_2[AsO_4]$

    *3.2.4a.2.1.1.1.5.3.1. Hydrates

*Joteite	$Ca_2CuAl(OH)_2[AsO_3(OH)]_2[AsO_4](H_2O)_5$

3.2.4a.2.1.1.1.5.4. Hydroarsenato-arsenates with $AsO_3OH : AsO_4 = 4 : 1$

    3.2.4a.2.1.1.1.5.4.1. Hydrates (neutral)

Mcnearite (x = 1.1)	$NaCa_5[AsO_3(OH)]_4[AsO_4] \cdot 4H_2O$

3.2.4a.2.1.1.1.6. Hydroarsenato-arsenato-phosphates
3.2.4a.2.1.1.1.6.1. Hydrates with $AsO_3OH : AsO_4 : PO_4 = 2 : 3 : 3$ (neutral)

Walentaite (x = 1.375)	$Ca_4Fe^{3+}_{12}[AsO_3(OH)]_4[AsO_4]_6[PO_4]_6 \cdot 28H_2O$

3.2.4a.2.1.1.1.6.2. Hydrates with $AsO_3OH : AsO_4 : PO_4 = 3 : 1 : 1$ (neutral)

Machatschkiite (x = 1,2)	$Ca_6[AsO_3(OH)]_3[AsO_4][PO_4] \cdot 15H_2O$

3.2.4a.2.1.1.1.7. Hydroarsenates   3.2.4a.2.1.1.1.7.1. Basic

Weilite (x = 2)	$Ca[AsO_3(OH)]$
*Švenekite	$Ca[AsO_2(OH)_2]_2$

    3.2.4a.2.1.1.1.7.2. Hydrates (neutral) (x=1)

Haidingerite family

Cobaltkoritnigite	$(Co,Zn)[AsO_3(OH)] \cdot H_2O$
*Burgessite	$Co_2(H_2O)_4[AsO_3(OH)]_2 \cdot H_2O$
Krautite	$Mn[AsO_3(OH)] \cdot H_2O$
Fluckite	$CaMn[AsO_3(OH)]_2 \cdot 2H_2O$
Haidingerite	$Ca[AsO_3(OH)] \cdot H_2O$

Pharmacolite family

Pharmacolite	$Ca[AsO_3(OH)] \cdot 2H_2O$
*Magnesiokoritnigite	$Mg[AsO_3(OH)] \cdot H_2O$
Brassite	$Mg[AsO_3(OH)] \cdot 4H_2O$
Rösslerite	$Mg[AsO_3(OH)] \cdot 7H_2O$

3.2.4a.2.1.1.1.6. Dihydroarsenates   3.2.4a.2.1.1.1.6.1.
    Hydrates (neutral) (x = 0,(3))

Kaatialaite	$Fe[H_2AsO_4]_3 \cdot 5H_2O$

3.2.4a.2.1.1.2. Orthoarsenates of Be → berylloarsenates   3.2.4a.2.1.1.2.1. Acid

Bergslagite	$Ca[Be(OH)AsO_4]^{\infty 2}$

3.2.4a.2.1.1.2.2. Hydrates
Bearsite $Be[Be(OH)AsO_4] 4H_2O$
*Okruschite $Ca_2Mn^{2+}{}_5Be_4(OH)_4[AsO_4]_6 \cdot 6H_2O$

3.2.4a.2.1.2. Orthoarsenates of f-cations
3.2.4a.2.1.2.1. Neutral (x = 1,5)
Chernovite-(Y) (compare with xenotime (group)) $Y[AsO_4]$
*Gasparite-(Ce) $(Ce,La,Nd)[AsO_4]$
3.2.4a.2.1.2.2. Basic
Arsenoflorencite-(Ce) $CeAl_3(OH)_6[AsO_4]_2$
*Arsenoflorencite-(La) $LaAl_3(OH)_6[AsO_4]_2$
*Graulichite-(Ce) $CeFe^{3+}{}_3(OH)_6[AsO_4]_2$
Retzian family
Retzian -(Ce) $Mn_2Ce(OH)_4[AsO_4]$
Retzian -(La) $Mn_2La(OH)_4[AsO_4]$
3.2.4a.2.1.2.2.1. Hydrates
*Goudeyite (x = 2,5) $(Al,Y)Cu^{2+}{}_6(OH)_6[AsO_4]_3 \cdot 3H_2O$
*Agardite-(Ce) $(Ce,Ca)Cu^{2+}{}_6(OH)_6[AsO_4]_3 \cdot 3H_2O$
*Agardite-(La) $(La,Ca)Cu^{2+}{}_6(OH)_6[AsO_4]_3 \cdot 3H_2O$
*Agardite-(Y) $(Y,Ca)Cu^{2+}{}_6(OH)_6[AsO_4]_3 \cdot 3H_2O$
*Agardite-(Nd) $NdCu^{2+}{}_6(OH)_6[AsO_4]_3 \cdot 3H_2O$

*3.2.4a.2.1.2.2.1.1. Hydroarsenato-orthoarsenates
*Plumboagardite $(Pb,REE,Ca)Cu^{2+}{}_6(OH)_6[HAsO_4][AsO_4]_3 \cdot 3H_2O$

*3.2.4a.2.1.2.2.1.2. Dihydroarsenates
*Vysokyite $U^{4+}[AsO_2(OH)_2]_4 \cdot H_2O$

3.2.4a.2.2. Subclass: Orthoarsenates of chalcophylic elements
3.2.4a.2.2.1. Orthoarsenates of Ag, Cu
3.2.4a.2.2.1.1. M^+ and M^{2+}
3.2.4a.2.2.1.1.1. Proper orthoarsenates
3.2.4a.2.2.1.1.1.1. Neutral
*Bradaczekite(x = 1,5) $NaCu_4[AsO_4]_3$
*Zincobradazekite $NaZn_2Cu_2[AsO_4]_3$
Johillerite (x=1,5) $NaCu(Mg,Zn)_3[AsO_4]_3$
*Nickenichite $Na_{0,8}Ca_{0,4}Cu_{0,4}(Mg,Fe^{3+})_3[AsO_4]_3$
*Hatertite $Na_2(Ca,Na)(Fe^{3+},Cu)_2[AsO_4]_3$
3.2.4a.2.2.1.1.1.1.1. Hydrates
Keyite (x = 1,5) $Cu^{2+}{}_3(Zn,Cu^{2+})_4Cd_2[(AsO_4]_6(H_2O)_2$
*Erikapohlite $Cu^{2+}{}_3(Zn,Cu^{2+},Mg)_4Ca_2[(AsO_4]_6 \cdot 2H_2O$
3.2.4a.2.2.1.1.1.1.2. Basic
Olivenite series (x=2)
Olivenite $Cu_2(OH)[AsO_4]$
*Zinkolivenite $CuZn(OH)[AsO_4]$
Conichalcite $CuCa(OH)[AsO_4]$
Cornwallite family (x=2,5)
Cornwallite $Cu_5(OH)_4[AsO_4]_2$
Cornubite $Cu_5(OH)_4[AsO_4]_2$
Clinoclase family (x=3)

Clinoclase (x=3)	$Cu_3(OH)_3[AsO_4]$
*Gilmarite	$Cu_3(OH)_3[AsO_4]$
Arhbarite	$Cu_2Mg(OH)_3[AsO_4]$

3.2.4a.2.2.1.1.1.2.1. Hydrates (basic)

*Rollandite (x = 1,5)	$Cu_3[AsO_4]_2·4H_2O$
*Rruffite	$Ca_2Cu^{2+}[AsO_4]_2·2H_2O$
*Lukrahnite (x = 1,75)	$Ca(Cu^{2+},Zn)(Fe^{3+},Zn)[AsO_4]_2(OH,·H_2O)_2$

Euchroite family (x=2)

Strashimirite	$Cu^{2+}_8(OH)_4[AsO_4]_4·5H_2O$
Euchroite	$Cu_2(OH)[AsO_4]·3H_2O$
*Guanacoite (x = 2,5)	$Cu_2Mg_2(Mg_{0.5}Cu_{0.5})(OH)_4[AsO_4]_2(H_2O)_4$
Philipsburgite (x=3)	$(Cu,Zn)_6(OH)_6[(AsO_4),(PO_4)]_2·H_2O$
(compare with kipushite (group))	
*Forêtite (x = 5)	$Cu_2Al_2(OH,O,·H_2O)_6[AsO_4]$

*3.2.4a.2.2.1.1.2. Orthoarsenato-oxides

*Ericlaxmanite	$Cu_4O[AsO_4]_2$
*Kozyrevskite	$Cu_4O[AsO_4]_2$
*Popovite	$Cu_5O_2[AsO_4]_2$
*Shchurovskyite	$Cu_6K_2CaO_2[AsO_4]_4$
*Dmisokolovite	$Cu_5K_3AlO_2[AsO_4]_4$
*Urusovite	$Cu^{2+}AlO[AsO_4]$
*Edtollite	$Cu_5K_2NaFe^{3+}O_2[AsO_4]_4$
*Melanarsite	$Cu_7K_3Fe^{3+}O_4[AsO_4]_4$

3.2.4a.2.2.1.1.3. Orthoarsenato-carbonates ($AsO_4 : CO_3 = 2 : 1$)

3.2.4a.2.2.1.1.3.1. Hydrates (basic)

Tyrolite (x=2)	$Ca_2Cu_9(OH)_8[AsO_4]_4[CO_3]·11H_2O$

*3.2.4a.2.2.1.1.4. Orthoarsenato-phosphates ($AsO_4 : PO_4 = 1 : 1$)

*Hermannroseite	$CaCu(OH)[(AsO_4),(PO_4)]$

3.2.4a.2.2.1.1.5. Orthoarsenato-sulfates
*3.2.4a.2.2.1.1.5.1. Orthoarsenato-sulfates with $AsO_4 : SO_4 = 1$
*3.2.4a.2.2.1.1.5.1.1. Orthoarsenato-oxido-sulfates with $AsO_4 : SO_4 = 1$

*Vasilseverginite	$Cu_9O_4[AsO_4]_2[SO_4]_2$

3.2.4a.2.2.1.1.5.2. Orthoarsenato-sulfates with $AsO_4 : SO_4 = 2$

3.2.4a.2.2.1.1.5.2.1. Hydrates (basic)

Parnauite (x=3)	$Cu_9(OH)_{10}[AsO_4]_2[SO_4]·7H_2O$

*3.2.4a.2.2.1.1.5.3. Orthoarsenato-sulfates with $AsO_4 : SO_4 = 4$

*3.2.4a.2.2.1.1.5.3.1. Hydrates

Leogangite	$Cu^{2+}_{10}(OH)_6[AsO_4]_4[SO_4]·8H_2O$

3.2.4a.2.2.1.1.6. Orthoarsenato-chlorides   3.2.4a.2.2.1.1.6.1. Hydrates

*3.2.4a.2.2.1.1.6.1.1. Neutral

*Lemanskiite	$NaCaCu^{2+}_5Cl[AsO_4]_4·5H_2O$
*Mahnertite	$(Na,Ca)Cu^{2+}_3Cl[AsO_4]_2·5H_2O$
*Zdenekite (x = 6,5)	$NaPbCu_5Cl[AsO_4]_4·5H_2O$

3.2.4a.2.2.1.1.6.1.2. Basic
Shubnikovite (x=1,(6)) $Ca_2Cu_8(OH)Cl[AsO_4]_6 \cdot 7H_2O$ (?)

*3.2.4a.2.2.1.1.6.2. Oxido-orthoarsenato-chlorides
*Coparsite (x = 4) $Cu_4O_2[(As,V)O_4]Cl$
*Arsmirandite $Na_{18}Cu_{12}Fe^{3+}O_8[AsO_4]_8Cl_5$

3.2.4a.2.2.1.1.7. Hydroarsenato-orthoarsenates
                                          3.2.4a.2.2.1.1.7.1. Hydrates (neutral)
*Pradeite $CoCu_4[AsO_3(OH)]_2[AsO_4]_2 \cdot 9H_2O$
Lindackerite (x=1) $Cu_5[AsO_3(OH)]_2[AsO_4]_2 \cdot 9H_2O$
*Hloušekite $(Ni,Co)Cu_4[AsO_3(OH)]_2[AsO_4]_2 \cdot 9H_2O$
*Ondrušite $CaCu_4[AsO_3(OH)]_2[AsO_4]_2 \cdot 10H_2O$
*Klajite $MnCu_4[AsO_3(OH)]_2[AsO_4]_2 \cdot 9\text{-}10H_2O$
*Zálesíite (x = 1) $CaCu_6\{[AsO_3(OH)][AsO_4]_2(OH)_6\} \cdot 3H_2O$
*Domerockite $Cu_4[AsO_3(OH)][AsO_4](OH)_3 \cdot H_2O$

3.2.4a.2.2.1.1.8.1.1. Hydroarsenato-orthoarsenato-oxides
*Braithwaiteite $NaCu^{2+}_5(Sb^{5+},Ti^{4+})[AsO_3(OH)]_2O_2[AsO_4]_4(H_2O)_8$

*3.2.4a.2.2.1.1.9. Hydroarsenates   *3.2.4a.2.2.1.1.9.1. Hydrates
*Pushcharovskite $Cu[AsO_3OH] \cdot 1,5H_2O$
*Yvonite $Cu[AsO_3OH] \cdot 2H_2O$

*3.2.4a.2.2.1.1.9.1. Hydroarsenato-oxides
*Lapeyreite $Cu^{2+}_3O[AsO_3(OH)]_2 \cdot 0.75H_2O$

3.2.4a.2.2.1.2. M^{2+} and M^3
*3.2.4a.2.2.1.2.1. Proper orthoarsenares   *3.2.4a.2.2.1.2.1.1. Basic
                                          3.2.4a.2.2.1.2.1.1.1. Hydrates
Arthurite (x=2) $CuFe^{3+}_2(OH)_2[AsO_4]_2 \cdot 4H_2O$
*Attikaite (x=2) $Ca_3Cu^{2+}_2Al_2(OH)_4[AsO_4]_4 \cdot 2H_2O$
  Chenevixite group (x=2,5)
  Luetheite $Cu_2Al_2(OH)_4[AsO_4]_2 \cdot H_2O$
  Chenevixite $Cu_2Fe^{3+}_2(OH)_4[AsO_4]_2 \cdot H_2O$
Ceruleite (x = 3,125) $Cu_2Al_7(OH)_{13}[AsO_4]_4 \cdot 11.5H_2O$
*Liroconite (x = 3,5) $Cu_2Al(OH)_4[AsO_4] \cdot 4H_2O$

*3.2.4a.2.2.1.2.1.1.1.1. Hydroarsenato-orthoarsenates
*Radovanite $Cu^{2+}_2Fe^{3+}[As^{5+}O_4][As^{3+}O_2OH]_2 \cdot H_2O$
*Segerstromite $Ca_3[As^{3+}(OH)_3]_2[As^{5+}O_4]_2$

3.2.4a.2.2.1.3. M^{2+}
3.2.4a.2.2.1.3.1. Proper orthoarsenato-phosphates
                                          3.2.4a.2.2.1.3.1.1. Neutral
Lammerite(x=1,5) $Cu_3[AsO_4]_2$
*Lammerite-β $Cu_3[AsO_4]_2$

3.2.4a.2.2.1.3.2. Orthoarsenato-sulfates with $AsO_4 : SO_4 = 1,(3)$
                                          3.2.4a.2.2.1.3.2.1. Hydrates (basic)

Chalcophyllite (x=3) $Cu_{18}Al_2(OH)_{24}[AsO_4]_4[SO_4]_3 \cdot 36H_2O$

3.2.4a.2.2.1.3.3. Orthoarsenato-chlorides
          3.2.4a.2.2.1.3.3.1. Hydrates
Lavendulan series (x=1,625)
Lavendulan $NaCaCu_5Cl[AsO_4]_4 \cdot 5H_2O$
Zinclavendulan $NaCa(Zn,Cu)_5Cl[AsO_4]_4 \cdot (4-5) \cdot H_2O$

*3.2.4a.2.2.1.3.4. Dihydroarsenato-orthoarsenates
          *3.2.4a.2.2.1.3.4.1. Hydrates (neutral)
*Andyrobertsite (x = 1,3) $KCdCu^{2+}_5[As(OH)_2O_2][AsO_4]_4(H_2O)_2$
*Calcioandyrobertsite $KCaCu^{2+}_5[As(OH)_2O_2][AsO_4]_4(H_2O)_2$

3.2.4a.2.2.2. Orthoarsetates of Zn
3.2.4a.2.2.2.1. M^{2+}
3.2.4a.2.2.2.1.1. Proper orthoarsenates
          3.2.4a.2.2.2.1.1.1. Neutral
Stranskiite (x=1,5) $Zn_2Cu[AsO_4]_2$
          3.2.4a.2.2.2.1.1.2. Basic
Adamite family (x=2)
Paradamite $Zn_2(OH)[AsO_4]$
Adamite $Zn_2(OH)[AsO_4]$
Cuproadamite $(Cu,Zn)_2(OH)[AsO_4]$
Austinite $CaZn(OH)[AsO_4]$
Nickelaustinite $Ca(Ni,Zn)(OH)[AsO_4]$
Cobaltaustinite $Ca(Co,Zn)(OH)[AsO_4$
*Pharmazincite $KZn[AsO_4]$
Chlorophoenicite family (x=5)
Magnesiochlorophoenicite $(Mg,Mn)_3Zn_2(OH,O)_6[AsO_4]$
Chlorophoenicite $(Mn,Mg)_3Zn_2(OH,O)_6[AsO_4]$
Theisite (x=5) $Cu_5Zn_5(OH)_{14}[AsO_4]_2$
          3.2.4a.2.2.2.1.1.3. Hydrates
          3.2.4a.2.2.2.1.1.3.1. Basic
Legrandite $Zn_2(OH)[AsO_4] \cdot H_2O$
*Ianbruceite $Zn_2(OH)(H_2O)[AsO_4](H_2O)_2$
Lotharmeyerite family (x=1,5)
Lotharmeyerite $Ca(Zn,Mn^{3+})_2[AsO_4]_2 \cdot 2H_2O$
*Manganlotharmeyerite $*Ca(Mn^{3+},Zn)_2[AsO_4]_2(OH)_2$
*Ferrilotharmeyerite $*CaFe^{3+}Zn[AsO_4]_2(OH) \cdot H_2O$
          3.2.4a.2.2.2.1.1.3.2. Neutral
Köttigite family(x=1,5)
Warikahnite $Zn_3[AsO_4]_2 H_2O$
Metaköttigite $(Zn,Fe)_3[AsO_4]_2 \cdot 8(H_2O,OH)$
Köttigite $Zn_3[AsO_4]_2 \cdot 8H_2O$
Prosperite $CaZn_2[AsO_4]_2 \cdot H_2O$
*Arsenohopeite (ortho.) $Zn_3[AsO_4]_2 \cdot H_2O$
*Davideloydite (tricl.) $Zn_3[AsO_4]_2 \cdot H_2O$

3.2.4a.2.2.2.1.2. Hydroarsenates 3.2.4a.2.2.2.1.2.1. Basic
          3.2.4a.2.2.2.1.2.2. Hydrates (basic)

Koritnigite (x=1) $Zn[AsO_3(OH)] \cdot H_2O$

*3.2.4a.2.2.2.1.3. Hydroarsenato-orthoarsenates
                              *3.2.4a.2.2.2.1.3.1. Hydrates
*Nyholmite (x = 1) $Cd_3Zn_2[AsO_3(OH)]_2[AsO_4]_2 \cdot H_2O$

3.2.4a.2.2.2.2. M^{2+} and M^{3+}
3.2.4a.2.2.2.2.1. Proper orthoarsenates
                              3.2.4a.2.2.2.2.1.1. Basic
*Wilhelmkleinite (x = 2) $ZnFe^{3+}_2(OH)_2[AsO_4]_2$
Gerdtremmelite (x=4) $ZnAl_2(OH)_5[AsO_4]$
                              3.2.4a.2.2.2.2.1.2. Hydrates
                              3.2.4a.2.2.2.2.1.2.1. Basic
Ojuelaite (x=2) $ZnFe^{3+}_2(OH)_2[AsO_4]_2 \cdot 4H_2O$
Mapimite (x=2,1(6)) $Zn_2Fe^{3+}_3(OH)_4[AsO_4]_3 \cdot 10H_2O$
                                   3.2.4a.2.2.2.2.1.2.2. Neutral
Fahleite (x=1,5) $CaZn_5Fe^{3+}_2[AsO_4]_6 \cdot 14H_2O$

3.2.4a.2.2.2.3. M^+ and M^{2+}
3.2.4a.2.2.2.3.1. Hydroarsenato-arsenatrs
                              3.2.4a.2.2.2.3.1.1. Neutral
O'Danielite (x=1,75) $Na(Zn,Mg)_3[HAsO_4]_2[AsO_4]$

3.2.4a.2.2.3. Orthoarsenates of Hg^+
3.2.4a.2.2.3.1. Proper orthoarsenates 3.2.4a.2.2.3.1.1. Neutral
Chursinite (x=1,5) $Hg^+Hg^{2+}[AsO_4]$

*3.2.4a.2.2.3.2. Orthoarsenato-chlorides
                              *3.2.4a.2.2.3.2.1. Neutral
*Kuznetsovite $Hg^+_2Hg^{2+}Cl[AsO_4]$

*3.2.4a.2.2.4 Orthoarsenates of In
*3.2.4a.2.2.4.1. Proper orthoarsenates
                              *3.2.4a.2.2.4.1.1. Hydrates
*Yanomamite $In[AsO_4] \cdot 2H_2O$

3.2.4a.2.2.5. Orthoarsenates of Pb
3.2.4a.2.2.5.1. M^{2+}
3.2.4a.2.2.5.1.1. Proper orthoarsenates
3.2.4a.2.2.5.1.1.1. Oxido-orthoarsenates
Jamesite (x=2,85) $Pb_2ZnFe^{3+}_2(Fe^{3+}_{2,8}Zn_{1,2})(OH)_8[(OH)_{1,2}O_{0,8}][AsO_4]_4$
                                   3.2.4a.2.2.5.1.1.2. Basic
Tsumcorite family(x=1,5)
Tsumcorite $Pb(Zn,Fe)_2(OH \cdot H_2O)_2[AsO_4]_2$
*Cobalttsumcorite $Pb(Co,Fe^{3+})_2(H_2O,OH)_2[AsO_4]_2$
*Nickeltsumcorite $Pb(Ni,Fe^{3+})_2(H_2O,OH)_2[AsO_4]_2$
Gartrellite $PbCuFe^{3+}(OH)[AsO_4]_2 \cdot H_2O$
*Zincgartrellite $Pb(Zn,Cu,Fe)_2(H_2O,OH)_2[AsO_4]_2$
Arsendecloizite family (x=2) (compare with decloizite (family))
Duftite $PbCu(OH)[AsO_4]$

Arsendecloizite	$PbZn(OH)[AsO_4]$
Gabrielsonite	$PbFe^{2+}(OH)[AsO_4]$
Carminite	$PbFe^{3+}_2(OH)_2[AsO_4]_2$
Bayldonite (x = 2)	$PbCu_3(OH)_2[AsO_4]_2$
*Segnitite (x = 2,75)	$PbFe^{3+}_3H(OH)_6[AsO_4]_2$

3.2.4a.2.2.5.1.1.3. Hydrates
3.2.4a.2.2.5.1.1.3.1. Basic

Mawbyite (x=1,5)	$Pb(Fe^{3+},Zn)_2[AsO_4]_2(OH,\cdot H_2O)_2$
*Longbanshuttanite	$Pb_2Mn_2Mg(OH)_4[AsO_4]_2\cdot 6H_2O$

3.2.4a.2.2.5.1.1.3.2. Neutral

Arsenbrackebuschite family

Arsenbrackebuschite (x=1,5)	$Pb_2(Fe,Zn)[AsO_4]_2\cdot H_2O$
*Feinglosite	$Pb_2(Zn,Fe)[(As,S)O_4]_2\cdot H_2O$
Thometzekite	$Pb(Cu,Zn)_2[AsO_4]_2\cdot 2H_2O$
Helmutwinklerite	$PbZn_2[AsO_4]_2\cdot 2H_2O$
*Rappoldite	$Pb(Co,Ni)_2[AsO_4]_2\cdot 2H_2O$

3.2.4a.2.2.5.1.2. Orthoarsenato-sulfates ($AsO_4 : SO_4 = 1$)
3.2.4a.2.2.5.1.2.1. Basic

Arsentsumebite (x=1,5)	$CuPb_2(OH)[AsO_4][SO_4]$

(compare with brackebuschite (group))

3.2.4a.2.2.5.1.3. Orthoarsenato-chromates ($AsO_4 : CrO_4 = 1$)
3.2.4a.2.2.5.1.3.1. Basic

Fornacite (x=1,5)	$Cu^{2+}Pb^{2+}_2(OH)[AsO_4][CrO_4]$
*Molybdofornacite	$CuPb_2(OH) [AsO_4][MoO_4]$

3.2.4a.2.2.5.1.4. Orthoarsenato-chlorides   3.2.4a.2.2.5.1.4.1. Neutral
Mimetite family (x = 1,(6)) (compare with apatite (gr.); pyromorphite (gr.); vanadinite (gr.))

Hedyphane	$Pb_3Ca_2Cl[AsO_4]_3$
Mimetite	$Pb_5Cl[AsO_4]_3$
*Clinomimetite synonym of Mimetite-M	$Pb_5Cl[AsO_4]_3$
*Vanackerite	$Pb_4CdCl[AsO_4]_3$
Nealite (x=2,5)	$Pb_4Fe^{2+}Cl_4[AsO_4]_2\cdot 2H_2O$

3.2.4a.2.2.5.1.5. Oxido-orthoarsenato-chlorides

Sahlinite (x = 2,3)	$Pb_{14}O_9Cl_4[AsO_4]_2$

3.2.4a.2.2.5.1.6. Hydroarsenates   3.2.4a.2.2.5.1.6.1. Neutral

Schultenite (x = 1)	$Pb[AsO_3(OH)]$

3.2.4a.2.2.5.2. M^{2+} and M^{3+}
3.2.4a.2.2.5.2.1. Proper oprthoarsenates
3.2.4a.2.2.5.2.1.1. Oxido-orthoarsenates
3.2.4a.2.2.5.2.1.2. Orthoarsenato-sulfates ($AsO_4 : SO_4 = 1$)
3.2.4a.2.2.5.2.1.2.1. Basic

Beudantite group (x=2,75)

Hidalgoite	$PbAl_3(OH)_6[AsO_4][SO_4]$
Beudantite	$PbFe^{3+}_3(OH)_6[AsO_4][SO_4]$

*Gallobeudantite $PbGa^{3+}_3(OH)_6[AsO_4][SO_4]$

3.2.4a.2.2.5.2.1.3. Hydroarsenato-arsenates 3.2.4a.2.2.5.2.1.3.1. Basic
Philipsbornite (x=2.75) $PbAl_3(OH)_6[AsO_3(OH)][AsO_4]$

3.2.4a.2.2.6. Orthoarsenates of Va cations
3.2.4a.2.2.6.1. Orthoarsenates of Bi^{3+} 3.2.4a.2.2.6.1.1. Neutral
Rooseveltite (x=1,5) α-$Bi[AsO_4]$
*Tetrarooseveltite β-$Bi[AsO_4]$

3.2.4a.2.2.6.2. Hydroxido-oxido-arsenates
Preisingerite (x=2,25) (compare with schumacherite (gr.)) $Bi_3(OH)O[AsO_4]_2$
Atelestite (x=3) $Bi_2(OH)O[AsO_4]$
*Arsenowaylandite (x = 3) $BiAl_3(OH)_6[AsO_4]_2$
                                          3.2.4a.2.2.6.1.2.1. Basic
Arsenobismite (x = 3) $Bi_2(OH)_3[AsO_4]$
*Neustädtelite (x = 3) $Bi_2Fe^{3+}Fe^{3+}O_2(OH)_2[AsO_4]_2$
*Cobaltneustädtelite $Bi_2Fe^{3+}Co^{2+}O(OH)_3[AsO_4]_2$
*Medenbachite $Bi_2Fe^{3+}Cu^{2+}O(OH)_3[AsO_4]_2$
                                          3.2.4a.2.2.6.1.2.2. Hydrates (basic)
Mixite (x=2,5) (compare with agardite (group) $Cu_6Bi(OH)_6[AsO_4]_3 \cdot 3H_2O$
Juanitaite $(Cu,Ca,Fe)_{10}Bi(OH)_{11}[AsO_4]_4 \cdot 2H_2O$
Schneebergite $BiCo_2(OH)[AsO_4]_2 \cdot H_2O_2$
Nickelschneebergite $BiNi_2(OH)[AsO_4]_2 \cdot H_2O_2$
*Bouazzerite $Bi_6(Mg,Co)_{11}Fe^{3+}_{14}O_{12}(OH)_4[AsO_4]_{18} \cdot 86H_2O$

3.2.4a.2.2.6.2. Orthoarsenates of Sb^{3+}
*3.2.4a.2.2.6.2.1. Proper orthoarsenates *3.2.4a.2.2.6.2.1.1. Hydrates
*Whitecapsite $H_{16}Fe^{2+}_5Fe^{3+}_{14}Sb^{3+}_6O_{16}[AsO_4]_{18} \cdot 120H_2O$

3.2.4a.2.2.6.2.2. Oxido-arsenates
Manganostibite (x=3,5) $Mn_7Sb(As,Si)O_{12} \rightarrow |^{(6)}(Mn_5Sb)_{\Sigma6}O_2||^{(4)}[Mn_2(As,Si)]_{\Sigma3}O_{10}|$

3.2.4a.2.2.6.3. Orthoarsenates of As^{3+}
3.2.4a.2.2.6.3.1 Proper orthoarsenates 3.2.4a.2.2.6.3.1.1. Oxido-arsenates
Aerugite (x=3,9) $Ni^{2+}_{18}As^{3+}O_{12}[AsO_4]_5$
Hematolite (x=8,25) $(Mn,Mg,Al)_{15}(As^{3+}O_3)(OH)_{23}[AsO_4]_2$
*Arakiite $(Zn,Mn^{2+})(Mn^{2+},Mg)_{12}(Fe,Al)_2(As^{3+}O_3)(OH)_{23}[AsO_4]_2$
                                          3.2.4a.2.2.6.3.1.2. Hydrates (basic)
Synadelphite (x=5,25) $(Mn,Mg,Ca,Pb)_9(As^{3+}O_3)(OH)_9[AsO_4]_2 \cdot 2H_2O$

3.2.4a.2.2.6.3.3. Orthoarsenato-chlorides   3.2.4a.2.2.6.3.2.1. Hydrates (basic)
Richelsdorfite (x=2,125) $Ca_2Cu^{2+}_5Sb^{5+}(OH)_6Cl[AsO_4]_4 \cdot 6H_2O$

*3.2.4a.3. Quasiclass: Pyroarsenates
*Petewilliamsite $(Ni,Co)_{30}[As_2O_7]_{15}$
                                          *3.2.4a.3.1.1. Basic
*Theoparacelsite $Cu^{2+}_3(OH)_2[As_2O_7]$

*3.2.4a.4. Quasiclass: Orthoarsenato-arsenites (basic)

*Carlfrancisite
$Mn^{2+}_3(Mn^{2+},Mg,Fe^{3+},Al)_{42}(OH)_{42}[As^{5+}O_4]_4[(Si,As^{5+})O_4]_6[(As^{5+},Si)O_4]_2[As^{3+}O_3)]_2$

*3.2.4б. **Class:** Arsenites
*3.2.4б.1. Subclass: Arsenites of cations with low FC
*3.2.4б.1.1. Arsenites of *s-*, *d$_s$-* and *p$_s$-* cations
*3.2.4б.1.1.1. Arsenito-oxides
*Fetiasite $Fe^{2+}Fe^{3+}_2O_2[As^{3+}_2O_5]$
3.2.4б.1.1.2.Arsenito-silicates   3.2.4б.1.1.2.1. Basic
*Ekatite $(Fe^{3+},Fe^{2+},Zn)_{12}(OH)_6[As^{3+}O_3]_6[(As^{3+}O_3),HO(SiO_3)]_2$

*3.2.4б.1.1.3. Арсенито-бораты (основные)
*Szklaryite $□Al_6BAs^{3+}_3O_{15}$

*3.2.4б.1.1.4. Arsenito-sulfates
*Tooeltite $Fe_6^{3+}(OH)_4[AsO_3]_4[SO_4]·4H_2O$

*3.2.4б.1.1.5. Arsenito-halogenides   *3.2.4б.1.1.5.1. Basic
*Georgiadesite (x = 4) $Pb_4(OH)Cl_4[As^{3+}O_3]$
*Unnamed $Pb_5Cl_7[As^{3+}O_3]$

3.2.5. **Class:** Sulfates
3.2.5.1.Subclass: Sulfates of cations with low FC
3.2.5.1.1. Sulfates of *s-*, *d$_s$-* and *p$_s$-* cations
3.2.5.1.1.1. Proper sulfates   3.2.5.1.1.1.1. Neutral
Millosevichite $(Al,Fe^{3+})_2[SO_4]_3$
*Mikasaite $(Fe^{3+},Al)_2[SO_4]_3$
*Perkovaite $Mg_3Ca_2[SO_4]_5$
Anhydrite $Ca[SO_4]$
  Barite group (compare with anglesite (group); hashemite (group))
  Celestite $Sr[SO_4]$
  Barite $Ba[SO_4]$
*Eldfellite $NaFe^{3+}[SO_4]_2$
*Steklite $KAl[SO_4]_2$
Yavapaiite $KFe^{3+}[SO_4]_2$
  Langbeinite group
  Langbeinite $K_2Mg_2[SO_4]_3$
  *Calciolangbeinite $K_2Ca_2[SO_4]_3$
  Manganolangbeinite $K_2Mn_2[SO_4]_3$
  Efremovite $(NH_4)_2Mg_2[SO_4]_3$
Glauberite $Na_2Ca[SO_4]_2$
Vanthoffite $Na_6Mg[SO_4]_4$
Thenardite $Na_2[SO_4]$
*Metathenardite $Na_2[SO_4]$
  Kalistrontite group
  Aphthitalite $(K,Na)_3Na[SO_4]_2$
  Kalistrontite $K_2Sr[SO_4]_2$
  Möhnite $(NH_4)K_2Na[SO_4]_2$
  Arcanite $K_2[SO_4]$
  Godovikovite group

Godovikovite	$NH_4Al[SO_4]_2$
*Pyracmonite	$(NH_4)_3Fe^{3+}[SO_4]_3$
*Aluminopyracmonite	$(NH_4)_3Al[SO_4]_3$
Sabieite	$NH_4Fe^{3+}[SO_4]_2$
Mascagnite	$(NH_4)_2[SO_4]$
*Bubnovaite	$K_2Na_8Ca[SO_4]_6$

*3.2.1.1.1.1. Acids

*Ivsite	$Na_3H[SO_4]_2$

3.2.5.1.1.1.2. Oxido-sulfates

Ye'elimite (x = 13)	$Ca_4Al_6O_{12}[SO_4]$

3.2.5.1.1.1.3. Basic and sulfato-halogenides

D'Ansite group (x = 1,15)

D'Ansite	$Na_{21}MgCl_3[SO_4]_{10}$
*D'Ansite-(Fe)	$Na_{21}FeCl_3[SO_4]_{10}$
*D'Ansite-(Mn)	$Na_{21}MnCl_3[SO_4]_{10}$
Cesanite (x = 1,1(6))	$Na_3Ca_2(OH)[SO_4]_3$
*Shuvalovite	$K_2NaCa_2F[SO_4]_3$
*Krasheninnikovite	$KNa_2CaMgF[SO_4]_3$
*Aiolosite	$Na_2(Na_2Bi)Cl[SO_4]_3$
*Kononovite	$NaMgF[SO_4]$

Alunite group (x = 1,25) (compare with argentojarosite; plumbojarosite)

Natroalunite	$NaAl_3(OH)_6[SO_4]_2$
Minamiite	$(Na,Ca,\square)Al_3(OH)_6[SO_4]_2$
*Huangite	$Ca_{0.5}Al_3(OH)_6[SO_4]_2$
Alunite	$KAl_3(OH)_6[SO_4]_2$
*Termessaite	$K_2AlF_3[SO_4]$
*Termessaite-(NH₄)	$(NH_4)_2AlF_3[SO_4]$
*Walhierite	$BaAl_6(OH)_{12}[SO_4]_4$
*Ammonioalunite	$(NH_4)Al_3(OH)_6[SO_4]_2$
*Adranosite-(Al)	$(NH_4)_4NaAl_2Cl(OH)_2[SO_4]_4$
*Adranosite-(Fe)	$(NH_4)_4NaFe_2Cl(OH)_2[SO_4]_4$
Schlossmacherite	$(H_3O)Al_3(OH)_6[SO_4]_2$
Natrojarosite	$NaFe^{3+}_3(OH)_6[SO_4]_2$
Jarosite	$KFe^{3+}_3(OH)_6[SO_4]_2$
Ammoniojarosite	$NH_4Fe^{3+}_3(OH)_6[SO_4]_2$
Hydronium-jarosite	$(H_3O)Fe^{3+}_3(OH)_6[SO_4]_2$
*Dorallcharite	$Tl_{0.8}K_{0.2}Fe^{3+}_3(OH)_6[SO_4]_2$

Sulphohalite polysomatic series n(Na₃X[SO₄]) with the proviso that X = Cl, F

Kogarkoite	$Na_3F[SO_4]$	(n=1)
Sulphohalite	$Na_6ClF[SO_4]_2$	(n=2)
Galeite	$Na_{15}ClF_4[SO_4]_5$	(n=5)
Schairerite	$Na_{21}ClF_6[SO_4]_7$	(n=7)

3.2.5.1.1.1.4. Hydrates

3.2.5.1.1.1.4.1. Basic (in that number oxido-sulfates

Metavoltine (x = 1,1(6))	$K_2Na_6Fe^{2+}Fe^{3+}_6O_2[SO_4]_{12}\cdot18H_2O$
*Alcaparrosaite (x = 1,25)	$K_3Ti^{4+}Fe^{3+}O[SO_4]_4\cdot2H_2O$

Copiapite group (x = 1,25)

Aluminocopiapite	$Fe^{3+}_4Al(OH)O[SO_4]_6\cdot20H_2O$
Ferricopiapite	$Fe^{3+}_4(Fe^{3+}_{2/3}\square_{1/3})(OH)_2[SO_4]_6\cdot20H_2O$
Magnesiocopiapite	$MgFe^{3+}_4(OH)_2[SO_4]_6\cdot20H_2O$

  *Botryogen $\qquad$ $MgFe^{3+}(OH)[SO_4]_2{\cdot}7H_2O$

  Copiapite $\qquad$ $(Fe^{2+},Mg)Fe^{3+}_4(OH)_2[SO_4]_6{\cdot}19H_2O$

  Calciocopiapite $\qquad$ $CaFe^{3+}_4(OH)_2[SO_4]_6{\cdot}19H_2O$

  *Volaschioite $\qquad$ $Fe^{3+}_4O_2(OH)_6[SO_4]{\cdot}2H_2O$

Sideronatrite family (x = 1,25)

  Metasideronatrite $\qquad$ $Na_2Fe^{3+}(OH)[SO_4]_2{\cdot}H_2O$

  Sideronatrite $\qquad$ $Na_2Fe^{3+}(OH)[SO_4]_2{\cdot}3H_2O$

  Clinoungemachite (x= 1,25) $\qquad$ $K_3Na_9Fe^{3+}(OH)_3[SO_4]_6{\cdot}9H_2O$

  Clairite (x = 1,375) $\qquad$ $(NH_4)_2Fe^{3+}_3(OH)_3[SO_4]_4{\cdot}3H_2O$

  *Caminite $\qquad$ $Mg_7(OH)_4[SO_4]_5{\cdot}H_2O$

Hohmannite family (x = 1,5)

  Metahohmannite $\qquad$ $Fe^{3+}(OH)[SO_4]{\cdot}1,5H_2O$

  Butlerite $\qquad$ $Fe^{3+}(OH)[SO_4]{\cdot}2H_2O$

  Parabutlerite $\qquad$ $Fe^{3+}(OH)[SO_4]{\cdot}2H_2O$

  Amarantite $\qquad$ $Fe^{3+}(OH)[SO_4]{\cdot}3H_2O$

  Hohmannite $\qquad$ $Fe^{3+}(OH)[SO_4]{\cdot}3,5H_2O$

Fibroferrite family (x = 1,5)

  *Riotintoite $\qquad$ $Al(OH)[SO_4]{\cdot}3H_2O$

  Jurbanite $\qquad$ $Al(OH)[SO_4]{\cdot}5H_2O$

  Rostite $\qquad$ $Al(OH,F)[SO_4]{\cdot}5H_2O$

  Fibroferrite $\qquad$ $Fe^{3+}(OH)[SO_4]{\cdot}5H_2O$

  Slavikite (x = 1,36) $\qquad$ $(H_3O)_3Mg_6Fe^{3+}_{15}(OH)_{18}[SO_4]_{21}{\cdot}98H_2O$

  Svyazhinite (x = 1,25) $\qquad$ $(Mg,Mn)(Al,Fe^{3+})F[SO_4]_2{\cdot}14H_2O$

  Uklonskovite (x = 2,5) $\qquad$ $NaMg(OH,F)[SO_4]{\cdot}2H_2O$

  *Kottenheimite $\qquad$ $Ca_6Si_2(OH)_{12}[SO_4]_4{\cdot}24H_2O$

  *Kottenheimite hexag. $\qquad$ $Ca_3Si(OH)_6[SO_4]_2{\cdot}12H_2O$

  *Laaherite mon. $\qquad$ $Ca_3Si(OH)_6[SO_4]_2{\cdot}12H_2O$

Aluminite family (x = 3)

  *Mangazeite $\qquad$ $Al_2(OH)_4[SO_4]{\cdot}3H_2O$

  Meta-aluminite $\qquad$ $Al_2(OH)_4[SO_4]{\cdot}5H_2O$

  Aluminite $\qquad$ $Al_2(OH)_4[SO_4]{\cdot}7H_2O$

Ettringite family (x = 3)

  Ettringite $\qquad$ $Ca_6Al_2(OH)_{12}[SO_4]_3{\cdot}26H_2O$

  Bentorite $\qquad$ $Ca_6(Cr,Al)_2(OH)_{12}[SO_4]_3{\cdot}26H_2O$

  Zaherite (x = 3,6) $\qquad$ $Al_{12}(OH)_{26}[SO_4]_5{\cdot}20H_2O$

Wermlandite family (x = 4.0-5.5)

  Mountkeithite (x = 4) $\qquad$ $Mg_{11}Fe_3^{3+}[SO_4]_{3.5}(OH)_{24}{\cdot}11H_2O$

  Wermlandite $\qquad$ $Mg_8Al_2(OH)_{18}[SO_4]_2{\cdot}12H_2O$

  Motukoreaite $\qquad$ $NaMg_6Al_3(OH)_{18}[SO_4]_2{\cdot}7H_2O$

Basaluminite family (x = 5-5.5)

  Basaluminite $\qquad$ $Al_4(OH)_{10}[SO_4]{\cdot}4H_2O$

  Felsöbányaite $\qquad$ $Al_4(OH)_{10}[SO_4]{\cdot}4H_2O$

  Hydrobasaluminite $\qquad$ $Al_4(OH)_{10}[SO_4]{\cdot}9H_2O$

  *Nikischerite(x = 5,5) $\quad$ $Fe^{2+}_6Al_3(OH)_{18}[Na(H_2O)_6][SO_4]_2{\cdot}6H_2O$

  Shigaite (x = 5,5) $\quad$ $*Mn^{2+}_6Al_3(OH)_{18}[Na(H_2O)_6][SO_4]_2{\cdot}6H_2O$

  Carrboydite (x = 6) $\quad$ $(Ni_{1-x}Al_x)[SO_4]_{x/2}(OH)_2{\cdot}nH_2O$ (x<0.5, n>3x/2)

  *Kuzelite (x = 7) $\qquad$ $Ca_4Al_2(OH)_{12}[SO_4]{\cdot}6H_2O$

Jamborite family (x = 9)

  Jamborite $\qquad$ $Ni^{2+}_{1-x}Co_x^{3+}(OH)_{2-x}[SO_4]_x{\cdot}nH_2O$ where [x≤1/3; n≤(1-x)]

  Honessite $\qquad$ $(Ni_{1-x}Fe_x^{3+})(OH)_2[SO_4]_{x/2}{\cdot}nH_2O$ $\qquad$ (x < 0.5, n < 3x/2)

Hydrohonessite $(Ni_{1-x}Fe_x^{3+})(OH)_2)[SO_4]_{x/2} \cdot nH_2O$ (x < 0.5, n > 3x/2)
Copper-aluminium analog of honessite
Copper-aluminium analog of hydrohonessite

3.2.5.1.1.1.4.2. Neutral (x = 1)

Alunogen family
Meta-alunogen $Al_4[SO_4]_6 \cdot 27H_2O$
Alunogen $Al_2[SO_4]_3 \cdot 17H_2O$
Coquimbite family
Lausenite $Fe^{3+}_2[SO_4]_3 \cdot 6H_2O$
Kornelite $Fe^{3+}_2[SO_4]_3 \cdot 7H_2O$
Paracoquimbite $Fe^{3+}_2[SO_4]_3 \cdot 9H_2O$
Coquimbite $Fe^{3+}_2[SO_4]_3 \cdot 9H_2O$
Quenstedtite $Fe^{3+}_2[SO_4]_3 \cdot 10H_2O$
*Aluminocoquimbite $AlFe^{3+}[SO_4]_3 \cdot 9H_2O$
Römerite $Fe^{2+}Fe^{3+}_2[SO_4]_4 \cdot 14H_2O$
  Halotrichite group
  Pickeringite $MgAl_2[SO_4]_4 \cdot 22H_2O$
  *Wupatkiite $(Co,Mg,Ni)Al_2[SO_4]_4 \cdot 22H_2O$
  Halotrichite $Fe^{2+}Al_2[SO_4]_4 \cdot 22H_2O$
  Apjohnite $Mn^{2+}Al_2[SO_4]_4 \cdot 22H_2O$
  Bilinite $Fe^{2+}Fe^{3+}_2[SO_4]_4 \cdot 22H_2O$
  Redingtonite $(Fe,Mg,Ni)(Cr,Al)_2[SO_4]_4 \cdot 22H_2O$
*Caichengyunite $Fe^{2+}_3Al_2[SO_4]_6 \cdot 30H_2O$
*Lanmuchangite $TlAl[SO_4]_2 \cdot 12H_2O$
Goldichite family
Krausite $KFe^{3+}[SO_4]_2 \cdot H_2O$
Goldichite $KFe^{3+}[SO_4]_2 \cdot 4H_2O$
Tamarugite family
Tamarugite $NaAl[SO_4]_2 \cdot 6H_2O$
Amarillite $NaFe^{3+}[SO_4]_2 \cdot 6H_2O$
  Mendozite group
  Mendozite $NaAl[SO_4]_2 \cdot 11H_2O$
  Kalinite $KAl[SO_4]_2 \cdot 11H_2O$
  Alum group
  Sodium alum $NaAl[SO_4]_2 \cdot 12H_2O$
  Potassium alum $KAl[SO_4]_2 \cdot 12H_2O$
  Tschermigite $NH_4Al[SO_4]_2 \cdot 12H_2O$
  Lonecreekite $NH_4Fe^{3+}[SO_4]_2 \cdot 12H_2O$
Voltaite $K_2Fe^{2+}_5Fe^{3+}_4[SO_4]_{12} \cdot 18H_2O$
*Ammoniovoltaite $(NH_4)_2Fe^{2+}_5Fe^{3+}_4[SO_4]_{12} \cdot 8H_2O$
*Ammoniomagnesiovoltaite $(NH_4)_2Mg_5Fe^{3+}_3Al[SO_4]_{12} \cdot 18H_2O$
*Zincovoltaite $K_2Zn_5Fe^{3+}_3Al[SO_4]_{12} \cdot 18H_2O$
*Pertlikite $K_2(Fe^{2+},Mg)_2(Mg,Fe^{3+})_4Fe^{3+}_2Al[SO_4]_{12} \cdot 18H_2O$
Ferrinatrite $Na_3Fe^{3+}[SO_4]_3 \cdot 3H_2O$
Gypsum family
Bassanite $Ca[SO_4] \cdot 0,5H_2O$
Gypsum $Ca[SO_4] \cdot 2H_2O$
*Omongwaite $Na_2Ca_5[SO_4]_6 \cdot 3H_2O$
Kieserite family (compare with gunningite (series))
Kieserite $Mg[SO_4] \cdot H_2O$

*Cobaltkieserite	$Co[SO_4] \cdot H_2O$
Dwornikite	$Ni[SO_4] \cdot H_2O$
Szomolnokite	$Fe[SO_4] \cdot H_2O$
Szmikite	$Mn[SO_4] \cdot H_2O$
Sanderite	$Mg[SO_4] \cdot 2H_2O$

Rozenite family (compare with boyleite (family))

Starkeyite	$Mg[SO_4] \cdot 4H_2O$
*β-starkeyite	$Mg[SO_4] \cdot 4H_2O$
*Cranswickite	$Mg[SO_4] \cdot 4H_2O$
Aplowite	$Co[SO_4] \cdot 4H_2O$
Rozenite	$Fe[SO_4] \cdot 4H_2O$

Pentahydrite group (compare with chalcanthite (group))

Pentahydrite	$Mg[SO_4] \cdot 5H_2O$
Siderotil	$Fe[SO_4] \cdot 5H_2O$
Jokokuite	$Mn[SO_4] \cdot 5H_2O$
*Chvaleticeite	$Mn[SO_4] \cdot 6H_2O$

Hexahydrite family(ср.вианкита (гр.))

Hexahydrite group

Hexahydrite	$Mg[SO_4] \cdot 6H_2O$
Retgersite	$Ni[SO_4] \cdot 6H_2O$

Moorhouseite group

Nickelhexahydrite	$Ni[SO_4] \cdot 6H_2O$
Moorhouseite	$Co[SO_4] \cdot 6H_2O$
Ferrohexahydrite	$Fe[SO_4] \cdot 6H_2O$

Epsomite group

Epsomite	$Mg[SO_4] \cdot 7H_2O$
Morenosite	$Ni[SO_4] \cdot 7H_2O$
Tauriscite	$Fe[SO_4] \cdot 7H_2O$

Melanterite group

Bieberite	$Co[SO_4] \cdot 7H_2O$
Melanterite	$(Fe,Mg)[SO_4] \cdot 7H_2O$
Mallardite	$Mn[SO_4] \cdot 7H_2O$

Polyhalite family

Görgeyite	$K_2Ca_5[SO_4]_6 \cdot H_2O$
Polyhalite	$K_2MgCa_2[SO_4]_4 \cdot 2H_2O$
Syngenite	$K_2Ca[SO_4]_2 \cdot H_2O$

Eugsterite family

Eugsterite	$Na_4Ca[SO_4]_3 \cdot 2H_2O$
Hydroglauberite	$Na_{10}Ca_3[SO_4]_8 \cdot 6H_2O$

Blödite family

Blödite	$Na_2Mg[SO_4]_2 \cdot 4H_2O$
*Cobaltoblödite	$Na_2Co[SO_4]_2 \cdot 4H_2O$
*Manganoblödite	$Na_2Mn[SO_4]_2 \cdot 4H_2O$
Nickelblödite	$Na_2(Ni,Mg)[SO_4]_2 \cdot 4H_2O$
Wattevillite	$Na_2Ca[SO_4]_2 \cdot 4H_2O$ (?)
Leonite	$K_2Mg[SO_4]_2 \cdot 4H_2O$
Löweite	$Na_{12}Mg_7[SO_4]_{13} \cdot 15H_2O$
Konyaite	$Na_2Mg[SO_4]_2 \cdot 5H_2O$

Picromerite group

Picromerite	$K_2Mg[SO_4]_2 \cdot 6H_2O$

*Nickelpicromerite	$K_2Ni[SO_4]_2 \cdot 6H_2O$
Boussingaultite	$(NH_4)_2Mg[SO_4]_2 \cdot 6H_2O$
Nickelboussingaultite	$(NH_4)_2(Ni,Mg)[SO_4]_2 \cdot 6H_2O$
Mohrite	$(NH_4)_2Fe^{2+}[SO_4]_2 \cdot 6H_2O$
*Mereiterite	$K_2Fe^{2+}[SO_4]_2 \cdot 4H_2O$
Mirabilite	$Na_2[SO_4] \cdot 10H_2O$
Koktaite	$(NH_4)_2Ca[SO_4]_2 \cdot H_2O$
Lecontite	$Na(NH_4,K)[SO_4] \cdot 2H_2O$

*3.2.5.1.1.2. Sulfato-borates
*Tatarinovite $Ca_3Al[SO_4](OH)_6[B(OH)_4] \cdot 12H_2O$

*3.2.5.1.1.3. Sulfato-orthophosphato-halogenides
                                          *3.2.5.1.1.3.1. Hydrates
*Arangasite $Al_2[SO_4][PO_4]F$

3.2.5.1.1.4. Sulfato-nitrates
3.2.5.1.1.4.1. Sulfato-nitrates with $SO_4 : NO_3 = 3$
                                          3.2.5.1.1.41.1. Hydrates (acid)
Ungemachite group
Ungemachite $K_3Na_8Fe^{3+}[SO_4]_6[NO_3]_2 \cdot 6H_2O$
Humberstonite $K_3Na_7Mg_2[SO_4]_6[NO_3]_2 \cdot 6H_2O$

*3.2.5.1.1.4.2. Sulfato-nitrates with $SO_4 : NO_3 = 2$
                                          3.2.5.1.1.4.2.1. Hydrates
*Witzkeite $Na_4K_4Ca[SO_4]_4[NO_3]_2 \cdot 2H_2O$

3.2.5.1.1.4.3. Sulfato-nitrates with $SO_4 : NO_3 = 1$
                                          3.2.5.1.1.4.3.1. Hydrates (basic)
Darapskite $Na_3[SO_4][NO_3] \cdot H_2O$

3.2.5.1.1.4.4. Sulfato-nitrates with $SO_4 : NO_3 = 0,1(6)$
                                          3.2.5.1.1.4.4.1. Hydrates (basic)
Mbobomkulite $(Ni,Cu^{2+})Al_4(OH)_{12}[NO_3,SO_4]_2 \cdot 3H_2O$

*3.2.5.1.1.5. Sulfato-halogenides	*3.2.5.1.1.5.1. Hydrates
*Khademite	$Al[SO_4]F \cdot 5H_2O$
*Vlodavetsite	$Ca_2Al[SO_4]_2F_2Cl \cdot 4H_2O$
*Vendidaite	$Al_2[SO_4](OH)_3Cl \cdot 6H_2O$
*Wilcoxite	$MgAl[SO_4]_2F \cdot 18H_2O$
Xitieshanite (x = 1,5)	$Fe^{3+}[SO_4]Cl \cdot 6H_2O$
Kainite (x = 1,5)	$KMg[SO_4]Cl \cdot 3H_2O$

3.2.5.1.1.6. Sulfato-iodates	3.2.5.1.1.6.1. Neutral
Hectorfloresite	$Na_9[SO_4][IO_3]$
	*3.2.5.1.1.6.1.1. Hydrates
*Fuenzalidaite	$K_6Na_4Na_6Mg_{10}[SO_4]_{12}[IO_3]_{12} \cdot 12H_2O$

| 3.2.5.1.1.7. Sulfato-fluoraluminates | 3.2.5.1.1.7.1. Hydrates (acid) |
| *Meniaulovite | $Ca_4AlSi[SO_4]F_{13} \cdot 12H_2O$ |

Creedite $Ca_3[SO_4][Al_2F_8(OH)_2]\cdot 2H_2O$

3.2.5.1.1.8. Acid sulfates (hydrosulfates)
3.2.5.1.1.8.1. Proper hydrosulfates   3.2.5.1.1.8.1.1. Neutral
Mercallite family
Mercallite $K[HSO_4]$
Misenite $K_8H_6[SO_4]_7$
                                            3.2.5.1.1.8.1.2. Hydrates
Matteuccite $Na[HSO_4]\cdot H_2O$

3.2.5.1.1.8.2. Hydrosulfato-sulfates   3.2.5.1.1.8.2.1. Neutral
Letovicite $(NH_4)_3[HSO_4][SO_4]$
                                            3.2.5.1.1.8.2.2. Hydrates
Rhomboclase $(H_3O)Fe^{3+}[SO_4]_2\cdot 3H_2O$
*Cossaite $(Mg_{0,5}\square)Al_6[HSO_4][SO_4]_6F_6\cdot 36H_2O$

3.2.5.1.1.9. Sulfates with unknown structure (sulfato-fluoraluminates ?)
Lannonite $HCa_4Mg_2Al_4[SO_4]_8F_9\cdot 32H_2O$

3.2.5.1.2. Sulfates of f-elements   *3.2.5.1.2.1. Hydrates
*Běhounekite $U[SO_4]_2(H_2O)_4$

3.2.5.1.2.2. Sulfato-fluoraluminates (0,5:1)   3.2.5.1.2.2.1. Hydrates (basic)
Chukhrovite group
*Chukhrovite-(Ca) $Ca_3Ca_{1,5}Al_2[SO_4]F_{13}\cdot 12H_2O$
Chukhrovite-(Ce) $Ca_3CeF(H_2O)_{10}[SO_4][AlF_6]_2$
*Chukhrovite-(Nd) $Ca_3NdF(H_2O)_{12}[SO_4][AlF_6]_2$
Chukhrovite-(Y) $Ca_3YF(H_2O)_{10}[SO_4][AlF_6]_2$

*3.2.5.1.2.3. Sulfato-oxalates
*Levinsonite-(Y) $(Y,Nd,Ce)Al[SO_4]_2(C_2O_4)\cdot 12H_2O$
*Zugshunstite-(Ce) $(Ce,Nd,La)Al[SO_4]_2(C_2O_4)\cdot 12H_2O$

3.2.5.2. Subclass: Sulfates of cations with middle FC
3.2.5.2.1. Sulfates of Zr   3.2.5.2.1.1. Hydrates (neutral)
Zircosulfate $Zr[SO_4]_2\cdot 4H_2O$

*3.2.5.2.2. Sulfates of Sn   *3.2.5.2.2.1. Hydrates (basic)
*Genplesite $Ca_3Sn(OH)_6[SO_4]_2\cdot 3H_2O$

3.2.5.2.3. Sulfates of Mn^{4+} 3.2.5.2.3.1. Hydrates (basic)
Despujolsite $Ca_3Mn^{4+}(OH)_6[SO_4]_2\cdot 3H_2O$

3.2.5.2.4. Sulfates of V^{4+} 3.2.5.2.4.1.Средние
*Pauflerite $\beta\text{-}VO[SO_4]$
                                            3.2.5.2.4.1.1. Hydrates (oxido)

Minasragrite family
Minasragrite $VO[SO_4]\cdot 5H_2O$
*Orthominasragrite $VO[SO_4]\cdot 5H_2O$
*Anorthominasragrite $VO[SO_4]\cdot 5H_2O$

Stanleyite	$VO[SO_4] \cdot 6H_2O$
*Bobjonesite	$VO[SO_4] (H_2O)_3$
*Karpovite	$Tl_2VO[SO_4]_2 (H_2O)$
*Evdokimovite	$Tl_4[VO]_3[SO_4]_5(H_2O)_5$

3.2.5.3. Sulfates of chalcophylic elements
3.2.5.3.1. Sulfates of Ag^+ 3.2.5.3.1.1. Basic
Argentojarosite (comp. with alunite (gr.) $AgFe^{3+}_3(OH)_6[SO_4]_2$

3.2.5.3.2. Sulfates of Cu   3.2.5.3.2.1. Neutral

Chalcocyanite (x = 1)	$Cu[SO_4]$
*Dravertite (x = 1)	$CuMg[SO_4]_2$
*Saranchinaite	$NaCu[SO_4]_2$

3.2.5.3.2.2. Oxido-sulfates

*Cryptochalcite (x = 1,2)	$K_2Cu_5O[SO_4]_5$
*Cesiodimite (x = 1,2)	$CsKCu_5O[SO_4]_5$
*Fedotovite (x = 1,(3))	$K_2Cu_3O[SO_4]_3$
*Wulffite (x = 1,5)	$K_3NaCu_4O_2[SO_4]_4$
*Parawulffite (x = 1,5)	$K_5Na_3Cu_8O_4[SO_4]_8$
Klyuchevskite (x = 1,5)	$K_3Cu_3Fe^{3+}O_2[SO_4]_4$
*Alumoklyuchevskite (x = 1,5)	$K_3Cu_3AlO_2[SO_4]_4$
*Eleomelanit(x = 1,5)	$(K_2Pb)Cu_4O_2[SO_4]_4$
Dolerophanite (x = 2)	$Cu_2O[SO_4]$

3.2.5.3.2.3. Basic and sulfato-halogenides

Piypite (x = 1,08)	$K_4(Na,Cu)Cu_4O_2Cl[SO_4]_4$
Kamchatkite (x = 1,75)	$KCu_3OCl[SO_4]_2$
Chlorothionite (x = 2)	$K_2CuCl_2[SO_4]$
Antlerite (x = 3)	$Cu_3(OH)_4[SO_4]$
Brochantite (x = 4)	$Cu_4(OH)_6[SO_4]$
*Grandviewite (x = 8.25))	$Cu_3Al_9(OH)_{29}[SO_4]_2$

3.2.5.3.2.4. Hydrates

Cuprocopiapite group
Cuprocopiapite (x = 1,1(6)) $Cu^{2+}Fe^{3+}_4(OH)_2[SO_4]_6 \cdot 20H_2O$
(compare with copiapite (group))
Aubertite family   (x = 1,25)

Magnesioaubertite	$(Mg,Cu)AlCl[SO_4]_2 \cdot 14H_2O$
Aubertite	$Cu^{2+}AlCl[SO_4]_2 \cdot 14H_2O$
Guildite	$Cu^{2+}Fe^{3+}(OH)[SO_4]_2 \cdot 4H_2O$
Natrochalcite (x = 1,25)	$NaCu^{2+}_2(OH)[SO_4]_2 \cdot H_2O$
*Kaliochalcite	$KCu_2(OH)[SO_4]_2(H_2O)$

Devilline family   (x = 2,5)

Devilline	$CaCu^{2+}_4(OH)_6[SO_4]_2 \cdot 3H_2O$
*Vonbezingite	$Ca_6Cu^{2+}_3(OH)_{12}[SO_4]_3 \cdot 2H_2O$
*Lautenthalite	$Cu^{2+}_4Pb(OH)_6[SO_4]_2 \cdot 3H_2O$
Campigliaite	$Cu^{2+}_4Mn^{2+}(OH)_6[SO_4]_2 \cdot 4H_2O$
*Niedermayrite	$Cu^{2+}_4Cd^{2+}(OH)_6[SO_4]_2 \cdot 4H_2O$
*Edwardsite	$Cu^{2+}_3Cd^{2+}_2(OH)_6[SO_4]_2 \cdot 4H_2O$

Langite family (x = 4)

| Posnjakite | $Cu_4(OH)_6[SO_4] \cdot H_2O$ |
| Langite | $Cu_4(OH)_6[SO_4] \cdot 2H_2O$ |

Wroewolfeite	$Cu_4(OH)_6[SO_4] \cdot 2H_2O$
*Kobyashevite	$Cu^{2+}_5(OH)_6[SO_4]_2 \cdot 2H_2O$
*Redgillite	$Cu^{2+}_6(OH)_{10}[SO_4] \cdot H_2O$
*Montetrisaite	$Cu^{2+}_6(OH)_{10}[SO_4] \cdot H_2O$
Chalcoalumite (x = 7)	$CuAl_4(OH)_{12}[SO_4] \cdot 3H_2O$

Cyanotrichite group (x = 7)

Cyanotrichite	$Cu^{2+}_4Al_2(OH)_{12}[SO_4] \cdot 2H_2O$
Carbonate-cyanotrichite	$Cu^{2+}_4Al_2(OH)_{12}[(CO_3),(SO_4)] \cdot 2H_2O$
Woodwardite	$(Cu^{2+},Al)_9(OH)_{18}[SO_4]_2 \cdot nH_2O$
*Hydrowoodwardite	$[Cu^{2+}_{1-x}Al_x(OH)_2][(SO_4)_{x/2}(H_2O)_n]$ x<0.5, n>3x/2n
*Camerolaite	$Cu_6Al_3(OH)_{18}(H_2O)_2[SO_4][Sb(OH)_6]$
Spangolite (x = 7,5)	$Cu^{2+}_6Al(OH)_{12}Cl[SO_4] \cdot 3H_2O$
Connellite (x = 19)	$Cu^{2+}_{19}(OH)_{32}Cl_4[SO_4] \cdot 3H_2O$

3.2.5.3.2.4.2. Neutral

Ransomite (x = 1)	$Cu^{2+}Fe^{3+}_2[SO_4]_4 \cdot 6H_2O$

Chalcanthite family (x = 1) (compare with pentahydrite; melanterite (group))

Bonattite	$Cu[SO_4] \cdot 3H_2O$
Chalcanthite	$Cu[SO_4] \cdot 5H_2O$
Boothite	$Cu[SO_4] \cdot 7H_2O$
Kröhnkite (x = 1)	$Na_2Cu[SO_4]_2 \cdot 2H_2O$
Leightonite (x = 1)	$K_2Ca_2Cu[SO_4]_4 \cdot 2H_2O$
Cyanochroite (x = 2)	$K_2Cu[SO_4]_2 \cdot 6H_2O$

(compare with picromerite (group))

*Alpersite	$(Mg,Cu)[SO_4] \cdot 7H_2O$

3.2.5.3.3. Sulfates of Hg^{2+} 3.2.5.3.3.1. Oxido (nitrido)-sulfates

Schuetteite family

Schuetteite	$Hg^{2+}_3O_2[SO_4]$
Gianellaite	$Hg^{2+}_4N_2[SO_4](H_2O)_x$

3.2.5.3.4. Sulfates of Zn

*3.2.5.3.4.1. Proper sulfates   *3.2.5.3.4.1.1. Neutral

Zincosite	$Zn[SO_4]$

3.2.5.3.4.1.2. Hydrates

Gunningite family (comp. with kieserite (family))

Gunningite	$Zn[SO_4] \cdot H_2O$
Poitevinite	$(Cu,Fe,Zn)[SO_4] \cdot H_2O$

Boyleite family (comp. with rozenite (gr.))

Boyleite	$Zn[SO_4] \cdot 4H_2O$
Ilesite	$(Mn,Zn)[SO_4] \cdot 4H_2O$
Bianchite	$Zn[SO_4] \cdot 6H_2O$

Goslarite family

Goslarite	$Zn[SO_4] \cdot 7H_2O$
Zinc-melanterite	$(Zn,Cu,Fe^{2+})[SO_4] \cdot 7H_2O$
*Changoite (x = 1)	$Na_2Zn[SO_4]_2 \cdot 4H_2O$
*Lishizhenite (x = 1)	$ZnFe^{3+}_2[SO_4]_4 \cdot 14H_2O$
Dietrichite (x = 1)	$ZnAl_2[SO_4]_4 \cdot 22H_2O$

3.2.5.3.4.1.3. Basic   3.2.5.3.4.1.3.1. Hedrates

Zincocopiapite (x = 1,1(6))	$ZnFe^{3+}_4(OH)_2[SO_4]_6 \cdot 18H_2O$
*Chaidamuite (x = 1,25)	$ZnFe^{3+}(OH)[SO_4]_2 \cdot 4H_2O$

Ktenasite family
Serpierite $Ca(Cu,Zn)_4(OH)_6[SO_4]_2 \cdot 3H_2O$
*Orthoserpierite $Ca(Cu,Zn)_4(OH)_6[SO_4]_2 \cdot 3H_2O$
Ktenasite $(Cu,Zn)_5(OH)_6[SO_4]_2 \cdot 6H_2O$
*Christelite $Zn_3Cu_2(OH)_6[SO_4]_2 \cdot 4H_2O$
Schulenbergite family
Schulenbergite $(x = 3,5)$ $(Cu,Zn)_7(OH)_{10}[(SO_4),(CO_3)]_2 \cdot 3H_2O$
Ramsbeckite $(x = 3,75)$ $(Cu,Zn)_{15}(OH)_{22}[SO_4]_4 \cdot 6H_2O$
*Osakaite $Zn_4(OH)_6[SO_4] \cdot 5H_2O$
*Lanshtainite$(x = 4)$ $Zn_4(OH)_6[SO_4] \cdot 3H_2O$
Namuwite $(x = 4)$ $(Zn,Cu)_4(OH)_6[SO_4] \cdot 4H_2O$
Zincaluminite $(x = 5,5)$ $(Zn,Al)_9(OH)_{18}[SO_4]_2 \cdot nH_2O$
Glaucocerinite $(x = 6,(3))$ $Zn_{1-x}Al_x(OH)_2[SO_4]_{x/2} \cdot nH_2O$ $(x<0.5, n>3x/2)$
*Natroglaucocerinite $[Zn_{8-x}Al_x(OH)_{16}][(SO_4)_{x/2+y/2}Na_y(H_2O)_6]$
*Zincowoodvardite $[Zn_{1-x}Al_x(OH)_2][(SO_4)_{x/2}(H_2O)_n]$ $(x<0.5, n>3x/2)$
*Kyrgyzstanite$(x = 7.0)$ $ZnAl_4(OH)_{12}[SO_4] \cdot 3H_2O$
Torreyite family $(x = 6,5)$
Torreyite $(Mg,Mn)_9Zn_4(OH)_{22}[SO_4]_2 \cdot 8H_2O$
Lawsonbauerite $(Mn,Mg)_9Zn_4(OH)_{22}[SO_4]_2 \cdot 8H_2O$
Mooreite family $(x = 7,5)$
Mooreite $Mg_9Zn_4Mn_2(OH)_{26}[SO_4]_2 \cdot 8H_2O$
*3.2.5.3.4.2. Sulfato-halogenides   *3.2.5.3.4.2.1. Neutral
*Belousovite $(x = 1)$ $KZnCl[SO_4]$
                                                         *3.2.5.3.4.2.2. Basic
*Gordaite $(x = 4,5)$ $NaZn_4(OH)_6Cl[SO_4] \cdot 6H_2O$
*Thèrésemagnanite $(x = 4,5)$ $NaCo_4(OH)_6Cl[SO_4] \cdot 6H_2O$
*Guarinoite $(x = 6)$ $(Zn,Co,Ni)_6(OH,Cl)_{10}[SO_4] \cdot 5H_2O$

*3.2.5.3.5. Sulfates of Tl^+
*Markhininite $Tl^+Bi^{3+}[SO_4]_2$

3.2.5.3.6. Sulfates of Tl^{3+} 3.2.5.3.6.1. Hydrates (neutral)
Monsmedite = voltaite with Tl

3.2.5.3.7. Sulfates of Pb^{2+} 3.2.5.3.7.1. Neutral
Anglesite   (comp. with barite (gr.)) $Pb[SO_4]$
Palmierite $(K,Na)_2Pb[SO_4]_2$

3.2.5.3.7.2. Basic and oxido-sulfates, sulfato-halogenides
Caracolite $(x = 1,1(6))$ $Na_3Pb_2Cl[SO_4]_3$
Wherryite $(x = 1,5)$ $Pb_7Cu_2(OH)_2[SO_4]_4[SiO_4]_2$
Linarite family $(x = 2)$
Lanarkite $Pb_2O[SO_4]$
Grandreefite $Pb_2F_2[SO_4]$
Linarite $PbCu^{2+}(OH)_2[SO_4]$
Plumbojarosite $(x = 2,5)$ $PbFe^{3+}_6(OH)_{12}[SO_4]_4$
Osarizawaite $(x = 2,5)$ $PbCuAl_2(OH)_6[SO_4]_2$
Chenite $(x = 2,5)$ $Pb_4Cu(OH)_6[SO_4]_2$
*Beaverite-(Zn) $(x = 2,5)$ $PbFe^{3+}_2Zn(OH)_6[SO_4]_2$
Beaverite $(x = 2,75)$ $Pb(Fe,Cu,Al)_3(OH, \cdot H_2O)_6[SO_4]_2$

*Krivovichevite (x = 4,5) $Pb_3AlOH)_7[SO_4]$
Elyite (x=5) $Pb_4Cu(OH)_8[SO_4]$
Pseudograndreefite (x = 6) $Pb_6F_{10}[SO_4]$
Sundiusite (x = 10) $Pb_{10}O_8Cl_2[SO_4]$
                                          *3.2.5.3.6.2.1. Hydrates
*Symesite $Pb_{10}O_7Cl_4[SO_4]·H_2O$

*3.2.5.3.7.3. Oxido-thiosulphates   *3.2.5.3.7.3.1. Basic
*Sidpietersite $Pb^{2+}_4(S^{6+}O_3S^{2-})O_2(OH)_2$
                                     *3.2.5.3.7.3.2. Hydrates
*Steverustite $Pb^{2+}_5Cu^+(S^{6+}O_3S^{2-})_3(OH)_5·2H_2O$

*3.2.5.3.7.4. Sulfato-arsenates   *3.2.5.3.7.4.1. Hydrates (basic)
*Mallestigite (x = 2,75) $Pb_3Sb(OH)_6[(SO_4),(AsO_4)]·3H_2O$

3.2.5.3.8.Sulfates of Ge^{4+} 3.2.5.3.8.1. Basic
Itoite (x = 2,5) $Pb_3^{(6)}Ge^{4+}(OH)_2O_2[SO_4]_2$
                                          3.2.5.3.8.2. Hydrates (basic)

 Fleischerite family (x= 2,5)
 Schaurteite $Ca_3^{(6)}Ge^{4+}(OH)_6[SO_4]_2·3H_2O$
 Fleischerite $Pb_3Ge^{4+}(OH)_6[SO_4]_2·3H_2O$

*3.2.5.3.8.2.1. Sulfato-carbonates
*Carraraite $Ca_3Ge(OH)_6[SO_4][CO_3]·12H_2O$

3.2.5.3.9. Sulfates of As^{3+}, Sb^{3+}, Bi^{3+} *3.2.5.9.1. Neutral
*Markhininite $Tl^+Bi^{3+}[SO_4]_2$
                                     *3.2.5.3.9.2. Basic and hydrates
*Riomarinaite (x = 1.5) $Bi^{3+}(OH)[SO_4]·H_2O$

*3.2.5.3.9.3. Oxido-sulfates   *3.2.5.3.9.3.1. Neutral
*Coquandite $Sb^{3+}_6O_8[SO_4]·H_2O$
                                     *3.2.5.3.9.3.1. Basic and hydrates
Klebelsbergite (x = 6) $Sb^{3+}_4O_4(OH)_2[SO_4]$
*Tavagnascoite $Bi_4O_4(OH)_2[SO_4]$
*Cannonite $Bi^{3+}_2O(OH)_2[SO_4]$
Peretaite (x = 3,5) $CaSb^{3+}_4O_4(OH)_2[SO_4]_2·2H_2O$

3.2.6. *Class:* Sulfites
3.2.6.1. Subclass: Sulfites of cations with low FC
3.2.6.1.1. Sulfites of *s-*, *d_s-* and *p_s-* cations
                                          3.2.6.1.1.1. Hydrates (neutral)
Hannebachite $Ca[SO_3]·0,5H_2O$
*Gravegliaite $Mn^{2+}[SO_3]·3H_2O$

*3.2.6.1.2. Sulfito-sulfates   *3.2.6.1.2.1. Hydrates
*Orschallite $Ca_3[SO_3]_2[SO_4]·12H_2O$
*Hielscherite $Ca_3Si(OH)_6[SO_3][SO_4]·11H_2O$

3.2.6.2. Subclass: Sulfites of chalcophylic cations   3.2.6.2.1. Neutral

Scotlandite $Pb[SO_3]$

3.2.6a. *Class:* Selenates
3.2.6a.1. Subclass: Selenates of chalcophylic cations (Pb $^{2+}$)
3.2.6a.1.1. Proper selenates
                                        3.2.6a.1.1.1. Neutral
Kerstenite $Pb[SeO_4]$ (?)
3.2.6a.1.2. Selnato-sulfates (1 : 1)   3.2.6a.1.2.1. Neutral
Olsacherite $Pb_2[SeO_4][SO_4]$

*3.2.6a.1.3. Selenato-iodates   3.2.6a.1.3.1. Proper
*3.2.6a.1.3.1.1. Complex   3.2.6a.1.3.1.1.1. Hydrates
*Carlosruizite $K_6Na_{10}Mg_{10}[IO_3]_{12}[SeO_4]_{12}·12H_2O$

3.2.6b. *Class:* Selenites
3.2.6b.1. Subclass: Selenites of catiobs with low FC
3.2.6b.1.1. Selenites of *s-, d$_s$-* and *p$_s$-* cations 3.2.6b.1.1.1. Hydrates (neutral)
*Nestolaite $Ca[SeO_3]·H_2O$
Mandarinoite $Fe^{3+}_2[Se^{4+}O_3]_3·6H_2O$
*Alfredopetrovite $Al^{3+}_2[Se^{4+}O_3]_3·6H_2O$
  Ahlfeldite group
  Ahlfeldite $Ni[Se^{4+}O_3]·2H_2O$
  Cobaltomenite $Co[Se^{4+}O_3]·2H_2O$

3.2.6b.2. Subclass : Selenites of chalcophylic elements
3.2.6b.2.1. Selenites of Cu^{2+} 3.2.6b.2.1.1. Hydrates (neutral)
Chalcomenite family
Chalcomenite $Cu[Se^{4+}O_3]·2H_2O$

*3.2.6б.2.1.2. Selenito-oxides   *3.2.6б.2.1.2.1. Basic and hydrates
*Favreauite $PbBiCu_6O_4(OH)[Se^{4+}O_3]_4·H_2O$

*3.2.6б.2.1.3. Selenito-sulfates   *3.2.6б.2.1.3.1. Basic and hydrates
*Pauladamsite $Cu_4(OH)_4[SeO_3][SO_4]·2H_2O$

*3.2.6б.2.1.4. Selenito-oxido-halogenides
*Ilinskite $NaCu_5O_2[Se^{4+}O_3]_2Cl_3$
*Georgbokiite $Cu_5O_2[SeO_3]_2Cl_2$
*Parageorgbokiite $β-Cu_5O_2[SeO_3]_2Cl_2$
*Nicksobolevite $Cu_7O_2[SeO_3]_2Cl_6$
*Burnsite $KCdCu_7O_2[SeO_3]_2Cl_9$
*Allochalcoselite $Cu^+Cu^{2+}_5PbO_2[SeO_3]_2Cl_5$
*Francisite $Cu^{2+}_3BiO_2[SeO_3]_2Cl$

3.2.6b.2.2. Selenites of Zn
*Zincomenite $Zn[SeO_3]$
                        3.2.6b.2.2.1. Basic selenites and selenito-halogenides
Sophiite = sofiite $Zn_2Cl_2[Se^{4+}O_3]$

*3.2.6б.2.2.2. Oxido-selenito-halogenides

*Chloromenite $Cu_9O_2[Se^{4+}O_3]_4Cl_6$
*Prewittite $KPb_{1.5}ZnCu^{2+}_6O_2[Se^{4+}O_3]_2Cl_{10}$

3.2.6b.2.3. Selenites of Pb^{2+}
3.2.6b.2.3.1. Proper selenites   3.2.6b.3.2.3.1.1. Neutral
Molybdomenite $Pb[Se^{4+}O_3]$

                      *3.2.6б.2.3.2. Oxido-selenites
*Plumboselite $Pb_3O_2[Se^{4+}O_3]$
                      3.2.6b.2.3.3. Selenito-selenates (1 :1)
                      3.2.6b.2.3.2.1. Basic
Schmiederite $Pb_2Cu^{2+}_2(OH)_4[Se^{4+}O_3][Se^{6+}O_4]$

*3.2.6б.2.3.4. Selenito-sulfates (1:1)   *3.2.6б.2.3.4.1. Основные
*Munakataite $Pb_2Cu^{2+}_2(OH)_4[Se^{4+}O_3][S^{6+}O_4]$

*3.2.6б.2.3.5. Selenito-chlorides   *3.2.6б.2.3.5.1. Средние
*Sarrabusite $Pb_5Cu^{2+}Cl_4[Se^{4+}O_3]_4$
                      *3.2.6b.2.3.5.2. Basic
*Unnamed $Pb_4Cu^{2+}Cl_3[Se^{4+}O_3]_3(OH)$
                      *3.2.6б.2.3.5.3. Hydrates
*Orlandite $Pb_3Cl_4[Se^{4+}O_3]·H_2O$

3.2. 7. *Class:* Chromates
3.2.7.1. Subclass: Chromates of cations with low FC
3.2.7.1.1. Chromates of *s-, d_s-* and *p_s-* cations
3.2.7.1.1.1. Proper chromates
3.2.7.1.1.1.1. Trichromates (bichromates)   3.2.7.1.1.1.1.1. Neutral
Lopezite $K_2[Cr_2O_7]$

3.2.7.1.1.1.2. Tetrachromates   3.2.7.1.1.2.1. Neutral
Chromatite $Ca[CrO_4]$
Hashemite (compare with barite (group)) $Ba[(Cr,S)O_4]$
Tarapacáite $K_2[CrO_4]$
                      *3.2.7.1.1.2.1.1. Hydrates
*Unnamed $Ca[CrO_4]·2H_2O$

3.2.7.1.1.2. Tetrachromato-iodates (0,5:1)   3.2.7.1.1.2.1. Neutral hydrates
Dietzeite $Ca_2[CrO_4][IO_3]_2·H_2O$

3.2.7.2. Subclass: Chromates of chalcophylic elements
3.2.7.2.1. Tetrachromates of Pb^{2+} 3.2.7.2.1.1. Neutral
Crocoite $Pb[CrO_4]$

*3.2.7.2.1.2. Oxido-tetrachromates
*Reynoldsite $Pb^{2+}_2Mn^{4+}_2O_5[CrO_4]$

*3.2.7.2.1.3. Tetrachromato-silicates   *3.2.7.2.1.3.1. Basic hydrates
*Maquartite $Pb^{2+}_3Cu^{2+}(OH)_4[CrO_4][SiO_3]·2H_2O$

*3.2.7.2.1.4. Тетрахроматы-галогениды
*Georgerobinsonite $Pb^{2+}_4(OH)_2FCl[CrO_4]_2$

3.2.7.2.2. Tetrachromates Pb^{4+} 3.2.7.2.2.1. Oxido-and tetrachromato-halogenides
Phoenicochroite (x = 4) $Pb_2O[CrO_4]$
Santanaite (x = 13) $Pb^{2+}_9Pb^{4+}_2O_{12}[CrO_4]$
Yedlinite $Pb^{2+}_2 Pb^{4+}_4(OH)_8O_2Cl_6[Cr^{6+}O_4]$

*3.2.7.2.3. Tetrachromates Hg^{2+}
*Edoylerite $Hg^{2+}_3[Cr^{6+}O_4]S_2$

*3.2.7.2.4. Tetrachromates Hg^+ и Hg^{2+} *3.2.7.2.4.1. Oxido-tetrachromates
*Wattersite $Hg^+_4Hg^{2+}Cr^{6+}O_6 \rightarrow Hg^+_4Hg^{2+}O_2[CrO_4]$

*3.2.7.2.5. Tetrachromates Bi^{3+}
*3.2.7.2.5.1. Oxido-tetrachromates
*Chrombismite $Bi^{3+}_{16}O_{23}[Cr^{6+}O_4]$

3.2.8. Class: Nitrates
3.2.8.1. Subclass: Nitrates of cations with low FC
3.2.8.1.1. Nitrates of s-, d_s- and p_s-cations
                                          3.2.8.1.1.1. Neutral
Nitratine $Na[NO_3]$
Nitrobarite $Ba[NO_3]_2$
Niter (saltpeter) $K[NO_3]$
*Nitrammite $(NH_4)[NO_3]$
*Gwihabaite $(NH_4,K)[NO_3]$
                                          3.2.8.1.1.2. Basic hydrates
Sveite $KAl_7(OH)_{16}Cl_2[NO_3]_4 \cdot 8H_2O$
                                          3.2.8.1.1.2.1.1. Neutral
Nitromagnesite $Mg[NO_3]_2 \cdot 6H_2O$
Nitrocalcite $Ca[NO_3]_2 \cdot 4H_2O$

3.2.8.2. Subclass: Nitrates of chalcophylic elements
3.2.8.2.1. Nitrates of Cu^{2+}
                                          3.2.8.2.1.1. Neutral
*Shilovite $Cu(NH_3)_4[NO_3]_2$
                                          3.2.8.2.1.2. Basic
3.2.8.2.1.1. Basic
Gerhardtite $Cu_2(OH)_3[NO_3]$
*Rouaite $Cu_2(OH)_3[NO_3]$
                                          3.2.8.2.1.3. Hydrates (basic)
Buttgenbachite $Cu_{19}(OH)_{32}Cl_4[NO_3]_2 \cdot 2H_2O$
Likasite $Cu_3(OH)_5 [NO_3] \cdot 2H_2O$

3.2.8a. *Class:* Iodates
3.2.8a.1. Subclass: Iodates of cations with low FC
3.2.8a.1.1. Iodates of s-, d_s- and p_s- cations
                                          3.2.8a.1.1.1. Neutral
Lautarite $Ca[IO_3]_2$

	3.2.8a.1.1.2. Hydrates (neutral)
Brüggenite	$Ca[IO_3]_2 \cdot H_2O$

*3.2.8a.1.2. Iodato-chromates	*3.2.8a.1.2.1. Hydrates
*Georgeericksenite	$Na_6CaMg[IO_3]_6[CrO_4]_2 \cdot (H_2O)_{12}$

3.2.8a.2. Subclass: Iodates of chalcophylic elements

3.2.8a.2.1. Iodates of Cu^{2+}	3.2.8a.2.1.1. Basic
Salesite	$Cu(OH)[IO_3]$
	3.2.8a.2.1.2. Hydrates
Bellingerite	$Cu_3[IO_3]_6 \cdot 2H_2O$

3.2.8a.2.2. Iodates of Pb^{2+}	3.2.8a.2.2.1. Iodato-halogenides
Seeligerite	$Pb_3OCl_3[IO_3]$

*3.2.8b. *Class:* Iodites
*3.2.8b.1. Subclass: Iodites of chalcophylic elements
*3.2.8b.1.1. Iodites Pb^{2+}
*3.2.8b.1.1.1. Hydroxido-oxido-iodito-chalogenides

Schwartzembergite	$Pb^{2+}_5I^{3+}O_6H_2Cl_3 \rightarrow Pb^{2+}_5(OH)_2O_2Cl^+_3[I^{3+}O_2]$

3.2.8c. *Class:* Rhodonates (tiocyanates)

4. TYPE: MINERALS WITH PRINCIPAL COVALENT-IONIC AND IONIC BOND – HALOGEN COMPAUNDS: HALOGENIDES (ISODESMICAL) → HALOGENSALTS (ANISODESMICAL)

4.1. SUBTYPE: HALOGENIDES (ISODESMICAL)
4.1.1.*Class*: Fluorides
4.1.1.1. Fluorides of *s*-, *d$_s$*- and *p$_s$*- cations
4.1.1.1.1. Fluorides of *s*-, *d$_s$*- and *p$_s$*- cations without Li and Be
4.1.1.1.1.1. Proper fluorides

4.1.1.1.1.1.1. Simple	4.1.1.1.1.1.1.1. Neutral
Sellaite	MgF_2
Fluorite group	
Fluorite	CaF_2
Frankdicksonite	BaF_2
*Strontiofluorite	SrF_2
Villiaumite group	
Villiaumite	NaF
Carobbiite	KF
*Oskarssonite	AlF_3
	*4.1.1.1.1.1.1.2. Hydrates
*Rosenbergite	$AlF[F_{0,5}(H_2O)_{0,5}]_4 \cdot H_2O$
4.1.1.1.1.1.2. Complex	4.1.1.1.1.1.2.1. Neutral
Neighborite	$NaMgF_3$
*Yakobssonite	$CaAlF_5$
*Karasugite	$SrCa[Al(F,OH)_7]$
*Calcioaravaipaite	$PbCa_2AlF_9$

4.1.1.1.1.2. Fluorido-hexafluoraluminates	4.1.1.1.1.2.1. Neutral

Weberite family

Weberite	$Na_2MgAlF_7 \rightarrow Na_2\{MgF[AlF_6]\}^{\infty 3}$
Usovite	$Ba_2CaMgAl_2F_{14} \rightarrow Ba_2\{CaMgF_2[AlF_6]_2\}^{\infty 3}$
*Bøgvadite	$Na_2SrBa_2Al_4F_{20} \rightarrow Na_2\{SrBa_2AlF_2[AlF_6]_3\}$
*Coulsellite	$CaNa_3Mg_3AlF_{14} \rightarrow Na_3\{CaMg_3F_8[AlF_6]\}$

4.1.1.1.1.2.2. Hydrates

Carlhintzeite	$Ca_2AlF_7 \cdot H_2O \rightarrow Ca\{CaF[AlF_6]\}^{\infty 3} \cdot H_2O$
*Leonardsenite	$MgAlF_5 \cdot 2H_2O$

*4.1.1.1.1.3. Fluorido-chlorides

*Rorisite	$CaFCl$
*Zhangpeishanite	$BaFCl$

4.1.1.1.2. Fluorides of Li
4.1.1.1.2.1. Simple   4.1.1.1.2.1.1. Neutral

Griceite	LiF

*4.1.1.1.2.2. Fluorido-hexafluoraluminates Li
*4.1.1.1.2.2.1. Complex   *4.1.1.1.2.2.1.1. Neutral

*Simmonsite	$Na_2Li[AlF_6]$

4.1.1.2. Fluorides of f- cations
4.1.1.2.1. Simple   4.1.1.2.1.1. Neutral
  Fluocerite series

Fluocerite-(Ce)	CeF_3
Fluocerite-(La)	LaF_3
*Waimirite-(Y)	YF_3

4.1.1.2.2. Complex   4.1.1.2.2.1. Neutral

Tveitite-(Y)	$(Y,Na)_6(Ca,Na,REE)_{12}(Ca,Na)F_{42}$
*Zajacite-(Ce) = gagarinite-(Ce)	$NaCaCeF_6$
Gagarinite-(Y)	$NaCaYF_6$
*Polezhaevaite-(Ce)	$NaSrCeF_6$

*4.1.1.2.3. Oxido-fluorides

*Håleniusite-(La)	$(La,Ce)OF$

*4.1.1.3. Fluorides of chalcophilic cations
*4.1.1.3.1. Fluorides of Ib-elements
*4.1.1.3.1.1. Fluorides of Cu^{2+}

*Khaidarkanite	$Cu_4Al_3(OH)_{14}F_3 \cdot 2H_2O$

*4.1.1.3 2. Fluorides of IVa-elements
*4.1.1.3.2.1. Fluorides Pb^{2+}
*4.1.1.3.2.1.1. Simple

*Fluorocronite	PbF_2

*4.1.1.3.2.1.1.1. Основные

*Artroeite	$PbAl(OH)_2F_3$

*4.1.1.3.2.1.2. Fluorido-chlorides

*Laurelite	$Pb_7F_{12}Cl_2$

*4.1.1.3.2.1.2.1. Fluorido-oxido-chlorides
*Rumseyite $[Pb_2OF]Cl$

4.1.2. *Class:* Chlorides and bromides
4.1.2.1. Chlorides of cations with low FC
4.1.2.1.1. Chlorides of *s-*, *d_s*- and *p_s*- cations
4.1.2.1.1.1. Simple
                                             4.1.2.1.1.1.1. Neutral
Molysite $FeCl_3$
Lawrencite $FeCl_2$
Scacchite $MnCl_2$
*Chloromagnesite (disputable) $MgCl_2$
*Chlorocalcite $KCaCl_3$
   Halite group
   Halite $NaCl$
   Sylvite KCl
Salammoniac NH_4Cl
                                                4.1.2.1.1.1.2. Basic
Kempite $Mn_2(OH)_3Cl$
*Hibbingite $\gamma\text{-}Fe^{2+}_2(OH)_3Cl$
                                                4.1.2.1.1.1.3. Hydrates
Hydrophilite = antarcticite or sinjarite $CaCl_2 \cdot 6H_2O$
*Ghiaraite $CaCl_2 \cdot 4H_2O$
                                             4.1.2.1.1.1.3.1. Basic
Cadwaladerite (x = 1,5) $Al(OH)_2Cl \cdot 4H_2O$
Korshunovskite (x = 2) $Mg_2(OH)_3Cl \cdot 3,5\text{-}4H_2O$
*Nepskoeite (x = 4) $Mg_4(OH)_7Cl \cdot 6H_2O$
 Chloraluminite family
 Chloraluminite $AlCl_3 \cdot 6H_2O \rightarrow [Al(H_2O)_6]Cl_3$
Roкühnite $FeCl_2 \cdot 2H_2O \rightarrow [^{(6)}FeCl_2(H_2O)_2]^\infty$
   Bischofite group
   Bischofite $MgCl_2 \cdot 6H_2O \rightarrow [Mg(H_2O)_6]Cl_2$
   Nickelbischofite $NiCl_2 \cdot 6H_2O \rightarrow [Ni(H_2O)_6]Cl_2$
Sinjarite $CaCl_2 \cdot 2H_2O \rightarrow [Ca(H_2O)_2Cl_2]^{\infty 2}$
Hydrohalite $NaCl \cdot 2H_2O$

4.1.2.1.1.2. Complex   4.1.2.1.1.2.1.Basic
*Koenenite $Na_4Mg_9Al_4(OH)_{22}Cl_{12}$
*Kuliginite (x = 2) $Fe^{2+}_3Mg(OH)_6Cl_2$
   Chlormagaluminite group   4.1.2.1.1.2.1.1. Hydrates
   Chlormagaluminite $(Mg,Fe^{2+})_4Al_2(OH)_{12}Cl_2 \cdot 2H_2O$
   Hydrocalumite $Ca_2Al(OH)_6(Cl,CO_3,OH) \cdot 2H_2O$
*Iowaite (x = 4) $Mg_6Fe^{3+}_2(OH)_{16}Cl_2 \cdot 4H_2O$
*Woodallite (x = 4) $Mg_6Cr_2(OH)_{16}Cl_2 \cdot 4H_2O$
*Kopeyskite $(NH_4)_2(Fe,Al,Ca,Mg)(Cl,OH)_5 \cdot H_2O$
                                             4.1.2.1.1.2.1.1.1. Neutral
Tachyhydrite $CaMg_2Cl_6 \cdot 12H_2O$
Carnallite $KMgCl_3 \cdot 6H_2O \rightarrow$
                                             $K[Mg(H_2O)_6]Cl_3$

4.1.2.2. Chlorides, bromides, iodides of chalcophylic elements
4.1.2.2.1. Chlorides, bromides, iodides of **Ib**-elements
4.1.2.2.1.1. Chlorides, bromides, iodides of Cu^+, Ag^+, Hg^+

4.1.2.2.1.1.1. Simple	4.1.2.2.1.1.1.1. Neutral
Nantokite group	
Nantokite	$CuCl$
Chlorargyrite group	
Chlorargyrite	$AgCl$
Bromargyrite	$AgBr$
Calomel	$HgCl$
*Kuzminite	$Hg_2(Br,Cl)_2$
4.1.2.2.1.1.2. Complex	4.1.2.2.1.1.2.1. Basic
Bideauxite (x = 1)	$Pb_2Ag^+(F,OH)_2Cl_3$

4.1.2.2.1.2. Chlorides of Cu^{2+}
4.1.2.2.1.2.1. Simple

	4.1.2.2.1.2.2. Neutral
Tolbachite	$CuCl_2$
*Sanguite	$KCuCl_3$
	4.1.2.2.1.2.1.2. Basic
*Belloite	$Cu(OH)Cl$
Atacamite family	
Atacamite	$Cu_2(OH)_3Cl$
*Clinoatacamite	$Cu_2(OH)_3Cl$
Botallackite	$Cu_2(OH)_3Cl$
*Fejerite	$Cu_4(OH)_6ClF$
Claringbullite	$Cu_4(OH)_6ClF$
	4.1.2.2.1.2.1.3. Hydrates
	4.1.2.2.1.2.1.3.1. Basic
*Bobkingite	$Cu_5(OH)_8Cl_2(H_2O)_2$
	4.1.2.2.1.2.1.3.1. Neutral
Eriochalcite	$CuCl_2 \cdot 2H_2O \rightarrow [CuCl_2(H_2O)_2]$
4.1.2.2.1.1.2. Complex	*4.1.2.2.1.1.2.1. Neutral
*Ammineite	$[Cu(NH_3)_2]Cl_2$
*Romanorlovite	$K_8Cu_6(OH)_3Cl_{17}$
	4.1.2.2.1.1.2.2. Основные
Сем. **паратакамита**(x = 2)	
Paratacamite	$Cu_3(Cu,Zn)(OH)_6Cl_2$
*Рагаатакамит-(Mg)	$Cu_3(Mg,Cu)(OH)_6Cl_2$
*Рагаатакамит-(Ni)	$Cu_3(Ni,Cu)(OH)_6Cl_2$
*Haydeeite	$Cu_3Mg(OH)_6Cl_2$
*Tondiite	$Cu_3Mg(OH)_6Cl_2$
*Gerbertsmithite	$Cu_3Zn(OH)_6Cl_2$
*Kapellasite	$Cu_3Zn(OH)_6Cl_2$
*Gillardite	$Cu_3Ni(OH)_6Cl_2$
*Leverettite	$Cu_3Co(OH)_6Cl_2$
Boleite family	
Boleite (x = 0,9)	$KPb^{2+}_{26}Ag^+_9Cu^{2+}_{24}(OH)_{48}Cl_{62}$

Pseudoboleite $Pb_{31}Cu^{2+}_{24}(OH)_{48}Cl_{62}$
Percylite = boleite + psevdoboleite (?) $PbCu(OH)_2Cl_2$ (?)
Diaboleite (x = 1,5) $\{Pb_2Cu(OH)_4Cl_2\}^{\infty 2}$
                                                      *4.1.2.2.1.1.2.2.1. Hydrates
Cumengite (x = 1) $Pb_{21}Cu_{20}(OH)_{40}Cl_{42} \cdot 6H_2O$
*Feodosiyite $Cu_{11}Mg_2(OH)_8Cl_{18} \cdot 16H_2O$
*Centennialite $CaCu_3(OH)_6Cl_2 \cdot nH_2O$ (n = 0.7)
*Avdoninite $K_2Cu_5(OH)_4Cl_8 \cdot 2H_2O$
*Dioskouriite $CaCu_4(OH)_4Cl_6 \cdot 4H_2O$
*Chrysothallite $K_6Cu_6Tl^{3+}(OH)_4Cl_{17} \cdot H_2O$
*Chanabayaite $Cu(N_3C_2H_2)(NH_3)Cl \cdot 0.25H_2O$

*4.1.2.2.1.1.3. Oxido-chlorides
*4.1.2.2.1.1.3.1. Basic
Chloroxiphite (x = 2) $Pb_3Cu(OH)_2O_2Cl_2$
*Fuettererite $Pb_3Cu^{2+}_6Te^{6+}O_6(OH)_7Cl_5$

4.1.2.2.2. Chlorides of **IIb**-elements
4.1.2.2.2.1. Chlorides of Zn   *4.1.2.2.2.1.1. Средние
*Flinteite K_2ZnCl_4
*Mellizinkalite $K_3Zn_2Cl_7$
*Amminite $[Zn(NH_3)_2]Cl_2$

                                                      4.1.2.2.2.1.2. Hydrates (basic)
*Cryobostryxite $ZnKCl_3 \cdot 2H_2O$
Simonkolleite $Zn_5(OH)_8Cl_2 \cdot H_2O \rightarrow [^{(4,6)}Zn_5(OH)_8]^{\infty 2}Cl_2\}^{\infty 2} \cdot H_2O$

*4.1.2.2.2.2. Chlorides of Hg
*4.1.2.2.2.2.1. Oxido-chlorides of Hg
*Hanawaltite $Hg^+_6Hg^{2+}Cl_2O_3$

*4.1.2.2.3. Хлориды **IIIa**-элементов
*4.1.2.2.3.1. Хлориды Tl   *4.1.2.2.3.1.1. Neutral
                                                      *4.1.2.2.3.1.1.1. Simple
*Lafossaite   TlCl
                                                      *4.1.2.2.3.1.1.2. Complex
*Steropesite Tl_3BiCl_6
*Hephaistosite $TlPb_2Cl_5$

4.1.2.2.4. Chlorides of **IVa**-elements
4.1.2.2.4.1. Chlorides of Pb^{2+}
4.1.2.2.4.1.1. Simple   4.1.2.2.4.1.1.1. Neutral
Cotunnite family
Matlokite   PbClF
Cotunnite $PbCl_2$
                                                      4.1.2.2.4.1.1.2. Basic and hydrates
Penfieldite (x = 0,(6)) $Pb_2(OH)Cl_3$
Fiedlerite (x = 0,75) $Pb_3F(OH)Cl_4 \cdot H_2O$
Laurionite family (x = 1)
Laurionite $Pb(OH)Cl$

Paralaurionite   Pb(OH)Cl

4.1.2.2.3.1.2. Complex   4.1.2.2.3.1.2.1. Neutral
*Challacolloite KPb_2Cl_5
Pseudocotunnite K_2PbCl_4 (?)
*Brontesite $(NH_4)_3PbCl_5$
                                        *4.1.2.2.4.1.2.2. Basic
*4.1.2.2.5. Oxido-chlorides Pb
*4.1.2.2.5.1. Simple   *4.1.2.2.5.1.1. Basic
*Blixite (x = 2)   *$Pb_8O_5(OH)_2Cl_4$

*4.1.2.2.5.2. Complex   *4.1.2.2.5.2.1. Basic
*Rickturneite $Pb_7O_4[Mg(OH)_4](OH)Cl_3$
*Hereroite $[Pb_{32}(O,\square)_{21}][AsO_4]_2[(Si,As,V,Mo)O_4]_2Cl_{10}$
                                        *4.1.2.2.5.2.1.1. Hydrates
*Vladkrivovichevite $[Pb_{32}O_{18}][Pb_4Mn_2O][BO_3]_8Cl_{14}\cdot2H_2O$

*4.1.2.2.6. Fluorides, chlorides IVa-elements
*4.1.2.2.6.1. Fluorides, chlorides Bi^{3+}
*4.1.2.2.6.1.1. Simple
*4.1.2.2.6.1.2. Comhlex
*Argesite $(NH_4)_7Bi^{3+}_3Cl_{16}$

4.1.2a. *Class*: Iodides
*Unnamed RhI_3
  Marshite series
  Marshite   CuI
  *Marshite cuprous   CuI
  Miersite   (Ag,Cu)I
Iodargyrite   AgI
*Iodargyrite 2H   AgI
Tocornalite   (Ag,Hg)I
*Coccinite HgI_2
*Moschelite Hg_2I_2

4.2. SUBTYPE: HALOGENOSALTS (ANISODESMICAL) (WITH
HEXACYANOFERRATES AND HEXATIOCYANATES, RHODONIDES)
*4.2a. QUASISUBTYPE*HALOGENOSALTS WITH d-CATION-COMPLEXFORMERS*
4.2a.1. *Class*: Chloroferrites and chlorocuprites (only s- cations and NH^+_4)
4.2a.1.1. Chlorooxidopolycuprites   4.2a.1.1.1. Neutral
Ponomarevite $K_4[Cu^{2+}_4OCl_{10}]$

4.2a.1.2. Hexachlorocuprites   4.2a.1.2.1. Neutral chloroaquacuprites*
Mitscherlichite $K_2[Cu^{2+}Cl_4(H_2O)_2]$

4.2a.1.3. Rhodanidocobaltites
4.2a.1.3.1. Tetrarhodanidocobaltites II

---

* Partical "aqua" in complex anion´s name of halogenosalt means, that at least one of its ligands is $\cdot H_2O$;
particles "oxo" and "hydroxo" means, that as ligands with halogen-ions appear O^{2-} - or OH^- -ions
respectively.

4.2a.1.3.1.1. Hydrates (neutral)

Julienite $Na_2[Co(SCN)_4] \cdot 8H_2O$

4.2a.2. *Class*: Hexachloroferrates and hexachloromanganates (only *s*- cations)
4.2a.2.1. Hexachloroferrates II and hexachloromanganates II

4.2a.2.1.1. Neutral

Rinneite group
Rinneite $K_3Na[Fe^{2+}Cl_6]$
*Saltonseaite $K_3Na[Mn^{2+}Cl_6]$
Chlormanganokalite $K_4[Mn^{2+}Cl_6]$
Douglasite $K_2[Fe^{2+}Cl_4(H_2O)_2]$

4.2a.2.2. Hexachloroferrates III
4.2a.2.2.1. Chloroaquaferrates III (neutral)

Kremersite group
Erythrosiderite $K_2[Fe^{3+}Cl_5(H_2O)]$
Kremersite $(NH_4,K)_2[Fe^{3+}Cl_5(H_2O)]$

4.2b. Quasisubtype*: Halogenosalts with *p*- anion-complexformers

4.2b.1. ***Class:*** Fluoroaluminates (only *s*- cations)
4.2b.1.1. Hexafluoropolyaluminates   4.2b.1.1.1. Neutral (*s*-cations)
Chiolite $Na_5[Al_3F_{14}]^{\infty 2}$
*Ralstonite $NaMg[Al_3F_8(OH)_4] \cdot H_2O$
Prosopite $Ca[Al_2F_4(OH)_4]$

Jarlite group
Calcjarlite $Na_2(Ca,Sr,Na,\square)_{14}Al_{12}Mg_2(F,OH)_{64}(OH)_4$
Jarlite $Na_2(Sr,Na)_{14}Mg_2Al_{12}F_{64}(OH,\cdot H_2O)_4$
*Jørgensenite $Na_2(Sr,Ba)_{14}Na_2Al_{12}F_{64}(OH,F)_4$

4.2b.1.2. Hexafluoroaluminates
4.2b.1.2.1. Hexafluoroaluminates of *s*-, *d*s- and *p*s-ations
4.2b.1.2.1.1. Hexafluoroaluminates of *s*-, *d*s- and *p*s-ations without Li and Be
4.2b.1.2.1.1.1. Proper hexafluoroaluminates

4.2b.1.2.1.1.1.1. Neutral

Cryolite family
Cryolite $Na_3[AlF_6]$
Elpasolite $K_2Na[AlF_6]$

4.2b.1.2.1.1.1.2. Hydrates

Thomsenolite family
Thomsenolite $NaCa[AlF_6] \cdot H_2O$
Pachnolite $NaCa[AlF_6] \cdot H_2O$

4.2b.1.2.1.1.2. Fluorohydroxyaluminates   4.2b.1.2.1.1.2.1. Hydrates
Yaroslavite $Ca_3[AlF_5(OH)]_2 \cdot H_2O$
*Ralstonite-like $Na_3CaMg_3\{AlF_{12}[(OH),O,F)]_2\}$

4.2b.1.2.1.1.3. Fluorohydroxaquaaluminates   4.2b.1.2.1.1.3.1. Neutral
Gearksutite family
Gearksutite $Ca[AlF_4(OH)(H_2O)]^{\infty 2}$

Tikhonenkovite $Sr[AlF_4(OH)(H_2O)]^{\infty 2}$
Acuminite $Sr[AlF_4(OH)(H_2O)]^{\infty 2}$

4.2b.1.2.1.2. Hexafluoroaluminates of Li   4.2b.1.2.1.2. Neutral
Cryolithionite family
Colquiriite $CaLi[AlF_6]$
Cryolithionite $Na_3Li_3[AlF_6]_2$

4.2b.2. *Class:* Fluoroborates (only *s*- cations)
                                          4.2b.2.1. Neutral
Ferruccite family
Ferruccite $Na[BF_4]$
Avogadrite $(K,Cs)[BF_4]$
*Barberiite $(NH_4)[BF_4]$

4.2b.3. *Class*: Fluorosilicates (only *s*- cations and NH^+_4)
                                          4.2b.3.1. Neutral
Malladrite family
Malladrite $Na_2[SiF_6]$
  Hieratite group
  Hieratite $K_2[SiF_6]$
  Cryptohalite $(NH_4)_2[SiF_6]$
  Bararite $(NH_4)_2[SiF_6]$
  *Demartinite $K_2[SiF_6]$
  *Heklaite $KNa[SiF_6]$
*Knasibfite $K_3Na_4[SiF_6]_3[BF_4]$

4.2b.4. *Class:* Chloroaluminates (only *s*- cations)
*4.2б.5. *Класс*: Hexachlorostannates
*Panichiite $(NH_4)_2[SnCl_6]$

5. TYPE: CARBON, ITS COMPOUNDS (WITHOUT CARBONATES) AND RELATED SUBSTENCES
5a. Quasitype*: Inorganic compaunds (withaut carbonates) and related substances
5a.1. SUBTYPE: NATIVE MINERALS
Native carbon family
Graphite   C
Lonsdaleite   C
Chaoite   C
Diamond   C
*Carbon cubic   C
Schungite   C
  Silicium group
  Silicium   Si
  Germanium   Ge

5a.2. SUBTYPE: MINERALS WITH PRICIPAL COVALENT AND METALLIC-COVALENT BOND - CARBIDES AND RELATED SUBSTANCES - SILICIDES, NITRIDES AND PHOSPHIDES
5a.2.1. *Class:* Carbides

5a.2.1.1. Carbides of **IV***a*-elements
  Moissanite-*6H*   SiC
  Moissanite -5H
  Moissanite -15R
  Moissanite -33R
  Moissanite –beta

5a.2.1.2. Carbides of cenosymmetrical *d*- elements
5a.2.1.2.1. Carbides of **VIII***b*- elements   Simple
Haxonite $(Fe,Ni)_{23}C_6$
  Cohenite group (?)
  Cohenite $(Fe,Ni,Co)_3C$
  Unnamed 448 $(Mn,Fe)_3(C,Si)$
Chalypite Fe_2C

*5a.2.1.2.2. Carbides of **VIII***b* и **VI***b*- elements
*Isovite $(Cr,Fe)_{23}C_6$

5a.2.1.2.2. Carbides of **VI***b*-elements   Simple
Unnamed 241 Cr_2C
Tongbaite Cr_3C_2

5a.2.1.2.3. Carbides of **IV***b*- elements   Simple
  Khamrabaevite series (?)
  Khamrabaevite $(Ti,V,Fe)C$
  Unnamed 330 $(V,Ti)C$

5a.2.1.3. Carbides of noncenosymmetrical *d*- elements
*5a.2.1.3.1. Carbides of **VII***b*- elements
*Carbide Mn Mn_3C

5a.2.1.3.2. Carbides of **VIII***b*-elements   Complex

5a.2.1.3.3. Carbides of **VI***b*-elements   Simple
Unnamed 290   WC

5a.2.1.3.4. Carbides of **V***b*-elements   Simple
Tantalcarbi $(Ta,Nb)C$
*Niobocarbide $(Nb,Ta)C$

5a.2.1a. *Class:* Silicides
5a.2.1a.1. Silicides of noncenosymmetrical *d*- elements
5a.2.1a.1.1. Silicides of **VIII***b*-elements Simple
*Palladosilicide Pd_2Si
Unnamed 449 (Mn,Si,Fe)
Suessite family (x = M : X = 3)
Suessite $(Fe,Ni)_3Si$
Gupeiite Fe_3Si
Unnamed 025 $(Cr,Fe)_3Si$
Perryite (x = 2.(6)) $(Ni,Fe)_8(Si,P)_3$

*Unnamed	Fe_2Si
*Hapkeite	Fe_2Si
*Mavlyanovite(x = 1.(6))	$(Mn,Fe)_5Si_3$
Xifengite (x = 1.(6))	Fe_5Si_3
*Naquite (tetr.) (x = 1)	$FeSi$
*Linzhiite	$FeSi_2$
Unnamed 424 (x = 0.43)	$FeSi_{2,3}$
*Luobusaite (x = 0.42)	$Fe_{0,84}Si_2$
	Complex
Unnamed 028 (x = 1)	$FeTiSi_2$
*Zangboite	$FeTiSi_2$

5a.2.1a.2. Silicides of *s*- elements
5a.2.1a.3 Silicides of **II***a*- elements   Simple
Unnamed 024 Mg_2Si

5a.2.2. **Class**: Nitrides
*5a.2.2.1. Nitrides *p*- elements
*Qingsongite   BN

5a.2.2.2. Nitrides (nitrido-oxides) **IV***a*- elements
                                                                                    Simple
Sinoite Si_2N_2O
*Nierite Si_3N_4

5a.2.2.2.1. Nitrides of cenosymmetrical *d*- elements
5a.2.2.2.1.1. Nitrides of **VIII***b*- elements   Simple
Roaldite Fe_4N
Siderazot Fe_5N_2

5a.2.2.2.1.2. Nitrides of **VI***b*- elements   Simple
Carlsbergite   CrN

5a.2.2.2.1.3. Nitrides of **IV***b*- elements   Simple
Osbornite   TiN

5a.2.2.2.2. Nitrides of noncenosymmetrical *d*- elements
5a.2.2.2.2.1. Nitrides of **II***b*- elements   Compounds inclusions
 Mosesite family
Mosesite $| Hg_2N |^{\infty 3} (Cl,SO_4,MoO_4,CO_3) \cdot H_2O$
Kleinite $| Hg_2N |^{\infty 3} (Cl,SO_4) \cdot (H_2O)_n$

5a.2.2a. *Class:* Phosphides
5a.2.2a.1. Phosphides of cenosymmetrical *d*- elements
5a.2.2a.1.1. Phosphides of **VIII***b*- elements   Simple
*Melliniite $(Ni,Fe)_4P$
*Murashkoite   FeP
*Zuktamrurite FeP_2
 Schreibersite series
 Rhabdite Ni_3P

Schreibersite	$(Fe,Ni)_3P$
*Nickelphosthide	$(Ni,Fe)_3P$
Barringerite	$(Fe,Ni)_2P$
*Allabogdanite	$(Ni,Fe)_2P$
*Transjordanite	Ni_2P

5a.2.2a.1.2. Phosphides of **IV***b*-elements.   Simple
Unnamed 027   TiP
                                                                                      *Complex
*Florenskyite   FeTiP
*Andreyivanovite   Fe(Cr,Fe,V,Ti)P

*5a.2.2a.1.3. Phosphides of **VI***b*- elements   Complex
*Monipite   MoNiP

5b. Quasitype*: Organic carbon compounds (minerals with principal van der Waals forces bond)
5b.1. SUBTYPE: SALTS OF ORGANIC ASIDS
5b.1.1. *Class:* Salts of benzopolycarbonic acids ($C_6H_{6-n}(COOH)_n$; n=6)
                                                    5b.1.1.1. Hydrates (neutral)
Mellite $Al_2[C_6(COO)_6]\cdot16H_2O$

5b.1.2. *Class:* Salts of citric acid (citrates)
                                                    5b.1.2.2.1. Hydrates
                                                    5b.1.2.2.1.1. Oxido-citrates
Pigotite $Al_4O_3[C_6H_5O_7]\cdot13H_2O$
                                                    5b.1.2.2.1.2. Neutral
Earlandite $Ca_3[C_6H_5O_7]_2\cdot4H_2O$

5b.1.3. *Class:* Salts of acetic acid (acetates)
*Ca-acetic $Ca[CH_3COO]\cdot H_2O$
*Hoganite $C_4H_8O_5Cu \rightarrow Cu[CH_3COO]_2\cdot H_2O$
*Paceite $C_8H_{24}O_{14}CaCu \rightarrow CaCu[CH_3COO]_4\cdot6H_2O$

5b.1.3.1. Acetato-chlorides   5b.1.3.1.1. Hydrates (neutral)
Calclacite $CaCl[CH_3COO]\cdot5H_2O$

5b.1.4. *Class:* Salts of oxalic acid (oxalates)
5b.1.4.1. Salts of oxalic acid (oxalates) *s-*, *d$_s$-*, *ps*-cations
                                                    *5b.1.4.1.1. Neutral
*Natroxalate $Na_2[C_2O_4]$
*Antipinite $KNa_3Cu_2[C_2O_4]_4$
                                                    5b.1.4.1.2. Hydrates (neutral)
Zhemchuzhnikovite series
Zhemchuzhnikovite $NaMg(Al,Fe^{3+})[C_2O_4]_3\cdot(8-9)H_2O$
Stepanovite $NaMgFe^{3+}[C_2O_4]_3\cdot(8-9)H_2O$
Minguzzite $K_3Fe^{3+}[C_2O_4]_3\cdot3H_2O$
Weddellite family
Glushinskite $Mg[C_2O_4]\cdot2H_2O$
Humboldtine $Fe^{2+}[C_2O_4]\cdot2H_2O$

Weddellite	$Ca[C_2O_4]\cdot 2H_2O$
*Caoxite	$Ca[C_2O_4]\cdot 3H_2O$
Whewellite	$Ca[C_2O_4]\cdot H_2O$
Oxammite	$(NH_4)_2[C_2O_4]\cdot H_2O$

*56.1.4.2. Salts of oxalic acid (oxalates) f- elements
*56.1.4.2.1. Hydrates
*Deveroite-(Ce) $Ce_2[C_2O_4]_3\cdot 10H_2O$

5b.1.4.3. Salts of chalcophylic elements (Cu^{2+}) 5b.1.4.1.1.1. Hydrates (neutral)
Moolooite	$Cu[C_2O_4]\cdot nH_2O$
Wheatleyite	$Na_2Cu^{2+}[C_2O_4]_2\cdot 2H_2O$
*Lindbergite	$Mn^{2+}[C_2O_4]\cdot 2H_2O$

*5b.1.4.3. Oxalato-chlorides
*Novgorodovaite $Ca_2[C_2O_4]Cl_2\cdot 2H_2O$

*5b.1.4.4. Oxalato-sulfates
*Coskrenite-(Ce) $(Ce,Nd,La)_2[SO_4]_2[C_2O_4]\cdot 8H_2O$

*5b.1.5. Salts of formic asids (formates)
| *Formicaite | $Ca[COOH]_2$ |
| *Dashkovaite | $Mg[COOH]_2\cdot 2H_2O$ |

*56.1.6. Salts of methylsulfonic acid CH_3SO_3H (methylsulfonates)
*Ernstburkeite $Mg(CH_3SO_3)_2\cdot 12H_2O$

5b.2. SUBTYPE: HYDROCARBONS AND RELATED COMPAUNDS
5b.2.1. *Class:* Hydrocarbons cyclic (in the order of decreasing x = H : C)
Evenkite family
Evenkite (n-tetracosene)	$C_{24}H_{48}$ (x = 2)
Fichtelite	$C_{19}H_{34}$ (x = 1,(8))
*Dinite	$C_{20}H_{36}$ (x = 1,8)
Hartite	$C_{20}H_{34}$ (x = 1,7)
Simonellite	$C_{19}H_{24}$ (x = 1,26)
(1,1-dimethyl-7-isopropyl-1,2,3,4-tetrahydrophenanthrene)	
Phylloretine	$C_{18}H_{18}$ (x = 1)
Kratochvilite	$C_{13}H_{10}$ (x = 0,77)
*Ravatite	$C_{14}H_{10}$ (x = 0,7)
Idrialite (dimethylbenzphenanthrene)	$C_{22}H_{14}$ (x = 0,64)
Karpatite = Carpathite	$C_{24}H_{12}$ (x = 0,5)

5b.2.2. *Class:* Hydrocarbons oxygenbearing (in the order of increasing O : C)
*Lidinite	$C_{27}H_{46}O$
Hoelite family	
Refikite (δ-13-dihydro-d-pimaric acid)	$C_{20}H_{32}O_2$
Hoelite (anthraquinone)	$C_{14}H_8O_2$
Flagstaffite (cis-terpin hydrate)	$C_{10}H_{18}(OH)_2\cdot H_2O \rightarrow C_{10}H_{22}O_3$
Sapperit (cellulose)	$C_6H_{10}O_5\cdot mH_2O$

5b.2.3. *Class:* Nitrogenbearing organic compounds

Abelsonite family

Abelsonite (nickel porphyrine)	$C_{31}H_{32}N_4Ni$
Kladnoite (phthalimide)	$C_6H_4(CO_2)NH$
Acetamide	CH_3CONH_2
Guanine	$C_5H_3(NH_2)N_4O$
Uricite (2,6,8-trihydroxypurine)	$C_5H_4N_4O_3$
*Tinnunculite	$C_5H_4N_4O_3 \cdot 2H_2O$
Urea	$CO(NH_2)_2$

5b.3. SUBTYPE: MIXTURES OF ORGANIC SUBSTANCES INCLUDING AMBER AND RELATED SUBSTANCES

Kerite

Asphalt

Ozokerite

Amber

References

1. Achmetov N. S. (1988). General and Inorganic Chemistry. 2-nd Ed., Moscow, Vystshaj Shkola, 640 p. (Russian)

2. Anthony J.W., Bideaux R.A., Bladh K.W., Nichols M.C. (1990). Handbook of mineralogy. Vol. I: Elements, sulfides, sulfosalts. Tucson, AZ, Mineral Data Publishing, , 588 p.

3. Atkins P.W., Beran J.A. (1992). Genetral Chemistry. 2-nd Ed. N.Y., Scientific American Books. 922 p.

4. Back E. (2014). Fleischers Glossary of mineral species, 11-th Ed. Tucson, AZ, The Mineralogical Record Inc., 420 p.

5. Belov N. V., Gododvikov A. A., Bakakin V. V. (1982). Essays on Theoretical Mineralogy. Moscow, Nauka, 208 p. (Russian)

6. Betekhtin A. G. (1950). Mineralogy. Moscow, Gosgeolizdat, 956 p. (Russian)

7. Bokiy G. B., Golubkova N. A. (1989). Introduction to Nomenclature of IUPAK. Moscow, Nauka, 184 p. (Russian)

8. Busev A. I., Efimov I. P. (1977). Definitions, Conceptions, Terms at Chemistry. 2-nd Ed., Moscow, Prosveshchenie, 224 p. (Russian)

9. CE - The Chemical Encyclopaedia. (1988). Moscow, Large Russian Encyclopaedia, vol. 1, 624 p. (Russian)

10. CE - The Chemical Encyclopaedia. (1992). Moscow, Large Russian Encyclopaedia, vol.3, 640 p. (Russian)

11. CED - The Chemical Encyclopaedic Dictionary. (1983). Moscow, Soviet Encyclopaedia, 792 p. (Russian)

12. Clark A.M.(1993).Hey's mineral index, 3-rd Ed. L., Chapman and Hall. 852 p.

13. Dana E.S. (1892). The system of mineralogy of James Dwight Dana 1837 - 1868. N.Y., John Wiley & Sons, Inc. L., Chapman & Hall, Ltd.

14. Fersman A.E. (1933). Geochemistry. Leningrad, Goschimtekhizdat, vol. 1, 328 p. (Russian)

15. Figurovskyi N.A. (1979). The essay of common history of chemistry. Evolution of classical chemistry at the XIX century. Moscow, Nauka, 478 p. (Russian)

16. Fleischer M. (1990). Glossary of mineral species. Transl. from english by Gorskaya M.G. edited by V.A. Frank-Kameneckyi. Moscow, MIR, 204 p. (Russian)

17. Fleischer M., Mandarino J.A. Glossary of mineral species, 6-th Ed. Tusson, AZ, The Mineralogical Record Inc., 1991, 256 p.

18. Godovikov A.A. (1979$_1$). Crystal Chemistry of Simple Substances. Novosibirsk, Nauka, 182 p. (Russian)

19. Godovikov A.A. (1979$_2$). Chemical Background of Mineral Classification. Moscow, Nedra, 302 p. (Russian)

20. Godovikov A.A. (1981$_1$). Remarks to Nomenclature's Rules of IUPAK on Chemistry (vol. 1, semi-vol. 1). Zhurnal Neorganicheskoi Khimii, 4, pp. 1159 - 1164. (Russian)

21. Godovikov A.A. (1981$_2$). Mendeleev Periodic System and Force Characteristics of Elements. Novosibirsk, Nauka, 96 p. (Russian)

22. Godovikov A.A. (1983$_1$). Mineralogy. Moscow, Nedra, 648 p. (Russian)

23. Godovikov A.A. (1983$_2$). Mixed Salts and their's Place at Minerals Systematic. Geologiya and Geofizika, N 12, pp.3 - 15. (Russian)

24. Godovikov A.A. (1984). Halogenosalts - Definition, Peculiarity of Structure, Place at Minerals Systematic, Common Features of Genesis. Geologiya and Geofizika, N 7, pp. 42 - 54. (Russian)

25. Godovikov A.A.(1988). About exposition "Chemical-Structural Systematic of Minerals". New data on minerals of SSSR, Moscow, Nauka, 35, pp. 78 - 104. (Russian)

26. Godovikov A.A. (1989). The connection between the elements properties and structure and properties of minerals. Moscow, Nauka, 118 p. (Russian)

27. Godovikov A.A. (1993). Using Electronegativities in the Classification of Minerals and Inorganic Substances. Zhurnal Neorganicheskoi Khimii, vol. 38, N 9, pp. 1468 - 1482. (Russian)

28. Godovikov A.A. (1994). The connection between the peculiarity of electron construction of elements, normal electrode potential and possibility of being of elements in native state. Zapiski Vsesoyuznogo Mineralogicheskogo Obshchestva, N 5, p. 47-50.

29. Godovikov A.A. (1997). Structural-chemical systematic of minerals. Moccow. Fersman Mineralogical Museum RAS.

30. Godovikov A.A., Nenasheva S.N. (2007). Structural-chemical systematic of minerals. 2-th Ed. Moccow. Fersman Mineralogical Museum RAS. Ocen Pictures Ltd. ISBN 5-900395-40-5.

31. Godovikov A.A., Harija Yu. The connection between the properties of elements and compounds; mineralogical-crystallochemical classification of elements. Jour. Fac. Sci., Hokkaido Univ., Ser. IV, vol. 22, no.2, 1987, pp. 357-385

32. Goldschmidt V.M. (1924). Beziehungen zwischen den geochemischen Verteilungsgesetzten und dem Bau der Atome. Vidensk.Selskab.Skrifter Chr.-Oslo. II, Nr. 4, SS. 1 - 37

33. Goldschmidt V.M. (1928). Ueber die Raumerfullung der Atome (Ionen) in Kristallen und ueber das Wesen der Lithosphare. Neues Jahrb.Min. Bd. LXII, A., SS. 1119 - 1130

34. Hoelzel A.R. (1992). Crystal data. Excerpt from systematics of minerals. Germany, 174 p.

35. Hume-Rothery W., Raynor G.V. (1956). The structure of metals and alloys. L., The Institut of metals, Monogr. & Rep. Ser., No 1. Translated from English. 1959. Moscow. Metallurgizdat, 392 p. (Russian).

36. Jarosch D., Zemann J. (1989). Yafsoanite: a garnet type calcium-tellurium(VI)-zink oxide. Miner.Petrol., Vol. 40, pp. 111 -116

37. Kim A.A., Zayakina N.V., Lavrentiev Yu.G. (1982). Yafsoanite $(Zn_{1.38} Ca_{1.36} Pb_{0.26})_3$ Te_1O_6 - the new mineral of tellurium. Zapiski Vsesoyuznogo Mineralogicheskogo Obshchestva, N 1, pp. 118 - 121. (Russian)

38. Klein C., Hurlbut C.S.Jr. Manual of mineralogy (after James D. Dana), 21-st Ed., N.Y., John Wiley and Sons, 1993, 684 p.

39. Kostov I. (1965). Geochemical principle at the classification of minerals. Problemy geochimii. Moscow., Izdatelstvo Akademii Nauk SSSR, pp. 457 - 462. (Russian).

40. Kostov I. (1968). Mineralogy. Edinburgh and London, Oliver and Boyd. Translated from English by I.Kostov. 1971. Moscow. MIR, 584 p. (Russian)

41. Kripyakevich P.I. (1977). The structural types of intermetallides compounds. Moscow, Nauka, 288 p. (Russian)

42. Krivovichev V.G. Mineralogical glossary. St. Petersburg: St. PetersburgUniv. Publ. House. 2009. 556 p. ISBN 978-5-288-04863-0. (Russian)

43. Lazarenko E.K. (1971). The course of mineralogy. Moscow, Vystshaj Shkola, 608 p. (Russian)

44. Lurie Yu.Yu. (1979). The dictionary of analytical chemistry. Moscow , Chimiya, 480 p. (Russian)

45. Liebau F. (1985). Structural chemistry of silicates. Berlin - Heidelberg -N.Y. - Tokyo, Springer-Verlag. Translated from English. 1988. Moscow. MIR, 410 p. (Russian)

46. Mandarino J.A., Back M.E. (2004). Fleischer's Glossary of mineral species, 9-th Ed. Tucson, AZ, The Mineralogical Record Inc., 309 p.

47. Massalskyi T.V. (1961). The intermediate phases and electronic structure. At the book: The phases theory in alloies . Moscow, Metallurgizdat, pp. 49 - 110. (Russian)

48. Mendeleev D.I. (1869). About atomic volume of simple bodies. At the book: Trudy vtorogo siezda russkikh estestvoispytateley. Part. 1, section. 2, pp. 62 - 71; at the book: D.I. Mendeleev. Essay, vol. II, Leningrad-Moscow, Izdatelstvo Akademii Nauk SSSR, 1934, pp. 20 -29.(Russian)

49. Meyer J.L. (1870). Die Natur der chemischen Elemente als Function ihrer Atomgewichte. Annalen der Chemie und Pharmacie, Bd. 7 (Suppl.), SS. 354 - 36458.

50 Mineralogy Database. www. Min date.org.

51. Nickel E.H., Nichols M.C. Mineral reference manual. N.Y., VNR, 1991, 250 p.

52. Nomenclature's Rules of IUPAK on Chemistry. (1979). Moscow, VINITI, vol. 1, semi-vol. 1. (Russian)

53. Pearson W.B. (1972). The crystal chemistry and physics of metals and alloys. N.Y. - L. - Sydney - Toronto, Wiley - Intersci. Translated from English. 1988. Moscow. MIR, 410 p. (Russian).

54. PED - The Physical Encyclopaedic Dictionary. (1983). Moscow, Soviet Encyclopaedia, 928 p. (Russian)

55. Povarennykh A.S. (1966). Crystal chemical classification of mineral species. Kiev, Naukova Dumka, 548 p. Transl. Povarennykh A.S. Crystal chemical classification of minerals. N.Y. - L., Plenum Press. 1972. V.1, 1 - 458 p., V.2, 459 - 766 p.

56 Ramdohr P., Strunz H. (1978). Klockmanns Lehrbuch der Mineralogie. 16.Auflage, ueberarbeitet und erweitert. Stuttgart, F.Enke Verlag, 876 s.

57. Roberts W.L., Campbell T.J., Rapp G.R.,Jr. Encyclopedia of minerals, 2-nd. Ed., N.Y., VNR, 1990, 978 p.

58. SCE - Short Chemical Encyclopaedia. (1964). Moscow, Soviet Encyclopaedia, vol. 3, 1112 p. (Russian)

59. SCE - Short Chemical Encyclopaedia. (1965). Moscow, Soviet Encyclopaedia, vol. 4, 1182 p. (Russian)

60. Semenov E.I. Systematic of minerals. (1991). Moscow, Nedra, 334 p. (Russian)

61. Shcherbina V.V. (1939). Geochemistry. Moscow-Leningrad, Izdatelstvo Akademii Nauk SSSR, 336 p. (Russian)

62. Shchukarev S.A. (1970 - 1974). Inorganic Chemistry. Moscow, Vysshaja Shkola, vol. 1, 352 p., vol. 2, 382 p. (Russian)

63. Strunz H. Mineralogische Tabellen. 5 Aufl.Leipzig, Akad. Verlagsges. Geest und Portig K.-G., 1970.

64. Umland J.B. (1993). General Chemistry. 5-th Ed. Minneapolis/St.Paul - N.Y. - L.A. - San Francisco, 1070 p.

65. Vernadskiy V. I. (1923). Minerals's Hystory of Crust. Pg., NHTI, N. 1, 208 p. (Russian)

66. Wells A.F. (1984). Structural inorganic chemistry. 5-th Edition. Oxford, Clarendon Press, 1382 p. Translated from English. 1987. Moscow. MIR, v. 1, 408 p., v. 2, 694 p., v.3, 564 p. (Russian).

67. Whitten K.W., Gailey K.D. (1984). General Chemistry with Qualitative Analysis. 2-nd Ed. Saunders Golden Sunburst Series. Philadelphia - N.Y. - Chicago - San Francisco - Montreal - Toronto - L. - Sydney - Tokyo - Mexico City - Rio de Janeiro - Madrid, Saunders College Publishing. 1032 p.

68. Zavaritskyi A.N. (1950). Introduction to petrochemistry of igneous mining rock. 2-nd Ed. Moscow- Leningrad. Izdatelstvo Akademii Nauk SSSR. 396 p. (Russian)

69. Zintl E., Kaiser H. (1933). Ueber die Fahigkeit der Elemente zur bildung negativer Ionen. Zt.anorgan.allgem.Chemie, Bd. 211, SS. 113 - 131

Index

© Springer Nature Switzerland AG 2020
A. A. Godovikov and S. N. Nenasheva, *Structural-Chemical
Systematics of Minerals*, https://doi.org/10.1007/978-3-319-72877-3

Printed by Printforce, the Netherlands